T0183421

Lecture Notes in Bioinformatics 12304

Subseries of Lecture Notes in Computer Science

More information about this series at http://www.springer.com/series/5381

Zhipeng Cai · Ion Mandoiu ·
Giri Narasimhan · Pavel Skums ·
Xuan Guo (Eds.)

Bioinformatics Research and Applications

16th International Symposium, ISBRA 2020
Moscow, Russia, December 1–4, 2020
Proceedings

 Springer

Editors
Zhipeng Cai (iD)
Department of Computer Science
Georgia State University
Atlanta, GA, USA

Giri Narasimhan (iD)
Florida International University
Miami, FL, USA

Xuan Guo (iD)
University of North Texas
Denton, TX, USA

Ion Mandoiu (iD)
University of Connecticut
Storrs Mansfield, CT, USA

Pavel Skums
Georgia State University
Atlanta, GA, USA

ISSN 0302-9743   ISSN 1611-3349   (electronic)
Lecture Notes in Bioinformatics
ISBN 978-3-030-57820-6   ISBN 978-3-030-57821-3   (eBook)
https://doi.org/10.1007/978-3-030-57821-3

LNCS Sublibrary: SL8 – Bioinformatics

This Springer imprint is published by the registered company Springer Nature Switzerland AG
The registered company address is: Gewerbestrasse 11, 6330 Cham, Switzerland

Preface

On behalf of the Program Committee, we would like to welcome you to the proceedings of the 16th edition of the International Symposium on Bioinformatics Research and Applications (ISBRA 2020), held in Moscow, Russia, December 1–4, 2020. The symposium provides a forum for the exchange of ideas and results among researchers, developers, and practitioners working on all aspects of bioinformatics and computational biology and their applications.

This year we received 131 submissions in response to the call for extended abstracts. The Program Committee decided to accept 41 of them for full publication in the proceedings and oral presentation at the symposium; a list of these contributions can be found in this front matter. The technical program also featured two keynote and three invited talks given by five distinguished speakers: Prof. Oxana V. Galzitskaya from the Russian Academy of Sciences, Russia; Prof. Phoebe Chen from La Trobe University, Australia; Prof. Nir Ben-Tal from Tel Aviv University, Israel; Prof. Cenk Sahinalp from National Cancer Institute, USA; and Prof. Srinivas Aluru from the Georgia Institute of Technology, USA.

We would like to thank the Program Committee members and the additional reviewers for volunteering their time to review and discuss symposium papers. We would like to extend special thanks to the steering and general chairs of the symposium for their leadership, and to the finance, publicity, workshops, local organization, and publications chairs for their hard work in making ISBRA 2020 a successful event. Last but not least, we would like to thank all authors for presenting their work at the symposium.

June 2020

Zhipeng Cai
Ion Mandoiu
Giri Narasimhan
Pavel Skums
Xuan Guo

Organization

Steering Committee

Yi Pan (Chair)	Georgia State University, USA
Dan Gusfield	University of California, Davis, USA
Ion Mandoiu	University of Connecticut, USA
Marie-France Sagot	Inria, France
Ying Xu	University of Georgia, USA
Aidong Zhang	University of Virginia, USA

General Chairs

Yuri Porozov	Sechenov University, Russia
Dongqing Wei	Shanghai Jiaotong University, China
Alex Zelikovsky	Georgia State University, USA

Local Chairs

Yuri Porozov	Sechenov University, Russia
Vladimir Poroikov	Institute of Biomedical Chemistry, Russia
Yuri Orlov	Sechenov University, Russia

Program Chairs

Zhipeng Cai	Georgia State University, USA
Ion Mandoiu	University of Connecticut, USA
Giri Narasimhan	Florida International University, USA
Pavel Skums	Georgia State University, USA

Publicity Chairs

Natalia Usova	Trialogue Ltd, Russia
Natalia Rusanova	Sechenov University, Russia
Jijun Tang	University of South Carolina, USA
Le Zhang	Sichuan University, China
Kiril Kuzmin	Georgia State University, USA

Program Committee

Kamal Al Nasr	Tennessee State University, USA
Max Alekseyev	George Washington University, USA
Nikita Alexeev	ITMO University, Russia

Mukul S. Bansal	University of Connecticut, USA
Paola Bonizzoni	Università di Milano-Bicocca, Italy
Anton Buzdin	Omicsway Corp., USA
Zhipeng Cai	Georgia State University, USA
Hongmin Cai	South China University of Technology, China
Xing Chen	China University of Mining and Technology, China
Xuefeng Cui	Tsinghua University, China
Ovidiu Daescu	The University of Texas at Dallas, USA
Lei Deng	Central South University, China
Pufeng Du	Tianjin University, China
Oliver Eulenstein	Iowa State University, USA
Oxana Galzitskaya	Institute of Protein Research, Russia
Xin Gao	King Abdullah University of Science and Technology, Saudi Arabia
Xuan Guo	University of North Texas, USA
Matthew Hayes	Xavier University of Louisiana, USA
Zengyou He	Dalian University of Technology, China
Steffen Heber	North Carolina State University, USA
Wooyoung Kim	University of Washington, USA
Danny Krizanc	Wesleyan University, USA
Xiujuan Lei	Shaanxi Normal University, China
Yuk Yee Leung	University of Pennsylvania, USA
Shuai Cheng Li	City University of Hong Kong, China
Yaohang Li	Old Dominion University, USA
Min Li	Central South University, China
Jing Li	Case Western Reserve University, USA
Xiaowen Liu	Indiana University – Purdue University Indianapolis, USA
Bingqiang Liu	Shandong University, China
Ion Mandoiu	University of Connecticut, USA
Serghei Mangul	University of Southern California, USA
Fenglou Mao	National Institute of Health, USA
Yury Orlovich	Belarus State University, Belarus
Andrei Paun	University of Bucharest, Romania
Nadia Pisanti	Università di Pisa, Italy
Yuri Porozov	Sechenov University, Russia
Russell Schwartz	Carnegie Mellon University, USA
Joao Setubal	University of São Paulo, Brazil
Xinghua Shi	The University of North Carolina at Charlotte, USA
Yi Shi	Shanghai Jiao Tong University, China
Pavel Skums	Georgia State University, USA
Ileana Streinu	Smith College, USA
Emily Chia-Yu Su	Taipei Medical University, China
Shiwei Sun	Chinese Academy of Sciences, China
Sing-Hoi Sze	Texas A&M University, USA
Weitian Tong	Eastern Michigan University, USA

Valentina Tozzini	CNR, Institute of Nanoscience, Italy
Gabriel Valiente	Universitat Politècnica de Catalunya, Spain
Alexandre Varnek	Université de Strasbourg, France
Jianxin Wang	Central South University, China
Seth Weinberg	Virginia Commonwealth University, USA
Fangxiang Wu	University of Saskatchewan, Canada
Yubao Wu	Georgia State University, USA
Yufeng Wu	University of Connecticut, USA
Zeng Xiangxiang	Xiamen University, USA
Alex Zelikovsky	Georgia State University, USA
Le Zhang	Sichuan University, China
Fa Zhang	Chinese Academy of Sciences, China
Xuegong Zhang	Tsinghua University, China
Shuigeng Zhou	Fudan University, China
Quan Zou	Tianjin University, China
Poroikov Vladimir	Institute of Biomedical Chemistry, Russia
Dmitry Osolodkin	Russian Academy of Sciences, Russia
Orlov Yuri	Russian Academy of Sciences, Russia

Additional Reviewers

Alexey Markin	Luca Denti
Alexey Sergushichev	Madalina Caea
Ana Rogoz	Manal Almaeen
Andrei Zugravu	Marian Lupascu
Arnav Jhala	Mengyao Wang
Arseny Shur	Mikhail Kolmogorov
Avin Chon	Ming Shi
Chong Chu	Ming Xiao
Cristian Stanescu	Mohammed Alser
Emanuel Nazare	Natalia Petukhova
Fanny Leung	Pavel Avdeyev
Fatima Zohra Smaili	Pijus Simonaitis
Gianluca Della Vedova	Renmin Han
Haoyang Li	Rocco Zaccagnino
Harry Taegyun Yang	Sergey Aganezov
Hind Aldabagh	Sergey Knyazev
Hossein Saghaian	Steluta Talpau
Igor Mandric	Tareq Alghamdi
Ilia Minkin	Viachaslau Tsyvina
Jaqueline Joice Brito	Yanting Luo
Kunihiko Sadakane	Yasir Alanazi
Lintai Da	Zehua Guo

Contents

Mitochondrial Haplogroup Assignment for High-Throughput Sequencing Data from Single Individual and Mixed DNA Samples

Fahad Alqahtani[1,2] and Ion I. Măndoiu[1(✉)]

[1] Computer Science and Engineering Department, University of Connecticut, Storrs, CT, USA
fahad.alqahtani@uconn.edu, ion.mandoiu@engr.uconn.edu
[2] National Center for Artificial Intelligence and Big Data Technology, King Abdulaziz City for Science and Technology, Riyadh, Saudi Arabia

Abstract. The inference of mitochondrial haplogroups is an important step in forensic analysis of DNA samples collected at a crime scene. In this paper we introduced efficient inference algorithms based on Jaccard similarity between variants called from high-throughput sequencing data of such DNA samples and mutations collected in public databases such as PhyloTree. Experimental results on real and simulated datasets show that our mutation analysis methods have accuracy comparable to that of state-of-the-art methods based on haplogroup frequency estimation for both single-individual samples and two-individual mixtures, with a much lower running time.

Keywords: Mitochondrial analysis · Haplogroup assignment · High-throughput sequencing · Forensic analysis

1 Introduction

Each human cell contains hundreds to thousands of mitochondria, each carrying a copy of the 16,569bp circular mitochondrial genome. Three main reasons have made mitochondrial DNA analysis an important tool for fields ranging from evolutionary anthropology [3] to medical genetics [6,12] and forensic science [1,4]. First, the high copy number makes it easier to recover mitochondrial DNA (mtDNA) compared to the nuclear DNA, which is present in only two copies per cell [9,14]. This is particularly important in applications such as crime scene or mass disaster investigations where only a limited amount of biological material may be available, and where sample degradation may render standard forensic tests based on nuclear DNA analysis unusable [20]. Second, mitochondrial DNA has a mutation rate about 10 times higher than the nuclear DNA, making it an information rich genetic marker. The higher mutation rate is due to the fact that mtDNA is subject to damage from reactive oxygen molecules released

© Springer Nature Switzerland AG 2020
Z. Cai et al. (Eds.): ISBRA 2020, LNBI 12304, pp. 1–12, 2020.
https://doi.org/10.1007/978-3-030-57821-3_1

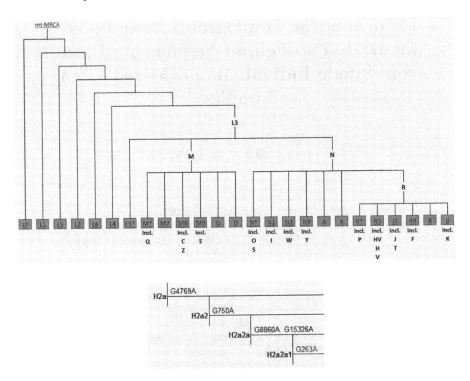

Fig. 1. Top level mtDNA haplogroups (top) and sample haplogroups with their mutations (bottom) from Build 17 of PhyloTree [25].

in mitochondria as by-product of energy metabolism. Finally, mitochondria are inherited maternally without undergoing recombination like the nuclear genome, which can simplify analysis, particularly for mixed samples [14].

Public databases have already amassed tens of thousands of such sequences collected from populations across the globe. Comprehensive phylogenetic analysis of these sequences has been used to infer the progressive accumulation of mutations in the mitochondrial genome during human evolution and track human migrations [31]. Combinations of these mutations, inherited as *haplotypes*, have also been used to trace back our most recent common matrilinear ancestor referred to as the "mitochondrial Eve" [15,29]. Last but not least, clustering of mitochondrial haplotypes has been used to define standardized *haplogroups* characterized by shared common mutations [29]. Due to lack of recombination, the evolutionary history of these haplogroups can be represented as a tree. The best curated haplogroup tree is PhyloTree [26], which currently catalogues over 5,400 haplogroups defined over some 4,500 different mutations (see Fig. 1).

Although many of the available mtDNA sequences have been generated using the classic Sanger sequencing technology, current mtDNA analyses are mainly performed using short reads generated by high-throughput sequencing technologies. Numerous bioinformatics tools have been developed to conduct mtDNA analysis of such short read data. The majority of these tools – including MitoSuite [11], HaploGrep [15], Haplogrep2 [33], mtDNA-Server [32], MToolBox [5], mtDNAmanager [16], MitoTool [8], Haplofind [29], Mit-o-matic [28], and Hi-MC [24] – take a reference-based approach, seeking to infer the haplotype (and assign a mitochondrial haplogroup) assuming that the DNA sample originates from a single individual. While these tools can be helpful for conducting population studies [14] or identifying mislabeled samples [18], they are not suitable for mtDNA analysis of mixed forensics samples that contain DNA from more than one individual, e.g., the victim and the crime perpetrator [30]. Even though the mtDNA haplotypes are not unique to the individual, mitochondrial analysis of mixed forensic samples is useful for including/excluding suspects in crime scene investigations since there is a large haplogroup diversity in human populations [10].

To the best of our knowledge, mixemt [30] is the only available bioinformatics tool that can assign haplogroups based on short reads generated from mixed DNA samples. By using expectation maximization (EM), mixemt estimates the relative contribution of each haplogroup in the mixture. To increase assignment accuracy, the EM algorithm of mixemt is combined with two heuristic filters. The first filter removes any haplogroup that has no support from short reads, while the second filter removes haplogroup mutations that are likely to be private or back mutations. Experiments with synthetic mixtures reported in [30] show that mixemt has high haplogroup assignment accuracy. More recently, mixemt has been used to infer mitochondrial haplogroup frequencies from short reads generated from urban sewer samples collected at tens of sites across the globe, and shown to generate estimates consistent with population studies based on sequencing randomly sampled individuals [22].

In this paper we propose new algorithms for haplogroup assignment from short sequencing reads generated from both single individual and mixed DNA samples. There are two types of prior information associated with haplogroups and available from resources such as PhyloTree [26]. First, each haplogroup has one or more complete mtDNA sequences collected from previous studies. These "exemplary" haplotypes can be leveraged to infer the *frequency* of each haplogroup from the short reads. Since many short reads are compatible with more than one of the existing haplotypes, an expectation maximization framework can be used to probabilistically allocate these reads and obtain maximum likelihood estimates for the frequency haplotypes (and hence the haplogroups) in the database. This is the primary approach taken by mixemt – the haplogroups with high estimated frequency are then deemed to be present in the sample, while the haplogroups with low frequency are deemed to be absent.

The second type of information captured by PhyloTree [26] are the mutations associated each branch of the haplogroup tree. Since each haplogroup corresponds to a node in the phylogenetic tree, haplogroups are naturally associated with the set of mutations accumulated on the path from the root to the respective tree node. As an alternative to the frequency estimation approach of mixemt, the short reads can be aligned to the reference mtDNA sequence and used to call the variants present in the sample. The set of detected variants can then be matched against the sets of mutations associated with each haplogroup, with the best match suggesting the haplogroup composition of the sample.

A priori it is unclear which of the two classes of approaches would yield better haplogroup assignment accuracy. The frequency estimation approach critically relies on having a good representation of the haplotype diversity in each haplogroup, and accuracy can be negatively impacted by lack of EM convergence to a global likelihood maximum due to the high similarity between haplogroups. In contrast, the accuracy of the mutation analysis approach depends on the haplogroup tree being annotated with all or nearly all of the shared mutations defining each haplogroup. High frequency of private and back mutations can negatively impact accuracy of this approach.

In this paper we show that an efficient implementation of the mutation analysis approach can match the accuracy of the state-of-the-art frequency based mixemt algorithm while running orders of magnitude faster. Specifically, our implementation of mutation-based analysis uses the SNVQ algorithm from [7] to identify from the short sequencing reads the mtDNA variants present in the sample. The SNVQ algorithm, originally developed for variant calling from RNA-Seq data, has been previously shown to be robust to large variations in sequencing depth (commonly observed in high-throughput mitogenome sequencing [7]) and allelic fraction (as may be expected for a mixed sample with skewed DNA contributions from different individuals). The set of variants called by SNVQ is then matched to the best set of mutations corresponding to single haplogroups or small collections of haplogroups using the classic Jaccard similarity measure. Exhaustively searching the space of small collections of haplogroups was deemed "computationally infeasible" in [30]. We show that for single individual samples finding the haplogroup with highest Jaccard similarity can be found substantially faster than running mixemt. For two individual mixtures, the pair of haplogroups with highest Jaccard similarity can be identified by exhaustive search within time comparable to that required by mixemt, and orders of magnitude faster when using advanced search algorithms [2].

The rest of the paper is organized as follows. In Sect. 2 we describe our mutation-based haplogroup assignment algorithms. In Sect. 3 we present experimental results comparing Jaccard similarity algorithms with mixemt on simulated and real sequencing data from single individuals and two-individual mixtures. Finally, in Sect. 4 we discuss ongoing and future work.

2 Methods

2.1 Algorithms for Single Individual Samples

In a preprocessing step, we generate the list of mutations for each haplogroup in PhyloTree (MToolBox [5] already includes a file with these lists). For a given sample, we start by mapping the input paired-end reads to the RSRS human mitogenome reference using hisat2 [13]. We next use SNVQ [7] to identify variants from the mapped data. In our brute-force implementation of the algorithm, referred to as *JaccardBF*, we compute the Jaccard coefficient between the set of SNVQ variants and *each* list of mutations associated with leaf haplogroups in PhyloTree. The Jaccard coefficient of two sets of variants is defined as the size of the intersection divided by the size of the union. The haplogroup with the highest Jaccard coefficient is then assigned as the haplogroup of the input data.

The brute-force algorithm can be substantially speeded up by using advanced indexing techniques. In Sect. 3 we report results using the "All-Pair-Binary" algorithm of [2], referred to as *JaccardAPB*, as implemented in the SetSimilaritySearch python library.

2.2   Algorithms for Two-Individual Mixtures

High-throughput reads are aligned to RSRS using hisat2 and then SNVQ is used to call variants as above. We experimented with several haplogroup assignment algorithms for two-individual mixtures. In the first, referred to as *JaccardBF2*, the Jaccard coefficient is computed using brute-force search for each leaf haplogroup, and the *top 2* haplogroups are assigned to the mixture. Unfortunately this algorithm has relatively low accuracy, mainly since the haplogroup with the second highest Jaccard similarity is most of the time a haplogroup closely related to the haplogroup with the highest similarity rather than the second haplogroup contributing to the mixture. To resolve this issue we experimented with computing the Jaccard coefficient between the set of SNVQ variants and all *pairs* of leaf haplogroups, with the output consisting of the pair with maximum Jaccard similarity. We implemented both brute-force and "All-Pair-Binary" indexing based implementations of this pair search algorithm, referred to as *JaccardBF_pair* and *JaccardAPB_pair*, respectively.

2.3   Algorithms for Mixtures of Unknown Size

When only an upper-bound k is known on the mixture size, the Jaccard coefficient can be computed against sets of mutations generated from unions of up to k leaf haplogroups. For mixtures of up to 2 individuals we report results using the "All-Pair-Binary" indexing based implementation, referred to as *JaccardAPB_1or2*.

3   Experimental Results

3.1   Datasets

Real Datasets. We downloaded all WGS datasets used in [26]. Specifically, whole-genome sequencing data for 20 different individuals with distinct haplogroups was downloaded from the 1000 Genomes project (1KGP). The 20 individuals come from two populations: British and Yoruba, with the Yoruba individuals sampled from two different locations (the United Kingdom, and Nigeria,

Table 1. Human WGS datasets for which ground truth haplogroups are Phylotree leaves. Percentage of mtDNA reads was estimated by mapping reads to the published 1KGP sequence, except for the datasets marked with "*" for which there is no 1KGP sequence and mapping was done against the RSRS reference.

Sample ID	Run ID	#Read pairs	#mtDNA pairs	%mtDNA	Haplogroup
HG00096	SRR062634	24,476,109	43,370	0.177	H16a1
HG00097	SRR741384	68,617,747	112,039	0.163	T2f1a1
HG00098*	ERR050087	20,892,714	37,602	0.180	J1b1a1a
HG00100	ERR156632	19,119,986	39,169	0.204	X2b8
HG00101	ERR229776	111,486,484	169,840	0.152	J1c3g
HG00102	ERR229775	109,055,650	217,187	0.199	H58a
HG00103	SRR062640	24,054,672	48,912	0.203	J1c3b2
HG00104*	SRR707166	58,982,989	94,242	0.159	U5a1b1g
NA19093	ERR229810	98,728,262	234,170	0.237	L2a1c5
NA19096	SRR741406	55,861,712	131,587	0.235	L2a1c3b2
NA19099	ERR001345	7,427,776	16,038	0.215	L2a1m1a
NA19102	SRR788622	15,134,619	28,239	0.186	L2a1a1
NA19107	ERR239591	9,217,863	13,297	0.144	L3b2a
NA19108	ERR034534	65,721,104	3,959	0.006	L2e1a

Table 2. Human WGS datasets for which ground truth haplogroups are Phylotree internal nodes.

Sample ID	Run ID	#Read pairs	#mtDNA pairs	%mtDNA	Haplogroup
HG00099	SRR741412	57,222,221	102,968	0.179	H1ae
HG00106	ERR162876	24,328,397	50,635	0.208	J2b1a
NA19092	SRR189830	125,888,789	337,350	0.268	L3e2a1b
NA19095	SRR741381	65,174,483	101,118	0.155	L2a1a2
NA19098	SRR493234	40,446,917	85,658	0.211	L3b1a
NA19113	SRR768183	48,428,152	62,412	0.128	L3e2b

respectively). The haplogroups of 14 of the 20 individuals correspond to leaves nodes in PhyloTree, while the haplogroups of the other 6 correspond to internal nodes. Accession numbers, basic sequencing statistics, and ground truth haplogroups for the 20 datasets are given in Tables 1 and 2.

Synthetic Datasets. For the synthetic datasets, we simulated reads using wgsim [17] based on exemplary sequences associated with leaf haplogroups in PhyloTree [27]. Of the 2,897 leaf haplogroups, 423 haplogroups have only one associated sequence, 2,454 haplogroups have two sequences, and 20 haplogroups have three or more sequences. For single individual experiments, we generated

Table 3. Experimental results on human WGS datasets for which the ground truth haplogroups are Phylotree leaves.

Sample ID	Ground truth	Mixemt		JaccardBF	
		Haplogroup	Time	Haplogroup	Time
HG00096	H16a1	**H16a1**	8,343	**H16a1**	264
HG00097	T2f1a1	**T2f1a1**	897	**T2f1a1**	546
HG00098	J1b1a1a	**J1b1a1a**	16,423	**J1b1a1a**	275
HG00100	X2b8	**X2b8**	12,477	**X2b8**	258
HG00101	J1c3g	**J1c3g**	61,091	**J1c3g**	2,523
HG00102	H58a	**H58a**	66,350	**H58a**	5,343
HG00103	J1c3b2	**J1c3b2**	29,733	**J1c3b2**	1,192
HG00104	U5a1b1g	**U5a1b1g**	27,067	**U5a1b1g**	5,628
NA19093	L2a1c5	**L2a1c5**	59,107	**L2a1c5**	4,345
NA19096	L2a1c3b2	**L2a1c3b2**	44,338	**L2a1c3b2**	1,054
NA19099	L2a1m1a	**L2a1m1a**	14,515	**L2a1m1a**	67
NA19102	L2a1a1	**L2a1a1**	13,642	**L2a1a1**	231
NA19107	L3b2a	**L3b2a**	8,607	**L3b2a**	166
NA19108	L2e1a	**L2e1a**	1,423	**L2e1a**	1,049

two sets of 10,000 simulated read pairs for each haplogroup, using different exemplary sequences as wgsim reference whenever possible, i.e., for all but the 423 haplogroups with a single associated sequence, for which the sole sequence was used to generate both sets of wgsim reads. For mixture experiments we similarly generated two groups of 2,897 two-individual mixtures by pairing each haplogroup with a second haplogroup selected uniformly at random from the remaining ones. Within each group, the reads were generated using wgsim and the first and the second PhyloTree sequence, respectively, except for haplogroups with a single PhyloTree sequence in which the sole sequence was used to generate both sets of wgsim reads. For each pair of haplogroups we generated 10,000 read pairs, with an equal number of read pairs from each haplogroup. We used default wgsim parameters for simulating reads, in particular the sequencing error rate was 1% and the mutation rate 0.001.

3.2   Results on Real Datasets

Tables 3 and 4 give the results obtained by mixemt and JaccardBF on the real datasets consisting of PhyloTree leaf and internal haplogroups, respectively. Both algorithms infer the expected haplogroup when the ground truth is a leaf PhyloTree node. For the six datasets in which the ground truth is an internal node of PhyloTree mixemt always infers the haplogroup correctly, while JaccardBF always infers a leaf haplogroup in the subtree rooted at the ground truth haplogroup. Despite using brute-force search to identify the best matching haplogroup, JacardBF is substantially faster (one order of magnitude or more) than mixemt.

Table 4. Experimental results on human WGS datasets for which the ground truth haplogroups are Phylotree internal nodes.

Sample ID	Ground truth	Mixemt		JaccardBF	
		Haplogroup	Time	Haplogroup	Time
HG00099	H1ae	**H1ae**	40,733	**H1ae1**	1,820
HG00106	J2b1a	**J2b1a**	24,614	**J2b1a5**	1,040
NA19092	L3e2a1b	**L3e2a1b**	137,218	**L3e2a1b1**	6,610
NA19095	L2a1a2	**L2a1a2**	94,921	**L2a1a2b**	1,529
NA19098	L3b1a	**L3b1a**	46,110	**L3b1a11**	650
NA19113	L3e2b	**L3e2b**	62,643	**L3e2b3**	822

Table 5. Experimental results on synthetic single individual datasets generated from the 2,897 leaf haplogroups in Phylotree.

	Mixemt		JaccardBF		JaccardAPB	
	Acc	Avg. time	Acc	Avg. time	Acc	Avg. time
Group1	99.275	7,251.490	99.379	83.780	99.413	0.041
Group2	99.448	7,185.373	99.517	81.428	99.620	0.043
Mean	99.361	7,218.432	99.448	82.604	99.517	0.042
Std. Dev	0.122	46.752	0.098	1.663	0.146	0.001

3.3   Accuracy Results for Single Individual Synthetic Datasets

The above results on real datasets already suggest that the mitochondrial haplogroup can be accurately inferred from WGS data. For a more comprehensive evaluation we simulated reads using exemplary sequences from all leaf haplogroups in PhyloTree. Table 5 gives the results of this comparison. Both mixem and Jaccard algorithms achieve over 99% accuracy on simulated datasets. As for real datasets, JaccardBF is more than one order of magnitude faster than mixemt. The indexing approach implemented in JaccardABP further reduces the running time needed to find the best matching haplogroup with no loss in accuracy.

3.4   Accuracy Results for Two-Individual Synthetic Mixtures

Table 6 gives experimental results on two-individual synthetic mixtures generated as described in Sect. 3.1. In these experiments we assume that it is *a priori* known that the mixture consists of two different haplogroups. Consistent with this assumption, the mixemt prediction is taken to be the two haplogroups with highest estimated frequencies (regardless of the magnitude of the estimated frequencies). Under this model, the accuracy of mixemt remains high but is slightly lower for mixtures than for single haplogroup samples, with an overall mean accuracy of 98.792% compared to 99.361%. JaccardBF2, which returns the two haplogroups with highest Jaccard similarity to the set of mutations called by SNVQ, performs quite poorly, with a mean accuracy of only 22.765%. The JaccardBF_pair algorithm, which returns the pair of haplogroups whose union has the highest Jaccard similarity to the set of mutations called by SNVQ, nearly matches the accuracy of mixemt (with a mean accuracy of 98.398%) with a lower running time. The running time is drastically reduced by indexing the haplogroups for Jaccard similarity searches, although the predefined threshold required for indexing (0.8 in our experiments) does lead to a small additional loss of accuracy (mean overall accuracy of 97.825% for JaccardAPB_pair).

Table 6. Experimental results on synthetic two-individual mixtures generated from the 2,897 leaf haplogroups in Phylotree.

	Mixemt		JaccardBF2		JaccardBF_pair		JaccardAPB_pair	
	Acc	Avg. time	Acc	Avg. time	Acc	Avg. time	Acc	Avg. time
Group1	98.619	4,890.769	22.540	83.116	98.343	1,224.589	97.480	2.101
Group2	98.964	5,273.326	22.989	80.440	98.452	1,484.743	98.171	2.315
Mean	98.792	5,082.048	22.765	81.778	98.398	1,354.666	97.825	2.208
Std. Dev	0.244	270.509	0.317	1.893	0.077	183.957	0.488	0.151

3.5   Accuracy Results for Unknown Mixture Size

In practical forensics applications there are scenarios in which the number of individuals contributing to a DNA mixture is not *a priori* known. In this case, joint inference of the number of individuals *and* their haplogroups is required. Although mitochondrial haplogroup inference with unknown number of contributors remains a direction of future research, in this section we report experimental results for the most restricted (but still practically relevant) such scenario, in which a mixture is *a priori* known to contain *at most two* haplogroups. Specifically, the 2,897 single individual synthetic datasets analyzed in Sect. 3.3 and the 2,897 two-individual synthetic datasets analyzed in Sect. 3.4 were reanalyzed using several joint inference algorithms. For mixemt, the joint inference was performed by using a 5% cutoff on the estimated haplogroup frequencies, while for JaccardAPB_1or2 the joint inference was performed by matching the set of SNVQ variants to the set of one or two haplogroups that has the highest Jaccard similarity. Table 7 reports the accuracy and runtime of the two methods. Overall, mixemt achieves a mean accuracy of 93.398%, with most of the errors due to the incorrect estimate of the number of individuals in the two-individual mixtures. In contrast, most of the JaccardAPB_1or2 errors are due to mis-classification of single individual samples as mixtures. Overall, JaccardAPB_1or2 achieves a mean accuracy of 96.538%.

Table 7. Experimental results for joint inference of mixture size and haplogroup composition.

	Mixemt		JaccardAPB_1or2	
	Acc	Avg. time	Acc	Avg. time
Group1 singles	99.275	7,251.490	94.028	1.4794
Group2 singles	99.448	7,185.373	96.548	2.098
Group1 pairs	83.914	4,890.769	97.376	1.468
Group2 pairs	90.956	5,273.326	98.205	2.244
Mean	93.398	6,150.240	96.539	1.822
Std. Dev	7.462	1,243.583	1.806	0.407

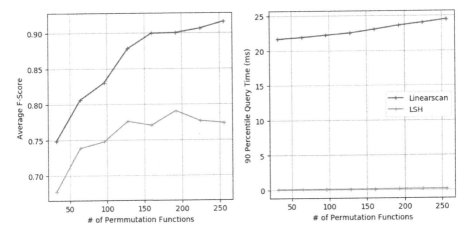

Fig. 2. Comparison of accuracy and running time needed to compute all sets with a Jaccard coefficient greater than 0.9 using MinHash sketches with varying number of hash functions from 2,897 randomly generated sets of average size 44.

4   Conclusions

In this paper we introduced efficient algorithms for mitochondrial haplogroup inference based on Jaccard similarity between variants called from high-throughput sequencing data and mutations annotated in public databases such as PhyloTree. Experimental results on real and simulated datasets show an accuracy comparable to that of previous state-of-the-art methods based on haplogroup frequency estimation for both single-individual samples and two-individual mixtures, with a much lower running time.

In ongoing work we are exploring methods for haplogroup inference of more complex DNA mixtures. Specifically, we are seeking to scale the mutation analysis approach to larger haplogroup mixtures by employing probabilistic techniques such as MinHash sketches and indexing for locality sensitive hashing (LSH) [23]. Implementations such as MinHashLSH [34] can be used to generate all haplogroups with a Jaccard similarity exceeding a given user threshold in sublinear time, resulting in dramatic speed-ups. However, MinHashLSH is an approximate algorithm, which may miss some of the haplogroups with high Jaccard similarity and may also generate false positives. The accuracy and runtime of MinHashLSH depend among other parameters on the number of hash functions, and the user can generally achieve higher precision and recall at the cost of increased running time (Fig. 2). Finally, we are exploring hybrid methods that combine mutation analysis with highly scalable frequency estimation algorithms such as IsoEM [19,21].

References

1. Amorim, A., Fernandes, T., Taveira, N.: Mitochondrial DNA in human identification: a review. PeerJ Prepr. **7**, e27500v1 (2019)
2. Bayardo, R.J., Ma, Y., Srikant, R.: Scaling up all pairs similarity search. In: Proceedings of the 16th International Conference on World Wide Web, pp. 131–140 (2007)
3. Blau, S., et al.: The contributions of anthropology and mitochondrial DNA analysis to the identification of the human skeletal remains of the Australian outlaw Edward 'Ned' Kelly. Forensic Sci. Int. **240**, e11–e21 (2014)
4. Budowle, B., Allard, M.W., Wilson, M.R., Chakraborty, R.: Forensics and mitochondrial DNA: applications, debates, and foundations. Annu. Rev. Genomics Hum. Genet. **4**(1), 119–141 (2003)
5. Calabrese, C., et al.: MToolBox: a highly automated pipeline for heteroplasmy annotation and prioritization analysis of human mitochondrial variants in high-throughput sequencing. Bioinformatics **30**(21), 3115–3117 (2014)
6. Chinnery, P.F., Howell, N., Andrews, R.M., Turnbull, D.M.: Clinical mitochondrial genetics. J. Med. Genet. **36**(6), 425–436 (1999)
7. Duitama, J., Srivastava, P.K., Măndoiu, I.I.: Towards accurate detection and genotyping of expressed variants from whole transcriptome sequencing data. BMC Genomic **13**(2), S6 (2012)
8. Fan, L., Yao, Y.G.: MitoTool: a web server for the analysis and retrieval of human mitochondrial DNA sequence variations. Mitochondrion **11**(2), 351–356 (2011)
9. Hahn, C., Bachmann, L., Chevreux, B.: Reconstructing mitochondrial genomes directly from genomic next-generation sequencing reads–a baiting and iterative mapping approach. Nucleic Acids Res. **41**(13), e129–e129 (2013)
10. Hu, N., Cong, B., Li, S., Ma, C., Fu, L., Zhang, X.: Current developments in forensic interpretation of mixed DNA samples. Biomed. Rep. **2**(3), 309–316 (2014)
11. Ishiya, K., Ueda, S.: MitoSuite: a graphical tool for human mitochondrial genome profiling in massive parallel sequencing. PeerJ **5**, e3406 (2017)
12. Johns, D.R.: Mitochondrial DNA and disease. N. Engl. J. Med. **333**(10), 638–644 (1995)
13. Kim, D., Langmead, B., Salzberg, S.: HISAT2: graph-based alignment of next-generation sequencing reads to a population of genomes (2017)
14. Kivisild, T.: Maternal ancestry and population history from whole mitochondrial genomes. Invest. Genet. **6**(1), 3 (2015)
15. Kloss-Brandstätter, A., et al.: HaploGrep: a fast and reliable algorithm for automatic classification of mitochondrial DNA haplogroups. Hum. Mutat. **32**(1), 25–32 (2011)

16. Lee, H.Y., Song, I., Ha, E., Cho, S.B., Yang, W.I., Shin, K.J.: mtDNAmanager: a web-based tool for the management and quality analysis of mitochondrial DNA control-region sequences. BMC Bioinform. **9**(1), 483 (2008)
17. Li, H.: Wgsim-read simulator for next generation sequencing. Github Repository (2011)
18. Luo, S., et al.: Biparental inheritance of mitochondrial DNA in humans. Proc. Nat. Acad. Sci. **115**(51), 13039–13044 (2018)
19. Mandric, I., Temate-Tiagueu, Y., Shcheglova, T., Al Seesi, S., Zelikovsky, A., Măndoiu, I.I.: Fast bootstrapping-based estimation of confidence intervals of expression levels and differential expression from rna-seq data. Bioinformatics **33**(20), 3302–3304 (2017)
20. Melton, T.W., Holland, C.W., Holland, M.D.: Forensic mitochondrial DNA analysis: current practice and future potential. Forensic Sci. Rev. **24**(2), 101–22 (2012)
21. Nicolae, M., Mangul, S., Măndoiu, I.I., Zelikovsky, A.: Estimation of alternative splicing isoform frequencies from RNA-Seq data. Algorithms Mol. Biol. **6**(1), 9 (2011)
22. Pipek, O.A., et al.: Worldwide human mitochondrial haplogroup distribution from urban sewage. Sci. Rep. **9**(1), 1–9 (2019)
23. Rajaraman, A., Ullman, J.D.: Mining of Massive Datasets. Cambridge University Press, Cambridge (2011)
24. Smieszek, S., et al.: HI-MC: a novel method for high-throughput mitochondrial haplogroup classification. PeerJ **6**, e5149 (2018)
25. Van Oven, M.: Phylotree. https://www.phylotree.org/. Accessed 7 Jan 2020
26. Van Oven, M.: PhyloTree build 17: growing the human mitochondrial DNA tree. Forensic Sci. Int.: Genet. Suppl. Ser. **5**, e392–e394 (2015)
27. Van Oven, M., Kayser, M.: Updated comprehensive phylogenetic tree of global human mitochondrial DNA variation. Hum. Mutat. **30**(2), E386–E394 (2009)
28. Vellarikkal, S.K., Dhiman, H., Joshi, K., Hasija, Y., Sivasubbu, S., Scaria, V.: mito-matic: a comprehensive computational pipeline for clinical evaluation of mitochondrial variations from next-generation sequencing datasets. Hum. Mutat. **36**(4), 419–424 (2015)
29. Vianello, D., Sevini, F., Castellani, G., Lomartire, L., Capri, M., Franceschi, C.: HAPLOFIND: a new method for high-throughput mtDNA haplogroup assignment. Hum. Mutat. **34**(9), 1189–1194 (2013)
30. Vohr, S.H., Gordon, R., Eizenga, J.M., Erlich, H.A., Calloway, C.D., Green, R.E.: A phylogenetic approach for haplotype analysis of sequence data from complex mitochondrial mixtures. Forensic Sci. Int.: Genet. **30**, 93–105 (2017)
31. Wallace, D.C., Chalkia, D.: Mitochondrial DNA genetics and the heteroplasmy conundrum in evolution and disease. Cold Spring Harb. Perspect. Biol. **5**(11), a021220 (2013)
32. Weissensteiner, H., et al.: mtDNA-Server: next-generation sequencing data analysis of human mitochondrial DNA in the cloud. Nucleic Acids Res. **44**(W1), W64–W69 (2016)
33. Weissensteiner, H., et al.: Haplogrep 2: mitochondrial haplogroup classification in the era of high-throughput sequencing. Nucleic Acids Res. **44**(W1), W58–W63 (2016)
34. Zhu, E.E.: Minhash lsh (2019). http://ekzhu.com/datasketch/index.html

Signet Ring Cell Detection with Classification Reinforcement Detection Network

Sai Wang[1,2], Caiyan Jia[1(✉)], Zhineng Chen[2], and Xieping Gao[3]

[1] School of Computer and Information Technology, Beijing Jiaotong University,
Beijing 100044, China
{wangsai18,cyjia}@bjtu.edu.cn
[2] Institute of Automation, Chinese Academy of Sciences, Beijing 100190, China
zhineng.chen@ia.ac.cn
[3] School of Medical Imaging and Examination, Xiangnan University,
Chenzhou 423043, China
xpgao@xtu.edu.cn

Abstract. Identifying signet ring cells on pathological images is an important clinical task that highly relevant to cancer grading and prognosis. However, it is challenging as the cells exhibit diverse visual appearance in the crowded cellular image. This task is also less studied by computational methods so far. This paper proposes a *Classification Reinforcement Detection Network* (CRDet) to alleviate the detection difficulties. CRDet is composed of a Cascade RCNN architecture and a dedicated devised *Classification Reinforcement Branch* (CRB), which consists of a dedicated context pool module and a corresponding feature enhancement classifier, aiming at extracting more comprehensive and discriminative features from the cell and its surrounding context. With the reinforced features, the small-sized cell can be well characterized, thus a better classification is expected. Experiments on a public signet ring cell dataset demonstrate the proposed CRDet achieves a better performance compared with popular CNN-based object detection models.

Keywords: Signet ring cell · Object detection · Deep learning

1 Introduction

Signet ring cell carcinoma (SRCC) is a highly malignant adenocarcinoma that commonly observed in the digestive system. Taking SRCC found in stomach as an example, it is a histological variant of gastric carcinoma generally occurred at a later stage, thus with a poor prognosis [14]. However, if early diagnosis of the tumor is achieved, i.e., finding signet ring cells timely, the prognosis could be improved greatly. It is the most practical way for detecting signet ring cells that pathologists examine the biopsy sample under the microscope, as the cell could only be visually observed at the cellular level. Nevertheless, the examination is time-consuming, laborious and highly dependent on the expertise of pathologists.

© Springer Nature Switzerland AG 2020
Z. Cai et al. (Eds.): ISBRA 2020, LNBI 12304, pp. 13–25, 2020.
https://doi.org/10.1007/978-3-030-57821-3_2

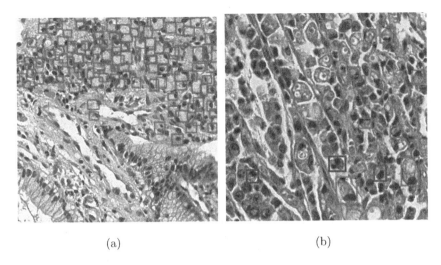

(a)   (b)

Fig. 1. Illustration of some key challenges for detecting signet ring cells in pathological images. The green and red regions represent annotated signet ring cells and signet ring cells missed by pathologist (i.e., unannotated), respectively. (a) Densely distributed and diverse visual appearance. (b) Incomplete annotation. (Color figure online)

The digitization of pathology, i.e., scanning the biopsy sample to the high-resolution whole slide image (WSI) via advanced equipment such as Philips Intel-liSite Pathology Solution, provide a unique opportunity to address the dilemma. It is promising to leverage computer-aided diagnosis to facilitate signet ring cell detection. However, it still remains difficult to develop an accurate method for this task, whose challenges mainly come from three aspects. First, the resolution of a WSI can up to $200,000 \times 100,000$ with millions of cells of various types, where a cell is typically small-sized, i.e., occupying dozens to hundreds pixels. Second, only a few signet ring cells are scattered on the image. They are unevenly distributed, and with diverse visual appearance in terms of morphology, scale, cytoplasmic ratio, nucleus layout, etc., as depicted in Fig. 1. Third and consequent, it is difficult for pathologists to enumerate all the signet ring cells even on a cropped image region. It is not rare that a signet ring cell looks like other cells without careful examination. To ensure the clinical significance, pathologists trend to follow a strict identification scheme that only marks the cells they are sure of. As a result, it leads to the problem of incomplete annotation, i.e., leaving many true signet ring cells unlabeled (the red rectangles in Fig. 1) that brings extra troubles to computational methods.

Despite with the difficulties, there are a few studies dedicated to signet ring cell detection recently. In [9], the authors presented a semi-supervised learning framework for the task, where a self-training and a cooperative-training methods were developed for a better use of labeled and unlabeled data. Convolutional neural network (CNN) based models (e.g., Faster RCNN [15], SSD [12], Cascade RCNN [1]), which shown competitive performance in cellular related applications such as cell detection [7] and nuclei segmentation [4,20], are also like to have decent performance in detecting signet ring cells. However, the models are mainly developed under the fully annotation setting. In contrast, signet ring cells are not only partially annotated, but have diverse visual appearance in morphology, scale, etc. Directly training models for this task seems not optimal, e.g., prone to cause false positive or miss true positive, which is not desired in clinical practice. Technically, the main reason for this is too many cells on an image. There are always some other types of cells have similar visual appearance with signet ring cells, or vice versa. Moreover, the cells are typically small but undergo a large down-sampling before feeding into both the classification and regression heads of a detection network, thus are not easy to distinguish. It is less studied that how to devise novel architectures, which not only are capable of alleviating the difficulties above, but also well inherit the advantages of traditional detection networks.

Motivated by the observations, in this paper we devise a novel architecture, called *Classification Reinforcement Detection Network* (CRDet), that aims to improve the performance of signet ring cell detection on the crowded cellular image. CRDet inherits the backbone of Cascade RCNN for its superiority in detecting densely distributed small objects, e.g., cells. Additionally, it devises a dedicated *Classification Reinforcement Branch* (CRB) for a more accurate signet ring cell identification. Specifically, the branch is composed of a *Context Pooling Module* (CPM) to explore visual appearance within and surrounding the cell for a comprehensive feature extraction, and a *Feature Enhancement Classifier* (FEC) that adopts the deconvolution and Squeeze-and-Excitation (SE) [6] operations for a richer and more discriminative feature representation. Besides, an agglomerative proposal sampling strategy is developed to effectively train the classifier. As a result, better classification results could be obtained even though the cells are small-sized and labeled incompletely. We have conducted extensive experiments on a public dataset containing over 10,000 labeled signet ring cells. The results show that both CPM and FEC are beneficial to the detection. CRDet achieves performance improvements of at least 4% in terms of F-measure compared with popular object detection models on this particular task.

2   Related Work

Signet ring cell is a special cell type. Despite studied relatively few, cell image analysis has been widely investigated in many other contexts for its essential role in medical problems. According to the studies carried out, we can broadly categorize existing related work into two branches, i.e., cell segmentation and abnormal cell detection.

2.1   Cell Segmentation

Cell segmentation aims to describe the contour of the cell. With the contour, it is beneficial to get some quantitative indicators, e.g., the density or morphology statistics of cells. Low-level feature based analysis (e.g., Otus thresholding, the watershed algorithm) dominates the segmentation in early years. The advent of CNN pushes a big step forward in this field recent years. Among them U-Net [16] and its variants [22,23] achieved promising results on challenging dataset, e.g., MoNuSeg for nuclei segmentation in histopathology images [8]. From these studies we can see that elegantly designed architectures are always helpful to cell segmentation.

2.2 Abnormal Cell Detection

Different from most existing studies that detect cells for tasks like cell counting [18], mitosis detection [3], etc., abnormal cell detection aims to identify cells that indicate the existence or severity of lesions. They are different from normal cells and usually few in number. Note that the term abnormal is highly relevant to the problem investigated thus quite different methods are developed in different contexts. For example in [19], the authors extracted histogram of gradient as feature and then used SVM as classifier to detect abnormal cervical cells in pap smear images. Yao et. al. [21] also developed a deep learning based cell subtype classification method for picking up tumor cells in histopathology images. The work most relevant to ours is [9], which uses a semi-supervised learning framework for signet ring cell detection. It used a self-learning mechanism to deal with the challenge of incomplete annotation and used a cooperative-training mechanism to explore the unlabeled regions. This work achieved good results, but it mainly focused on how to better use both labeled and unlabeled data. While in our study, we only utilize the labeled data and emphasize on devising a dedicated classification branch with comprehensive and discriminative feature representation. The reinforced feature is expected to benefit signet ring cell detection.

Note that there were also a few studies that achieved better detection performance by reinforcing the classification branch. For example, *decoupled classification Reinforcement network* (DCR) [2] proposed to use a separate classification network in parallel with the localization network. It achieved a better performance by reducing the rate of false positive compared with methods without this design. However, only a ResNet [5] was directly employed as the classification branch. While our work takes a more careful design of the classification branch from both surrounding context utilization and feature reinforcement, which will be elaborated in the next section.

3   Method

3.1   Background

This section provides a brief overview of Cascade RCNN which is used in the proposed CRDet as backbone. Cascade RCNN inherits from Faster RCNN and further proposes a multi-stage architecture for a more accurate object detection. Given a number of region proposals extracted by the Region Proposal Network

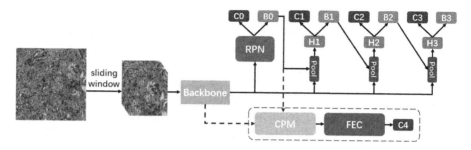

Fig. 2. The architecture of CRDet. The whole network excluding the dashed rectangle part is Cascade RCNN, while the dashed rectangle illustrates the proposed CRB composed of a CPM and a FEC for comprehensive and discriminative feature representation. Dotted line indicates that the sampled region proposals and feature maps are feeding into CPM. "H", "B" and "C" means the detection head, outputted bounding boxes and classification head, respectively.

(RPN), a sequence of detection branches trained with increasing IoU thresholds are stacked. Through ROI pooling, all proposals are resized to 7×7 at each branch for further classification and bounding box regression. Compared with other detection networks, Cascade RCNN uses an iterative mechanism to refine the detected bounding boxes, thus better suppresses false positives, i.e., visually similar non-object background is better identified.

For the case of signet ring cell detection, there are a large number of cells distributed densely on an image, while signet ring cells also have diverse visual appearance that is easy to cause false positive or miss true positive. Therefore, we choose Cascade RCNN as the base model and further devise a reinforced classification branch for better handling the wrongly detected cases.

3.2 Architecture

Overview architecture of the proposed CRDet is illustrated in Fig. 2. The network is composed of two components: the base Cascade RCNN detector and the proposed CRB. As a dedicated branch aims at improving the detection performance, CRB consists of a CPM for cell and its surrounding context utilization, and a FEC for deconvolutional-based feature enhancement and classification. Similar to [2], we place CPM behind Stage-1 of ResNet50 (i.e., the first convolution group in ResNet series [5]) to receive coarsely processed feature maps, which mainly consist of low-level image features such as textures and edges at a relatively high resolution. Meanwhile, region proposals generated by RPN are delivered to CPM to localize cells in the feature maps. With these inputs, CPM uses two ROI poolings, each with a different scale factor, to extract comprehensive features within and surrounding the proposal. Then, FEC employs deconvolution and SE operations to gradually increase the richness and discrimination of feature maps, where a classifier is followed to make predictions on top of the reinforced feature. At last, predictions of different classification heads are merged to generate the final probability of a region proposal being classified as a signet ring cell or not.

Context Pool Module. As mentioned, the signet ring cell is small-sized and with diverse visual appearance. When using traditional CNN-based detectors, the image would experience a sharp down-sampling before reaching ROI pooling. For our task, we argue that using conventional ROI pooling only for feature extraction is not sufficient for two reasons. First, the bounding box generated by RPN is also small-sized. Therefore feature associated with the box is limited. It has the risk of inadequate to distinguish signet ring cells from other cells, resulting in more false positives or missed cells. Second, the bounding box is a rough estimate, and may only cover a part of a cell. Using it directly for ROI pooling would like to weaken the feature representation, which also increase the detection difficulties.

With these in mind, CPM is developed to extract more comprehensive features to facilitate the cell identification. CPM gets feature maps from shallow convolutional layers. It contains local details that are beneficial to describing small objects. The detail structure of CPM is depicted in Fig. 3(a). As can be seen, CPM has two ROI poolings, one the same as conventional ROI pooling while the other is carried out on an enlarged ROI which is the original one multiplied by a scale factor α. It aims at extracting feature also from the surrounding context. We argue that this is meaningful as a proper enlargement not only makes more relevant feature accessible, it also effectively compensates the problem of inaccurate ROI localization. Note that the pooling is carried out channel-by-channel thus each operation outputs feature maps of $7 \times 7 \times 256$. We concatenate the two separately pooled feature maps, and halve its dimension through a 1×1 convolution. Then the resulted feature maps are forward to the subsequent FEC.

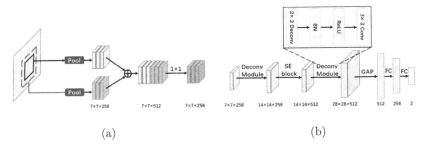

(a)   (b)

Fig. 3. The architecture of CPM and FEC. (a) Context pool module. (b) Feature enhancement classifier. The dashed line shows details of the Deconv Module.

Feature Enhancement Classifier. With the comprehensive feature representation, we devise a dedicated FEC to further enrich the feature as well as its discrimination, and then use it to classify the cells. As seen in Fig. 3(b), FEC is actually a special classification network. Unlike traditional CNN gradually down-sampling the input, FEC first experiences two proposed deconv modules and one SE block to enlarge the feature maps meanwhile selectively weighting their channels. It increases the spatial resolution from 7×7 to 28×28 thus more rich and discriminative feature is obtained. Then, global average pooling [10] is applied to associate the feature maps with two fully-connected layers, followed by a 2-class classification layer that decides whether a proposal is signet ring cell or not.

As for the deconv module, it first enlarges the feature maps by 2 times by a deconvolution operation, followed by a batch normalization and a rectified linear unit, which provides standardization and nonlinearity that are critical to a CNN. In the following, a 3×3 convolution is applied to reduce the dimension of feature maps while further increasing their semantic level. The SE block is placed between the two deconv modules to increase the feature selectivity. The whole classifier is lightweight while still discriminative. It is particularly suitable for classifying objects that are small-sized but with a large quantity, e.g., densely distributed cells.

Agglomerative Proposal Sampling. We also develop an agglomerative proposal sampling strategy analogous to OHEM [17] to effectively train the classifier. Since we can obtain the category (i.e., signet ring cell or not) of a proposal from RPN by checking the ground truth, we first rank those correctly classified ones in a descending order by their classification scores. Then, the top 24 proposals are picked out for training. Since the proposals are correctly classified with high confidence, we named them "simple samples". Meanwhile, we also sort those wrongly classified proposals in a descending order by their classification scores. Also, 24 false positives are picked out as "hard samples". Considering the memory overhead, those 48 "simple samples" and "hard samples" are fetched for more effectively classifier training at each epoch. Compared with the conventional random sampling, the proposed strategy is simple but able to guide the training process more targeted, thus increases the discrimination of the classifier.

Loss Function. As for the base detector, we use smooth L1 loss and cross-entropy loss for bounding box localization and classification, respectively. The loss of base detector is defined as L_{det}. The CRB carries out only the cell classification task. Its loss can be defined as the cross entropy over 2 object classes, i.e., classifying as signet ring cell or not. Denote the set of selected "simple samples" and "hard samples" as Ω, loss function for this classification can be formulated as:

$$L_{cls} = \frac{1}{|\Omega|} \sum_{x \in \Omega} L_{CRB}(p_x, p_x^*)$$

$$= \frac{1}{|\Omega|} \sum_{x \in \Omega} (-p_x^* \log p_x - (1 - p_x^*) \log(1 - p_x)) \qquad (1)$$

where $|\cdot|$ is the number of samples in the set, and $L_{CRB}(p_x, p_x^*)$ represents the cross entropy loss between the prediction result p_x and its labeled categories p_x^*. Therefore the overall loss function can be written as:

$$L = L_{det} + \lambda_{cls} L_{cls} \qquad (2)$$

where λ_{cls} is a hyper-parameter. Since the number of samples are typically smaller than the number of proposals, λ_{cls} is empirically set to 2 in all the experiments to balance L_{det} and L_{cls}.

4   Experiment

To evaluate our CRDet, we conduct extensive experiments on the recently released signet ring cell dataset [9], which is the first public dataset on this task. Details of the experimental setting and results are described as follows.

4.1   Dataset

The signet ring cell dataset contains 687 Hematoxylin and Eosin (H&E) stained pathological WSIs scanned at 40X magnification, which are collected from 155 patients from 2 organs, i.e., gastric mucosa and intestine. The dataset is divided into non-overlapping training and testing sets. The training set contains 99 patients' 460 images, where 77 of them from 20 patients contain signet ring cells. Since the WSI is too large, 455 image regions of size $2{,}000 \times 2{,}000$ are cropped from the WSIs, of which 77 positive images labeled with 12,381 signet ring cells and the rest negative images do not label any signet ring cell. All the cells are labeled by experienced pathologists in the form of tight bounding box. Due to overcrowding and occlusion of cells, it can only guarantee that the labeled cells are indeed signet ring cells, but there are some unannotated signet ring cells in positive images. Since an image of $2{,}000 \times 2{,}000$ is still too large for model training, we split the image to patches of 512×512 in a sliding-window manner. The stride is set to 400. Note that the setting would lead to certain overlapping between neighbor patches.

Since only the training set is public available, we use the training set only as our entire dataset and randomly split it to 7:1:2, respective for model training, validation and test in our experiments. As a result, there are 11821, 1433 and 2609 signet ring cells in the three partitions.

4.2   Implement Detail

Since cell would not exhibit a wide range of aspect ratios and scales as natural objects, we redesign the default anchor. Specifically, we calculate the distributions of both aspect ratio and scale based on the 11821 training cells at first. With the intention of covering the vast majority of cells, we then set the aspect ratios to $\{0.7, 1.0, 1.3\}$ and the scales to $\{32, 64, 112, 128, 256\}$ among all feature maps in FPN uniformly. The scale factor in CPM is empirically set to 1.2 in all experiments. Meanwhile, we keep other hyper-parameters the same as in Cascade RCNN.

We employ ResNet50 with FPN as our backbone (in Cascade RCNN part) as it well balances the performance and computational complexity. We use parameters pre-trained on ImageNet dataset to initialize the backbone, while parameters associated to the remaining part of CRDet are randomly initialized. Our model is trained on 2 Nvidia TITAN Xp GPUs with batch size of 4 images for 30 epochs. We optimize our model using SGD. The initial learning rate is set to 1×10^{-2} and is divided by 10 at 12 and 24 epochs. We also use the warm up and synchronized batchnorm mechanism [13]. No other data augmentation is employed except for the standard random image flipping with ratio 0.5. During the inference, the scores of all classification heads in Cascade RCNN are averaged at first. It then is merged with the classification score of CRB by element-wise product to generate the final classification score. When all sliced patches corresponding to an original image have been predicted, we add the prediction results to their respective offsets in the original image and apply an NMS with IoU threshold 0.5 to remove duplicate boxes and get the final results.

4.3   Ablation Study

To better understand CRDet, we adopt Cascade RCNN as the baseline and execute controlled experiments to evaluate the effectiveness of the two proposed sub-modules, i.e., CPM and FEC, as follows.

Table 1 lists the results of different models on training images. It is seen that when equipped the baseline with FEC, an improvement of 2.2% in terms of F-measure is obtained. It is also observed that the improvement is the combination of a relatively large improved precision and a slightly reduced recall. FEC can increase the discrimination of feature thus reduces the number of false positives at little cost. With CPM further equipped, an additional improvement of 2%

Table 1. The performance of baseline equipped with CPM and FEC.

Method	Recall	Precision	F-measure
baseline	0.736	0.368	0.491
baseline + FEC	0.722	**0.398**	0.513
CRDNet: baseline + CPM + FEC	**0.742**	0.396	**0.516**

on recall is observed. It implies that the exploration of cell surrounding context is beneficial to more effective feature extraction, and reducing the missing rate. Employing both modules leads to an improvement of 2.5% in terms of F-measure compared with the baseline.

4.4   Comparison with State-of-the-art Detectors

We compare CRDet with popular CNN-based object detection models including SSD [12], RetinaNet [11], Faster RCNN [15] and Cascade RCNN [1]. The experiments aims to evaluate how well CRDet tests against state-of-the-art detection models in this particular task, as signet ring cell detection is an emerging research topic with few methods dedicated to it now. The results are given in Table 2. Note that the compared methods are all supervised while [9] is a no open-sourced semi-supervised method, So we do not implement and compare to it.

As can be seen, the proposed CRDet achieves prominent results of 51.6% and 53.9% in terms of F-measure with different backbones, both of which surpasses all the compared models. As anticipated, the one-stage detectors (i.e., SSD and RetinaNet) exhibit the commonly observed low precision problem. With cell proposals given by two-stage detectors (i.e., the RCNN series, CRDet), the extracted feature is more targeted thus the precision improves a lot. However, it is observed that the feature discrimination is still the bottleneck such that a more powerful backbone (i.e., ResNet101) could lead to an improvement of 2% to 3% in terms of precision, mainly attributed to reducing false positives. The proposed CRDet also takes feature reinforcement as the main objective. By incorporating CRB on top of Cascade RCNN, it gains a better feature representation thus further reduces false positives, where improvements of at least 4% in terms of F-measure are consistently observed. It is also shown in Fig. 4. Although the signet ring cells are densely distributed and with diverse visual appearance, CRDet can still produce accurate bounding boxes to localize them.

Table 2. The performance of different methods on signet ring cell detection.

Method	Backbone	Recall	Precision	F-measure
SSD [12]	VGG16	**0.819**	0.043	0.082
RetinaNet [11]	ResNet50	0.670	0.212	0.322
Faster RCNN [15]	ResNet50	0.737	0.331	0.456
Cascade RCNN [1]	ResNet50	0.736	0.368	0.491
	ResNet101	0.731	0.394	0.512
CRDNet	ResNet50	<u>0.742</u>	<u>0.396</u>	<u>0.516</u>
	ResNet101	0.726	**0.428**	**0.539**

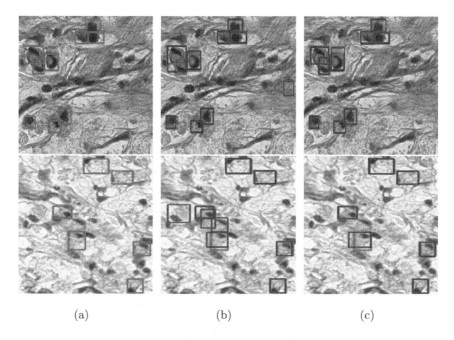

(a)   (b)   (c)

Fig. 4. Qualitative results of CRDNet and Cascade RCNN. The green and red boxes represent ground truth and detection results respectively. (a) Ground Truth. (b) Cascade RCNN. (c) CRDet. (Color figure online)

5   Conclusion

In this paper, we have presented an end-to-end trainable network named CRDet for signet ring cell detection. With the dedicated CRB, the cell could be effectively identified from challenging pathological images. The experiments conducted on a public dataset basically validate our proposal, where performance improvements are consistently observed. We note that the improvement mainly comes from reducing false positives rather than increasing the recall, which is also an important aspect to clinical practice. Thus, future work includes the incorporation of more features to improve the recall. Moreover, we are also interested in leveraging unlabeled data to further boost the detection performance.

Acknowledgements. This work was supported by the Natural Science Foundation of China (No. 61972333, 61876016, 61772526) and the Fundamental Research Funds for the Central Universities (2019JBZ110).

References

1. Cai, Z., Vasconcelos, N.: Cascade R-CNN: delving into high quality object detection. In: Proceedings of the IEEE Conference on Computer Vision and Pattern Recognition, pp. 6154–6162 (2018)

2. Cheng, B., Wei, Y., Shi, H., Feris, R., Xiong, J., Huang, T.: Decoupled classification refinement: Hard false positive suppression for object detection. arXiv preprint arXiv:1810.04002 (2018)
3. Cireşan, D.C., Giusti, A., Gambardella, L.M., Schmidhuber, J.: Mitosis detection in breast cancer histology images with deep neural networks. In: Mori, K., Sakuma, I., Sato, Y., Barillot, C., Navab, N. (eds.) MICCAI 2013. LNCS, vol. 8150, pp. 411–418. Springer, Heidelberg (2013). https://doi.org/10.1007/978-3-642-40763-5_51
4. He, K., Gkioxari, G., Dollár, P., Girshick, R.: Mask R-CNN. In: Proceedings of the IEEE International Conference on Computer Vision, pp. 2961–2969 (2017)
5. He, K., Zhang, X., Ren, S., Sun, J.: Deep residual learning for image recognition. In: Proceedings of the IEEE Conference on Computer Vision and Pattern Recognition, pp. 770–778 (2016)
6. Hu, J., Shen, L., Sun, G.: Squeeze-and-excitation networks. In: Proceedings of the IEEE Conference on Computer Vision and Pattern Recognition, pp. 7132–7141 (2018)
7. Hung, J., Carpenter, A.: Applying faster R-CNN for object detection on malaria images. In: Proceedings of the IEEE Conference on Computer Vision and Pattern Recognition Workshops, pp. 56–61 (2017)
8. Kumar, N., Verma, R., Sharma, S., Bhargava, S., Vahadane, A., Sethi, A.: A dataset and a technique for generalized nuclear segmentation for computational pathology. IEEE Trans. Med. Imaging $36(7)$, 1550–1560 (2017)
9. Li, J., et al.: Signet ring cell detection with a semi-supervised learning framework. In: Chung, A.C.S., Gee, J.C., Yushkevich, P.A., Bao, S. (eds.) IPMI 2019. LNCS, vol. 11492, pp. 842–854. Springer, Cham (2019). https://doi.org/10.1007/978-3-030-20351-1_66
10. Lin, M., Chen, Q., Yan, S.: Network in network. arXiv preprint arXiv:1312.4400 (2013)
11. Lin, T.Y., Goyal, P., Girshick, R., He, K., Dollár, P.: Focal loss for dense object detection. In: Proceedings of the IEEE International Conference on Computer Vision, pp. 2980–2988 (2017)
12. Liu, W., et al.: SSD: single shot multibox detector. In: Leibe, B., Matas, J., Sebe, N., Welling, M. (eds.) ECCV 2016. LNCS, vol. 9905, pp. 21–37. Springer, Cham (2016). https://doi.org/10.1007/978-3-319-46448-0_2
13. Peng, C., et al.: Megdet: a large mini-batch object detector. In: Proceedings of the IEEE Conference on Computer Vision and Pattern Recognition, pp. 6181–6189 (2018)
14. Pernot, S., Voron, T., Perkins, G., Lagorce-Pages, C., Berger, A., Taieb, J.: Signet-ring cell carcinoma of the stomach: impact on prognosis and specific therapeutic challenge. World J. Gastroenterol.: WJG $21(40)$, 11428 (2015)
15. Ren, S., He, K., Girshick, R., Sun, J.: Faster R-CNN: towards real-time object detection with region proposal networks. In: Advances in Neural Information Processing Systems, pp. 91–99 (2015)
16. Ronneberger, O., Fischer, P., Brox, T.: U-Net: Convolutional networks for biomedical image segmentation. In: Navab, N., Hornegger, J., Wells, W.M., Frangi, A.F. (eds.) MICCAI 2015. LNCS, vol. 9351, pp. 234–241. Springer, Cham (2015). https://doi.org/10.1007/978-3-319-24574-4_28
17. Shrivastava, A., Gupta, A., Girshick, R.: Training region-based object detectors with online hard example mining. In: Proceedings of the IEEE Conference on Computer Vision and Pattern Recognition, pp. 761–769 (2016)

18. Sirinukunwattana, K., Raza, S.E.A., Tsang, Y.W., Snead, D.R., Cree, I.A., Rajpoot, N.M.: Locality sensitive deep learning for detection and classification of nuclei in routine colon cancer histology images. IEEE Trans. Med. Imaging **35**(5), 1196–1206 (2016)
19. Sophea, P., Handayani, D.O.D., Boursier, P.: Abnormal cervical cell detection using hog descriptor and SVM classifier. In: 2018 Fourth International Conference on Advances in Computing, Communication & Automation, pp. 1–6. IEEE (2018)
20. Wang, E.K., et al.: Multi-path dilated residual network for nuclei segmentation and detection. Cells **8**(5), 499 (2019)
21. Yao, J., Wang, S., Zhu, X., Huang, J.: Imaging biomarker discovery for lung cancer survival prediction. In: Ourselin, S., Joskowicz, L., Sabuncu, M.R., Unal, G., Wells, W. (eds.) MICCAI 2016. LNCS, vol. 9901, pp. 649–657. Springer, Cham (2016). https://doi.org/10.1007/978-3-319-46723-8_75
22. Zhou, Z., Rahman Siddiquee, M.M., Tajbakhsh, N., Liang, J.: UNet++: a nested U-Net architecture for medical image segmentation. In: Stoyanov, D., et al. (eds.) DLMIA/ML-CDS -2018. LNCS, vol. 11045, pp. 3–11. Springer, Cham (2018). https://doi.org/10.1007/978-3-030-00889-5_1
23. Zhu, Y., Chen, Z., Zhao, S., Xie, H., Guo, W., Zhang, Y.: ACE-Net: biomedical image segmentation with augmented contracting and expansive paths. In: Shen, D., et al. (eds.) MICCAI 2019. LNCS, vol. 11764, pp. 712–720. Springer, Cham (2019). https://doi.org/10.1007/978-3-030-32239-7_79

SPOC: Identification of Drug Targets in Biological Networks via Set Preference Output Control

Hao Gao[1], Min Li[1($\boxtimes$)], and Fang-Xiang Wu[2($\boxtimes$)]

[1] School of Computer Science and Engineering, Central South University, Changsha 410083, People's Republic of China
limin@csu.edu.cn
[2] Division of Biomedical Engineering and Department of Mechanical Engineering, University of Saskatchewan, Saskatoon, SK S7N5A9, Canada
faw341@mail.usask.ca

Abstract. Biological networks describe the relationships among molecular elements and help in the deep understanding of the biological mechanisms and functions. One of the common problems is to identify the set of biomolecules that could be targeted by drugs to drive the state transition of the cells from disease states to health states called desired states as the realization of the therapy of complex diseases. Most previous studies based on the output control determine the set of steering nodes without considering available biological information. In this study, we propose a strategy by using the additionally available information like the FDA-approved drug targets to restrict the range for choosing steering nodes in output control instead, where we call it the Set Preference Output Control (SPOC) problem. A graphic-theoretic algorithm is proposed to approximately tackle it by using the Maximum Weighted Complete Matching (MWCM). The computation experiment results from two biological networks illustrate that our proposed SPOC strategy outperforms the full control and output control strategies to identify drug targets. Finally, the case studies further demonstrate the role of the combination therapy in two biological networks, which reveals that our proposed SPOC strategy is potentially applicable for more complicated cases.

Keywords: Set Preference Output Control · Drug targets · Biological networks · Maximum Weighted Completed Matching

1 Background

In the last decade, network medicine has gradually formed an efficient framework to systematically discover the underlined mechanisms of complex diseases and

This work was supported by Natural Sciences and Engineering Research Council of Canada (NSERC), the National Natural Science Foundation of China [61772552], and the Program of Independent Exploration Innovation in Central South University (2019zzts959), the Fundamental Research Funds for the Central Universities, CSU (2282019SYLB004).

© Springer Nature Switzerland AG 2020
Z. Cai et al. (Eds.): ISBRA 2020, LNBI 12304, pp. 26–37, 2020.
https://doi.org/10.1007/978-3-030-57821-3_3

then provide potential strategies to target therapy of a particular disease [1,2]. Based on the data-driven construction algorithms, different types of complex biological networks like the protein-protein interaction networks, the gene regulatory networks, and the human brain networks have been generally studied to reveal the pivotal components that might be linked to the particular biological processes. For example, centrality-based methods [3] have been applied to the protein-protein interaction networks to identify the essential proteins the single deletion of which leads to lethality or infertility [4].

Though the-state-of-art machine learning methods can outperform them in some fields [5], the advantages of the network-based methods are to explore the original dynamics of the biological models and give the inherent explanation of biological mechanisms so that we can finally control the transition of the state of complex biological networks [6,7]. In recent years, full control(FC) [8] and output control(OC) [9,10] have been one type of the fundamental methods to analyze complex diseases in the area of network biology [11–13]. Owing to the demand for real biological problems, additional information is used to improve the results when exploring biological networks. Wu *et al.* have added the drug target information which ranks the possibility and efficiency of the genes targeted by drugs to the MSS algorithm to promote the accuracy of the detection of drug targets [14]. Guo *et al.* have referred to the personalized mutated genes as the constrained set of nodes for choosing personalized driver genes in the constrained output control(COC) [12,15].

But there are two fundamental problems when using the previous methods to some extreme cases. As for the MSS, the number of steering nodes identified in large gene regulatory networks accounts for the 80% of all nodes [16]. It's not practical in clinical experiments. Another one is that the restricted condition in COC is too strict in the identification of drug targets. Due to the incomplete discovery of drugs not like the case in mutated genes, many biological networks might be impossible to be controlled by existing drugs and other target-based treatments.

Hence, we propose the Set Preference Output Control(SPOC) strategy which is a relaxed condition in output control so that the identification of the set of steering nodes can be enriched with as many drug targets as possible. Additionally, we provide a theoretical analysis between SPOC and maximum weighted complete matching (MWCM). Finally, the new strategy is applied to MAPK signaling network and ER+ breast cancer signaling network to further illustrate the advantages of our SPOC strategy.

2   Problem Formulation

2.1   Dynamics Model of Biological Networks

As previous studies [6–15], a regulatory network with n biomolecules can be described by the following linear time-invariant dynamic model [17]:

$$\begin{cases} \dot{x}(t) = Ax(t) + Bu(t) \\ y(t) = Cx(t) \end{cases} \tag{1}$$

where the state transition matrix $A \in \mathbf{R}^{n \times n}$ describes the interactions between biomolecules, where $a_{ij} \neq 0$ if biomolecule j regulates biomolecule i. The input matrix $B \in \mathbf{R}^{n \times m}$ represents the connections between biomolecules and drugs, where $b_{ij} \neq 0$ if biomolecule i is targeted by drug j. The output matrix $C \in \mathbf{R}^{s \times n}$ specifies the set of biomolecules that define the disease genotype. $x(t) = \{x_1(t), x_2(t), \ldots, x_n(t)\}^T$ represents the state of n biomolecules in networks at time t. $y(t) = \{y_1(t), y_2(t), \ldots, y_s(t)\}^T$ represents observed states of the genotype. $u(t) = \{u_1(t), u_2(t), \ldots, u_m(t)\}^T$ represents the drugs regulating the biomolecules directly.

Generally, such a system can be represented as a digraph $G(V, E)$. The set of nodes in the digraph can be represented as the nodes $V = V_A \cup V_B$, where V_A is the set of biomolecules and V_B represents the set of control signals (drugs). The set of edges can be represented as $E = E_A \cup E_B$, which is defined as follows,

$$E_A : v_j \rightarrow v_i \qquad if \ A_{ij} \neq 0$$

and

$$E_B : u_j \rightarrow v_i \qquad if \ B_{ij} \neq 0$$

In practices, the nonzero values in matrices A and B are difficult to be determined so that the structural systems are the framework to study the controllability of biological networks [6,7], which we are adopting in this study.

2.2   Problem Description

Previous output (target) control (OC) [9,10] can approximately identify a set of genes in networks without any biological constraints. However, the identified steering nodes should be enriched in the drug-target genes to practically drive the disease states to healthy states. Hence, here we first define the Set Preference Output Control (SPOC) strategy that preferentially determines the minimum set of steering nodes from a predefined set of genes to fully control the set of biomolecules that define the disease genotype. The problem could be mathematically formulated as follows,

Problem 1: *Given a directed network with system matrix $A \in \{0,1\}^{n \times n}$, input matrix $B \in \{0,1\}^{n \times m}$ and output matrix $C \in \{0,1\}^{s \times n}$, SPOC is to control the set of output nodes $V_O = \{v_{o_1}, \ldots, v_{o_s}\}$ by a minimum set of steering nodes containing as many as possible preferential steering nodes $V_P = \{v_{p_1}, \ldots, v_{p_l}\}$.*

The above formulated problem is different from Constrained Output Control (COC) proposed by Guo et al. [12,15] in which the steering nodes in COC could only be selected from the constrained set, while the steering nodes in SPOC could be selected from nodes outsides the preferential set. For example, if a disease gene in a gene regulatory network is a source node whose in-degree is 0 and couldn't be targeted by any drug, the COC would fail to deal with this situation because the disease gene is inaccessible and uncontrollable [18]. Hence, we relax the constrained set in COC to the preferential set (Fig. 1).

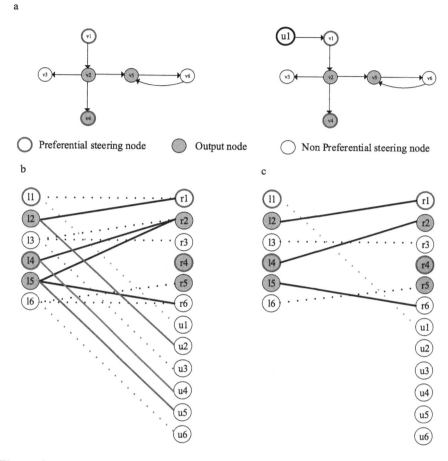

Fig. 1. Overview of SPOC strategy. a. The left network shows the definition of the original network including the preferential steering nodes, output nodes, and non-preferential nodes. The right network indicates that node $\{v1\}$ is chosen from the preferential set acted as the steering node to control all the output nodes. b. It represents the process of the construction of a weighted bipartite graph. The weight of the black solid line, black dot line, red solid line, red dot line, blue solid line and blue dot line are 2, 1, $\epsilon + 1$, ϵ, $\gamma + 1$ and γ, respectively. c. After removing the unmatched edges, the left matched edges indicate the result of the MWCM. Cycle $\{v5, v6\}$ is accessible, indicating that no additional nodes are needed. (Color figure online)

3   Methods

Inspired by the algorithm of output control [9], we first define a bipartite graph $G(V_l, V_r \cup V_U, E, W)$ where $V_l = \{l_1, l_2, \ldots, l_n\}$, $V_l = \{r_1, r_2, \ldots, r_n\}$, and $V_U = \{u_1, u_2, \ldots, u_m\}$ based on the structural system $G(V, E)$. We add a self-loop to each node if the node $v_i \in \overline{V_O}, \overline{V_O} = V - V_O$ because those nodes with a self-regulation need no additional signals to control in any cases. Here, we can construct the weighted bipartite graph as follows,

$$w_{l_i r_j} = \begin{cases} 2 & a_{ji} \neq 0 \text{ and } v_i \in V_O \\ 1 & a_{ji} \neq 0 \text{ and } v_i \in \overline{V_O} \end{cases} \tag{2}$$

For each $v_i \in V_P$

$$w_{l_i u_i} = \begin{cases} \epsilon + 1 & v_i \in V_O \\ \epsilon & v_i \in \overline{V_O} \end{cases} \tag{3}$$

Otherwise, for each $v_i \in \overline{V_P}, \overline{V_P} = V - V_P$,

$$w_{l_i u_i} = \begin{cases} \gamma + 1 & v_i \in V_O \\ \gamma & v_i \in \overline{V_O} \end{cases} \tag{4}$$

where $\epsilon = \frac{1}{n}$ and $\gamma < \epsilon$.

Next, we demonstrate two definitions related to the problem.

Definition 1. (Complete matching) Complete matching is a matching where all nodes in the V_l are matched.

Definition 2. (Maximum Weight Complete Matching (MWCM) in Bipartite) Given a bipartite graph $G = (V, E)$ with structural system (A, B) and weight functions, find a complete matching M maximizing the sum of weights.

After the construction of the weighted bipartite graph, we can prove that the steering nodes selected by the maximum weight complete matching (MWCM) algorithm [19,20] have the minimum number of nodes matching to control signals and the maximum preferential nodes matching to control signals.

Theorem 1. *The number of nodes matching to control signals is the minimum.*

Proof. Consider the nodes belonging to the preferential set, if an output node is determined as the steering node, the contribution would be $\epsilon + 1 = 2 + (\epsilon - 1)$. If a non-output node was determined as the steering node, the contribution would be $\epsilon = 1 + (\epsilon - 1)$. The case in non-preferential set is similar. Given a set of nodes $V'_U \subseteq V_l$ matching to control signals is determined by the MWCM, the maximum score can be calculated as follows,

$$Scroe_{V'_U} = (|\overline{V_O}| + 2|V_O|) + \epsilon|V'_U \cap V_P| + \gamma|V'_U \cap \overline{V_P}| - |V'_U| \tag{5}$$

If another set of nodes $V''_U \subseteq V_l$ matching to control signals is determined by a complete matching and sufficient to $|V''_U| + 1 \leq |V'_U|$,

$$Scroe_{V''_U} = (|\overline{V_O}| + 2|V_O|) + \epsilon|V''_U \cap V_P| + \gamma|V''_U \cap \overline{V_P}| - |V''_U| \tag{6}$$

Owing to the fact that $0 < \epsilon|V'_U \cap V_P| + \gamma|V'_U \cap \overline{V_P}|, \epsilon|V''_U \cap V_P| + \gamma|V''_U \cap \overline{V_P}| < 1$,

$$Scroe_{V''_U} - Scroe_{V'_U} > |V'_U| - |V''_U| - 1 \geq 0 \tag{7}$$

This result contradicts with the assumption that $Scroe_{V'_U}$ is the maximum. Hence, the number of nodes $|V'_U|$ matching to control signals is the minimum.

Based on Theorem 1, $\epsilon|V_U \cap V_P| + \gamma|V_U \cap \overline{V_P}|$ is the maximum which is only correlated to the score of nodes in V_l because $|V_O|$ is constant and $|V_U|$ is the minimum.

Theorem 2. *The number of preferential nodes matching to control signals is the maximum.*

Proof. Assuming that there are two set of nodes $V'_U \subseteq V_l$ and $V''_U \subseteq V_l$ matching to control signals, V'_U is determined by the MWCM, while V''_U can be obtained by exchanging an augmenting path from node $v_m \in V'_U \cap \overline{V_P}$ to node $v_n \in V'_U \cap V_P$, indicating that $|V'_U| = |V''_U|$, $|V''_U \cap V_P| = |V'_U \cap V_P| + 1$, and $|V''_U \cap \overline{V_P}| = |V'_U \cap \overline{V_P}| - 1$.

$$Scroe_{V''_U} = \epsilon(|V'_U \cap V_P| + 1) + \gamma(|V'_U \cap V_{nP}| - 1) + (|\overline{V_O}| + 2|V_O|) - |V''_U|$$
$$= Scroe_{V'_U} + \epsilon - \gamma > Scroe_{V'_U} \tag{8}$$

This result contradicts with the assumption that $Scroe_{V'_U}$ is the maximum. Hence The number of preferential nodes $V'_U \cap V_P$ matching to control signals is the maximum.

After the MWCM, the nodes in V_l matching to the set of control signals are selected as the first part of steering nodes. In the meantime, the cycles covering all output nodes could be detected by strong connected components(SCC) algorithm to identify the set of inaccessible nodes as the second part of steering nodes. The time complexity of MWCM and SCC are $O(n^3)$ and $O(n)$, respectively, where n is the number of nodes in biological networks.

4 Datasets

4.1 MAPK Signaling Network

The Mitogen-Activated Protein Kinase (MAPK) signaling network determine the cancer cell fate through closely regulating diverse biological activities including cell cycle, apoptosis, and differentiation. It contains three pivotal components ERK, JNK, and p38 defining states of the MAPK signaling network. To deeply understand the mechanism of cancer cell fate decision, the MAPK signaling network was constructed as a comprehensive map of the chemical reactions in cell [21], which consists of 46 genes and 92 edges. In the network, the state of 7 genes can directly affect the phenotypes, which are regarded as the output nodes. Genes MYC, p70, and p21 are considered to regulate the proliferation of the cell, while genes FOXO3, p53, ERK, and BCL2 are found to be associated with cell apoptosis. We find that 5 nodes (BCL2, FGFR3, EGFR, JUN, and PDK1) are the approved drug targets in DrugBank database [22], which are the set of preferential nodes. The set of nodes have the ability to affect the cell fate according to previous studies [23–27] (Fig. 2).

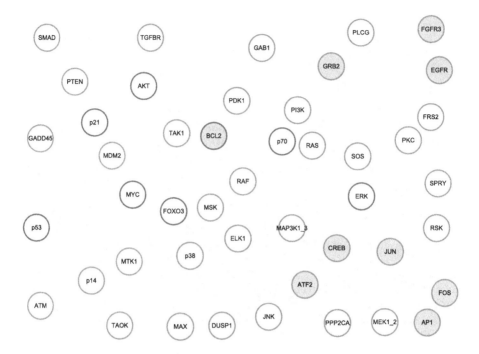

Fig. 2. MAPK signaling Network. The nodes marked with red color are the set of output nodes. The nodes filled with gray color are the set of preferential nodes. (Color figure online)

4.2 ER+ Breast Cancer Signaling Network

The ER+ breast cancer signaling network was constructed on the basis of the literature of ER+, HER2+, and PIK3CA-mutant breast cancers to deeply comprehend the response of variable drug targets [28]. It contains 49 nodes consisting of proteins and transcripts and 83 edges consisting of transcriptional regulation and signaling process, which involves the main signaling pathways of the breast cancer. Six nodes are mapped to control apoptosis(BIM, MCL1, BCL2, and BAD) and proliferation (translation and E2F), which are the set of output nodes. There are 10 nodes directly targeted by 7 drugs: Alpelisib, Everolimus, Fulvestrant, Ipatasertib, Neratinib, Palbociclib, and Trametinib. In the network, all drug targets are considered as the preferential set of steering nodes.

5   Results

5.1   Measurements

In this study, we use two metrics to evaluate the result of three network control scenarios: full control (FC), output control (OC) and set preference output control (SPOC). The first metric is defined as the percentage of identified steering nodes in the biological networks.

$$P_{SN} = \frac{the\ number\ of\ steering\ nodes}{the\ number\ of\ all\ nodes} \tag{9}$$

The lower score P_{SN} indicates that the biological networks are easier to be controlled. Owing to the non-uniqueness of network control scenarios, we randomly relabel the network nodes for 1000 times in this study to generate different control configurations. Another one is defined as the average proportion of identified drug targets in all control configurations.

$$P_{DT} = \frac{the\ average\ number\ of\ drug\ targets}{the\ number\ of\ steering\ nodes} \tag{10}$$

The higher score P_{DT} demonstrates that the biological networks are with a higher probability to be controlled by the existing drugs.

5.2 Results in Two Biological Networks

In this section, we discuss the results of controllability analysis in three scenarios: full control (FC), output control (OC) and set preference output control (SPOC) on MAPK signaling network and ER+ breast cancer signaling network.

In MAPK signaling network (as shown in Fig. 3a), the number of identified steering nodes by FC is 23.9% (11/46), which is around 3 times larger than the one 8.7% (4/46) by OC and SPOC. The results show the advantage of OC and SPOC compared with FC in terms of the number of steering nodes. Then the average proportion of drug targets in the 1000 times identified steering nodes through FC, OC and SPOC are 5.9%, 11.2%, and 100%, respectively. The present studies reveal that our algorithm can promote the enrichment of steering nodes to the identification of drug targets, which is coherent to our objective.

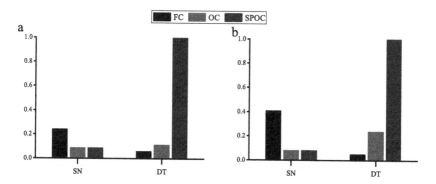

Fig. 3. The P_{SN} and P_{DT} of three methods in two biological networks. a. MAPK signaling network. b. ER+ breast cancer network.

We also study the set of identified steering nodes on ER+ breast cancer signaling network to further evaluate our proposed algorithm (as shown in Fig. 3b).

Similar results can be found in this network. As for the three control scenarios: FC, OC, and SPOC, the number of identified steering nodes are 40.8%(20/49), 8.2%(4/49), and 8.2%(4/49), while the average proportion of drug targets are 5%, 23.7%, and 100%, respectively. Similar results further demonstrate the advantages of our proposed algorithm.

5.3   Case Studies

According to previous results of SPOC, each control configuration is chosen from the predefined set of nodes so that we analyze the binary patterns of 1000 sets of steering nodes. The results of two biological networks are visualized through the heatmap as shown in Fig. 4. The elements, except for diagonal elements, indicate the frequency of two nodes in all control configurations. The diagonal elements represent the frequency of steering nodes.

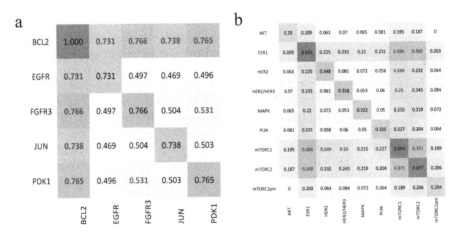

Fig. 4. The frequency of paired steering nodes. a. MAPK signaling network. b. ER+ breast cancer network.

MAPK Signaling Network. As shown in Fig. 4a, all the preferential nodes appear in at least 731 out of the 1000 control configurations, in which gene BCL2 appears in all control configurations. Then we choose the top5 high score combinations of genes {(BCL2, PDK1), (BCL2,FGFR3), (BCL2, JUN), (BCL2,EGFR), (FGFR3, PDK1)}, most of which contain the gene BCL2 which is a popular target to cure cancers [23]. In this study, we conduct a case study to evaluate whether the combination of genes could regulate the state of cancer cells. The venetoclax treatment [29] can not only inhibit the BCL2 family members to activate the apoptosis signaling but also regulate the PDK1 to affect cellular metabolism. The inhibition of EGFR using gefitinib can be strengthened to induce the apoptosis when the BCL2 is silent in non-small cell lung cancer in the meantime [30]. Hence, our proposed algorithm can potentially identify the combination of genes to the target cancer.

ER+ Breast Cancer Signaling Network. For this network, we can find nodes ESR1, mTORC1 and mTORC2 are with higher probability to be the steering nodes than other nodes in Fig. 4b. Then Top5 combinations of nodes {(mTORC1, ESR1), (mTORC1, mTORC2), (mTORC2, ESR1), (mTORC1, HER2/HER3), (mTORC2, HER2)} with high score are selected for the further analysis. After mapping to drugs, {(Everolimus), (Everolimus, Fulvestrant), (Everolimus, Neratinib)} can be found to cure breast cancer. Though Everolimus has been found to be a drug to treat breast cancer [31], the combination therapy is still a great challenge. Here using our algorithm, we can find two useful combinations. Recently, a randomized phase II trial test has proved that Everolimus can enhance the treatment of Fulvestrant [32]. Moreover, phase I study of the combination of Neratinib and Everolimus is ongoing [33], which indirectly indicate the potential of their synergistic function in treating advanced cancers.

6   Conclusions

In this paper, we have proposed a relaxed condition for output control and an associated algorithm to make it applicable for more general situations especially for the detection of drug targets. Besides the theoretical analysis of SPOC, we have compared our algorithm with two existing methods FC and OC. The results illustrate that our algorithm can do better than two existing methods from the integrated view of the number of steering nodes and its enrichment of drug targets. Furthermore, we find that our algorithm can provide a new avenue to identify the combination therapy through known treatments based on two case studies.

SPOC is a new framework based on the output control to analyze biological networks. However, the algorithm is determined by not only the definition of the state nodes which dominate the state transition but also the collection of the preferential nodes which contain the objectives. Hence, we need to get a deeper understanding of the biological problems so that we could use the available information to accurately apply our algorithm.

References

1. Barabási, A.L., et al.: Network Science. Cambridge University Press, Cambridge (2016)
2. Barabási, A.L., Gulbahce, N., Loscalzo, J.: Network medicine: a network-based approach to human disease. Nat. Rev. Genet. **12**(1), 56 (2011)
3. Jeong, H., Mason, S.P., Barabási, A.L., Oltvai, Z.N.: Lethality and centrality in protein networks. Nature **411**(6833), 41 (2001)
4. Winzeler, E.A., et al.: Functional characterization of the S. cerevisiae genome by gene deletion and parallel analysis. Science **285**(5429), 901–906 (1999)
5. Zeng, M., Zhang, F., Wu, F.X., Li, Y., Wang, J., Li, M.: Protein-protein interaction site prediction through combining local and global features with deep neural networks. Bioinformatics **36**(4), 1114–1120 (2020)

6. Li, M., Gao, H., Wang, J., Wu, F.X.: Control principles for complex biological networks. Brief. Bioinform. **20**(6), 2253–2266 (2019)

7. Guo, W.F., Zhang, S.W., Zeng, T., Akutsu, T., Chen, L.: Network control principles for identifying personalized driver genes in cancer. Brief. Bioinform. (2019)

8. Liu, Y.Y., Slotine, J.J., Barabási, A.L.: Controllability of complex networks. Nature **473**(7346), 167 (2011)

9. Wu, L., Shen, Y., Li, M., Wu, F.X.: Network output controllability-based method for drug target identification. IEEE Trans. Nanobiosci. **14**(2), 184–191 (2015)

10. Gao, J., Liu, Y.Y., D'souza, R.M., Barabási, A.L.: Target control of complex networks. Nat. Commun. **5**, 5415 (2014)

11. Wu, L., Li, M., Wang, J., Wu, F.X.: Minimum steering node set of complex networks and its applications to biomolecular networks. IET Syst. Biol. **10**(3), 116–123 (2016)

12. Guo, W.F., et al.: Discovering personalized driver mutation profiles of single samples in cancer by network control strategy. Bioinformatics **34**(11), 1893–1903 (2018)

13. Hu, Y., et al.: Optimal control nodes in disease-perturbed networks as targets for combination therapy. Nat. Commun. **10**(1), 2180 (2019)

14. Wu, L., Tang, L., Li, M., Wang, J., Wu, F.X.: Biomolecular network controllability with drug binding information. IEEE Trans. Nanobiosci. **16**(5), 326–332 (2017)

15. Guo, W.F., et al.: Constrained target controllability of complex networks. J. Stat. Mech: Theory Exp. **2017**(6), 063402 (2017)

16. Müller, F.J., Schuppert, A.: Few inputs can reprogram biological networks. Nature **478**(7369), E4 (2011)

17. Kailath, T.: Linear Systems, vol. 156. Prentice-Hall, Englewood Cliffs (1980)

18. Lin, C.T.: Structural controllability. IEEE Trans. Autom. Control **19**(3), 201–208 (1974)

19. Jungnickel, D.: Graphs, Networks and Algorithms. Springer, Heidelberg (2005). https://doi.org/10.1007/b138283

20. Crouse, D.F.: On implementing 2D rectangular assignment algorithms. IEEE Trans. Aerosp. Electron. Syst. **52**(4), 1679–1696 (2016)

21. Grieco, L., Calzone, L., Bernard-Pierrot, I., Radvanyi, F., Kahn-Perles, B., Thieffry, D.: Integrative modelling of the influence of mapk network on cancer cell fate decision. PLoS Comput. Biol. **9**(10), e1003286 (2013)

22. Wishart, D.S., et al.: Drugbank: a knowledgebase for drugs, drug actions and drug targets. Nucleic Acids Res. **36**(suppl–1), D901–D906 (2007)

23. Frenzel, A., Grespi, F., Chmelewskij, W., Villunger, A.: Bcl2 family proteins in carcinogenesis and the treatment of cancer. Apoptosis **14**(4), 584–596 (2009). https://doi.org/10.1007/s10495-008-0300-z

24. Lafitte, M., et al.: FGFR3 has tumor suppressor properties in cells with epithelial phenotype. Mol. Cancer **12**(1), 83 (2013)

25. Sigismund, S., Avanzato, D., Lanzetti, L.: Emerging functions of the EGFR in cancer. Mol. Oncol. **12**(1), 3–20 (2018)

26. Vleugel, M.M., Greijer, A.E., Bos, R., van der Wall, E., van Diest, P.J.: c-jun activation is associated with proliferation and angiogenesis in invasive breast cancer. Hum. Pathol. **37**(6), 668–674 (2006)

27. Raimondi, C., Falasca, M.: Targeting PDK1 in cancer. Curr. Med. Chem. **18**(18), 2763–2769 (2011)

28. Gómez Tejeda Zañudo, J., Scaltriti, M., Albert, R.: A network modeling approach to elucidate drug resistance mechanisms and predict combinatorial drug treatments in breast cancer. Cancer Converg. **1**(1), 1–25 (2017). https://doi.org/10.1186/s41236-017-0007-6

29. Eischen, C., Adams, C., Clark-Garvey, S., Porcu, P.: Targeting the Bcl-2 family in B cell lymphoma. Front. Oncol. **8**, 636 (2018)
30. Zou, M., et al.: Knockdown of the Bcl-2 gene increases sensitivity to EGFR tyrosine kinase inhibitors in the H1975 lung cancer cell line harboring T790m mutation. Int. J. Oncol. **42**(6), 2094–2102 (2013)
31. Royce, M.E., Osman, D.: Everolimus in the treatment of metastatic breast cancer. Breast Cancer: Basic Clin. Res. **9**, BCBCR–S29268 (2015)
32. Kornblum, N., et al.: Randomized phase ii trial of fulvestrant plus everolimus or placebo in postmenopausal women with hormone receptor-positive, human epidermal growth factor receptor 2-negative metastatic breast cancer resistant to aromatase inhibitor therapy: results of pre0102. ASCO (2018)
33. Piha-Paul, S.A., et al.: Phase i study of the pan-HER inhibitor neratinib given in combination with everolimus, palbociclib or trametinib in advanced cancer subjects with EGFR mutation/amplification, HER2 mutation/amplification or HER3/4 mutation (2018)

Identification of a Novel Compound Heterozygous Variant in *NBAS* Causing Bone Fragility by the Type of Osteogenesis Imperfecta

D. A. Petukhova[1]([✉]) [iD], E. E. Gurinova[2] [iD], A. L. Sukhomyasova[1,2] [iD],
and N. R. Maksimova[1] [iD]

[1] Research Laboratory "Molecular Medicine and Human Genetics", Medical Institute,
M. K. Ammosov North-Eastern Federal University, 677013 Yakutsk, Russia
petukhovadial@gmail.com
[2] Medical Genetic Center, Republican Hospital, №1 – «National Medical Center»,
677019 Yakutsk, Russia

Abstract. Biallelic mutations in the *NBAS* gene have been reported to cause three different clinical signs: short stature with optic nerve atrophy and Pelger-Huët anomaly (SOPH) syndrome, infantile liver failure syndrome 2 (ILFS2) and a combined severe phenotype including both SOPH and ILFS2 features. Here, we describe a case of a 6-year-old Yakut girl who presented with clinical signs of SOPH syndrome, acute liver failure (ALF) and bone fragility by the type of osteogenesis imperfecta (OI). Targeted panel sequencing for 494 genes of connective tissue diseases of the patient revealed that he carried novel compound heterozygous missense mutation in *NBAS*, c.2535G>T (p.Trp845Cys), c.5741G>A (p.Arg1914His). Mutation affect evolutionarily conserved amino acid residues and predicted to be highly damaging. Timely health care of patients with such a set of clinical spectrum of SOPH syndrome, ALF and bone fragility by the type of OI can contribute to establishment coordinated multispecialty management of the patient focusing on the health problems issues through childhood.

Keywords: *NBAS* · SOPH syndrome · Infantile liver failure syndrome 2 · Osteogenesis imperfecta

1 Introduction

NBAS contains 52 exons, spans 420 kb and is mapped to chromosome 2 p.24.3 [1]. Recent studies has shown that there are three main subgroups of phenotypes associated with biallelic variants of *NBAS* gene: 1) short stature, optic nerve atrophy and Pelger-Huët anomaly (SOPH; MIM# 614800), which involves skeletal, ocular and immune system in the absence of frank hepatic involvement [2]. The cause of this syndrome is the homozygous missense mutation in *NBAS* p.5741G>A (p.Arg1914His), which has only been reported in an isolated Russian Yakut population (31 families). Clinical features included postnatal growth retardation, senile face, reduced skin turgor and elasticity, osteoporosis, optic atrophy with loss of visual acuity and color vision, underlobulated

© Springer Nature Switzerland AG 2020
Z. Cai et al. (Eds.): ISBRA 2020, LNBI 12304, pp. 38–43, 2020.
https://doi.org/10.1007/978-3-030-57821-3_4

neutrophils, and normal intelligence; 2) infantile liver failure syndrome 2 (ILFS2; MIM# 616483), an autosomal recessive condition characterized by isolated hepatic involvement with recurrent episodes of acute liver failure (ALF) during intercurrent febrile illness. *Haack et al.* identified homozygous mutations in *NBAS* gene in a cohort of 11 individuals from ten German families with ILFS2 [3]. It has been proposed that the sensitivity to fever in these patients may be due to thermal susceptibility of the syntaxin complex 18; 3) Further reports have suggested a multi-system phenotype [4–7].

To date, combined severe phenotype including such of clinical spectrum as SOPH, ILFS2 and bone fragility have been reported in one case in two patients, whose parents was of North-European, Northern Spanish and Italian origin [8]. Osteogenesis imperfecta (OI) is a congenital disorder characterized by low bone mass and increased bone fragility, affects 1 in 15,000 live births. Early diagnosis is important, as therapeutic advances can lead to improved clinical outcome and patient benefit.

The NBAS protein is involved in Golgi-to-endoplasmic reticulum (ER) retrograde transport [7] and is considered to be a component of the SNAREs (Soluble N-Ethylmaleimide-sensitive Factor (NSF) Attachment Protein Receptors). Essentially every step of membrane transport is carried out by a pair of different SNARE proteins (v-SNARE and t-SNARE). The SNARE proteins mediate intracellular transport of vesicles, such as ER to Golgi and Golgi to plasma membranes and are conserved from yeast to human. The NBAS protein interacts with t-SNARE p31 directly and with other proteins, forming complex syntaxin 18 [9]. In the cells of patients, a reduction in NBAS is accompanied by a decrease in p31, supporting an important function for NBAS in the SNARE complex [3]. Although these intracellular events are known, the specific mechanisms by which NBAS contributes to liver disease, bone fragility is not fully understood.

Because early health care for patients with severe combined syndrome can effectively improve the course of liver disease and bone fragility with the *NBAS* mutation, it is important for differentially diagnose *NBAS* mutation-based disease.

Here, we describe a case of a 6-year-old Yakut girl who presented with clinical signs of SOPH syndrome, ILFS2 and bone fragility by the type of OI.

2   Materials and Methods

The present study was approved by the local ethics committee of the M.K. Ammosov North-Eastern Federal University (Yakutsk, Russia). Patient was recruited into a research to study bone fragility by the type of OI.

Deoxyribonucleic acid (DNA) extracted by phenol-chloroform extraction from peripheral blood of our patient with informed consent of his parents. Targeted panel sequencing for 494 genes of connective tissue diseases was performed on our patient. The average depth in the target region was $144.5\times$.

All variants were annotated using the dbSNP (http://www.ncbi.nlm.nih.gov/SNP/), HapMap (http://hapmap.ncbi.nlm.nih.gov/), HGMD (http://www.hgmd.org/), and the 1000 Genome (http://www.1000genomes.org/).

To predict the protein-damaging effects of the suspected pathogenic variations in *NBAS*, three different software programs were used: Polyphen-2 (http://genetics.bwh.har

vard.edu/pph2/), SIFT (https://sift.bii.a-star.edu.sg/), MutationTaster (http://www.mut ationtaster.org/).

Evolutionary conservation of the sequences and structures of the proteins and nucleic acids was assessed using MUSCLE Sequence Alignment Tool (https://www. ebi.ac.uk/Tools/msa/muscle/). Conservation across species indicates that a sequence has been maintained by evolution despite speciation. The human NBAS protein sequence was aligned with the following vertebrate species: *P.troglodytes*, *M.mulatta*, *C.lupus*, *B.taurus*, *M.musculus*, *G.gallus*, *X.tropicalis*, *D.rerio*. Homologous sequences are obtained from HomoloGene (https://www.ncbi.nlm.nih.gov/homologene/).

3   Results

3.1   Clinical Status

The patient was a 6-year-old girl born from non-consanguineous, healthy parents of Yakut origin. The pregnancy was threatened interruption. Delivery at term, physiological. Her birth weight was 2250 g and body height was 45 cm. Apgar scores were 8/8 points. On the 2nd day after birth, a fracture of the middle third of the right femur was diagnosed with a shift. Genealogy history of burden due to OI was not identified. Clinical examination of the patient at the age of 6 years showed that the patient has a dysplastic physique, with nanism (growth at the age of 2 years - 68 cm, at 4 years - 73 cm, at 6 years - 77 cm), defeat of the optic nerve, connective tissue by type of SOPH syndrome. Also observed developmental delay, ALF and immunodeficiency. According to X-ray, there is a shortening of the right thigh due to a consolidating fracture of the middle third of the diaphysis of the right femur, deformation and contracture of the left elbow joint.

3.2   Clinical Significance of Variants

For determination clinical significance of genetic variants was used a set of criteria. First, the allele frequency of variants should be less than 0.05 in dbSNP, Exome Aggregation Consortium (ExAC), 1000 Genomes Project, Exome Sequencing Project (ESP6500) databases. Second, variants were considered as disease mutation, if they were predicted as loss-of-function variants (stop-gain, frameshift, canonical splicing variants, insertion/deletion). Third, the variant was novel and forecasted to be damaging. If these set of criteria were met, then the mutation was considered a disease mutation.

The following *NBAS* variants were detected: c.5741G>A (p.Arg1914His) in 45 exon, and c.2535G>T (p.Trp845Cys) in 23 exon (Table 1). The variant c.2535G>T is absent in dbSNP, ExAC, 1000 Genomes Project, ESP6500 databases. This variant predicted harmful by SIFT (score 0.000), PolyPhen-2 (HumDiv score 1.000, HumVar score 0.998), MutationTaster (score 1.000). This information identified these two sites as pathogenic.

3.3   Mutation Analysis

Molecular genetic study revealed the known heterozygous mutation c.5741G>A (p.Arg1914His) in 45 exon of NBAS. This mutation is located in the C-terminal region

Table 1. *In silico* variant predictions

Variation	SIFT prediction	PolyPhen-2	MutationTaster	Region of NBAS
c.5741G>A (p.Arg1914His)	Damaging	Possibly damaging	Disease causing	C-terminal
c.2535G>T (p.Trp845Cys)	Damaging	Possibly damaging	Disease causing	Seq39

of NBAS protein and associated with SOPH syndrome on a homozygous state (Fig. 1A). Also, in 23 exon of *NBAS* was identified a previously unreported on a public databases missense heterozygous variant c.2535G>T (p.Trp845Cys). This variant is located in the region encoding the Sec39 domain of the NBAS. The Seq39 region is a necessary part of Golgi-ER retrograde transport and the mutation was located in this evolutionary conserved region. MUSCLE Sequence Alignment Tool shows that the Trp845 is highly conserved among vertebrates (Fig. 1B).

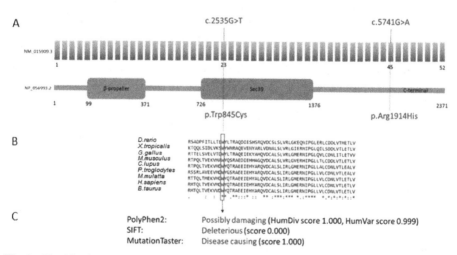

Fig. 1. Identification of a compound heterozygous mutation in *NBAS* gene in Yakut patient (A) The exon structure of the *NBAS* gene and a schematic representation of the NBAS protein with known domains are shown. (B) Multiple sequence alignment of the NBAS protein among vertebrates. (C) *In silico* single nucleotide variant predictions by PolyPhen-2, SIFT, MutationTaster bioinformatics tools.

4   Discussion

Recent advances technologies now provide rapid and cost-effective analysis of the causative mutations of human disorders. High throughput sequencing combined with capture techniques are increasingly used for routine diagnosis of Mendelian disease.

OI is a rare genetic disorder characterized by low bone mass, decreased bone strength, increased bone fragility, and shortened stature. Bone fragility has not been previously reported as a feature associated with variants in *NBAS*, except for the study of *Balasubramanian et al.* [5]. The specific mechanisms by which NBAS contributes to bone fragility, are not fully understood.

In this study, we have identified a novel compound heterozygous variant in NBAS (2535G>T (p.Trp845Cys); c.5741G>A (p.Arg1914His) by target NGS sequencing in 6-year-old Yakut girl with bone fragility by the type of OI.

Based on the localization of missense variants and in-frame deletions, *Staufner et al.* highlighted three clinical subgroups that differ significantly regarding main clinical features and are directly related to the affected region of the NBAS protein: β-propeller (combined phenotype), Sec39 (infantile liver failure syndrome type 2/ILFS2), and C-terminal (short stature, optic atrophy, and Pelger–Huët anomaly/SOPH) [10]. The high prevalence of various multisystemic symptoms associated with variants affecting the C-terminal part of NBAS point to important functions of this part of the protein, which so far has no defined domains and for which no reliable 3D modeling is currently possible. However, further studies are needed to improve our knowledge about the tertiary structure of NBAS, domain-specific protein functions, and potential sites of interaction.

Timely health care of patients with such a set of clinical spectrum of SOPH syndrome, ILFS2 and bone fragility can contribute to establishment coordinated multispecialty management of the patient focusing on the health problems issues through childhood.

This study increases the number of *NBAS* mutation associated with the disease and will facilitate further studies of the syndrome.

Acknowledgments. The authors are grateful to all participants of this study.

Funding
The study was supported by the Ministry Education and Science of Russian Federation (Project No. FSRG-2020-0014 "Genomics of Arctic: epidemiology, hereditary and pathology").

Conflict of interest
The authors declare that they have no competing interests.

References

1. Scott, D.K., Board, J.R., Lu, X., Pearson, A.D.J., Kenyon, R.M., Lunec, J.: The neuroblastoma amplified gene. Gene **307**, 1–11 (2003)
2. Maksimova, N., et al.: Neuroblastoma amplified sequence gene is associated with a novel short stature syndrome characterised by optic nerve atrophy and Pelger-Huet anomaly. J. Med. Genet. **47**, 538–548 (2010). https://doi.org/10.1136/jmg.2009.074815
3. Haack, T.B., et al.: Biallelic mutations in NBAS cause recurrent acute liver failure with onset in infancy. Am. J. Hum. Genet. **97**, 163–169 (2015). https://doi.org/10.1016/j.ajhg.2015.05.009
4. Kortüm, F., et al.: Acute liver failure meets SOPH syndrome: a case report on an intermediate phenotype. Pediatrics **139**, e20160550 (2017). https://doi.org/10.1542/peds.2016-0550
5. Segarra, N.G., et al.: NBAS mutations cause a multisystem disorder involving bone, connective tissue, liver, immune system, and retina. Am. J. Med. Genet. **167**, 2902–2912 (2015). https://doi.org/10.1002/ajmg.a.37338

6. Capo-Chichi, J.-M., et al.: Neuroblastoma Amplified Sequence (NBAS) mutation in recurrent acute liver failure: confirmatory report in a sibship with very early onset, osteoporosis and developmental delay. Eur. J. Med. Genet. **58**, 637–641 (2015). https://doi.org/10.1016/j.ejmg.2015.11.005

7. Staufner, C., et al.: Recurrent acute liver failure due to NBAS deficiency: phenotypic spectrum, disease mechanisms, and therapeutic concepts. J. Inherit. Metab. Dis. **39**(1), 3–16 (2015). https://doi.org/10.1007/s10545-015-9896-7

8. Balasubramanian, M., et al.: Compound heterozygous variants in NBAS as a cause of atypical osteogenesis imperfecta. Bone **94**, 65–74 (2017). https://doi.org/10.1016/j.bone.2016.10.023

9. Aoki, T., et al.: Identification of the neuroblastoma-amplified gene product as a component of the syntaxin 18 complex implicated in golgi-to-endoplasmic reticulum retrograde transport. MBoC **20**, 2639–2649 (2009). https://doi.org/10.1091/mbc.e08-11-1104

10. Staufner, C., et al.: Defining clinical subgroups and genotype–phenotype correlations in NBAS-associated disease across 110 patients. Genet. Med. **22**, 610–621 (2020). https://doi.org/10.1038/s41436-019-0698-4

Isoform-Disease Association Prediction by Data Fusion

Qiuyue Huang[1], Jun Wang[1], Xiangliang Zhang[2], and Guoxian Yu[1(✉)]

[1] College of Computer and Information Science, Southwest University,
Chongqing 400715, China
gxyu@swu.edu.cn
[2] CEMSE, King Abdullah University of Science and Technology,
Thuwal, Saudi Arabia

Abstract. Alternative splicing enables a gene spliced into different isoforms, which are closely related with diverse developmental abnormalities. Identifying the isoform-disease associations helps to uncover the underlying pathology of various complex diseases, and to develop precise treatments and drugs for these diseases. Although many approaches have been proposed for predicting gene-disease associations and isoform functions, few efforts have been made toward predicting isoform-disease associations in large-scale, the main bottleneck is the lack of ground-truth isoform-disease associations. To bridge this gap, we propose a multi-instance learning inspired computational approach called IDAPred to fuse genomics and transcriptomics data for isoform-disease association prediction. Given the bag-instance relationship between gene and its spliced isoforms, IDAPred introduces a dispatch and aggregation term to dispatch gene-disease associations to individual isoforms, and reversely aggregate these dispatched associations to affiliated genes. Next, it fuses different genomics and transcriptomics data to replenish gene-disease associations and to induce a linear classifier for predicting isoform-disease associations in a coherent way. In addition, to alleviate the bias toward observed gene-disease associations, it adds a regularization term to differentiate the currently observed associations from the unobserved (potential) ones. Experimental results show that IDAPred significantly outperforms the related state-of-the-art methods.

Keywords: Isoform-disease association · Alternative splicing · Data fusion · Multi-instance learning

1 Introduction

Deciphering human diseases and the pathology is one of key fundamental tasks in life science [4]. Thousands of genes have been identified as associated with a variety of diseases. Identifying gene-disease associations (GDA) contributes to decipher the pathology, which helps us to find new strategies and drugs to treat diverse complex diseases. Many computational solutions have been developed

© Springer Nature Switzerland AG 2020
Z. Cai et al. (Eds.): ISBRA 2020, LNBI 12304, pp. 44–55, 2020.
https://doi.org/10.1007/978-3-030-57821-3_5

to predict GDAs in large-scale, such as network propagation [32,35], literature mining [23], clustering analysis [31], data fusion [23], matrix completion [18], deep learning-based methods [16] and so on.

A single gene can produce multiple isoforms by alternative splicing, which greatly increases the transcriptome and proteome complexity [29]. More than 95% multi-exon genes in human genome undergo alternative splicing [20,33]. In practice, a gene can be associated with diverse diseases mainly owing to its abnormally spliced isoforms [29]. Increasing studies confirm that alternative splicing is associated with diverse complex diseases, such as autism spectrum disorders [28], ischemic human heart disease [19], and Alzheimer disease [10]. Neagoe *et al.* [19] observed that a titin isoform switch in chronically ischemic human hearts with 47:53 average N2BA-to-N2B ratio in severely diseased coronary artery disease transplanted hearts, and 32:68 in nonischemic transplants. Long-term titin modifications can damage the ability of the heart. Apolipoprotein E (apoE) is localized in the senile plaques, congophilic angiopathy, and neurofibrillary tangles of Alzheimer disease. Strittmatter *et al.* [30] compared the difference of binding of synthetic amyloid beta (beta/A4) peptide to apoE4 and apoE3, which are two common isoforms of apoE, and observed that apoE4 is associated with the increased susceptibility to disease. The results show that the pathogenesis of Alzheimer disease may be related to different bindings in apoE. Sanan *et al.* [24] observed the apoE4 isoform binds to a beta peptide more rapidly than apoE3. Holtzman *et al.* [10] found the expression of apoE3 and apoE4 in APPV717F transgenic (TG), no apoE mice resulted in fibrillar amyloid-β deposits and neuritic plaques by 15 months of age and substantially (>10-fold) more fibrillar deposits were observed in apoE4-expressing APPV717F TG mice. Lundberg *et al.* [15] demonstrated that FOXP3 in CD4+ T cells is associated with coronary artery disease and alternative splicing of FOXP3 is decreased in coronary artery disease.

Existing isoform-disease associations (IDAs) are mainly detected by wetlab experiments (*i.e.*, gel electrophoresis and immunoblotting). To the best of authors knowledge, there is *no computational solution* for predicting IDAs at a large-scale. The main bottleneck is that there is *no public database* that stores sufficient IDAs, which are required for typical machine learning methods to induce a reliable classifier for predicting IDAs. In fact, such lack also exists in functional analysis of isoforms [13]. To bypass this issue, some researchers take a gene as a bag and its spliced isoforms as instances of that bag, and adapt multiple instance learning (MIL) [2,17] to distribute the readily available functional annotations of a gene to its isoforms [3,6,14,26,34,40].

Based on the accumulated GDAs in public databases (*i.e.*, DisGeNET [22], OMIM (www.omim.org)) and inspired by the MIL-based isoform function prediction solutions, we kickoff a *novel* task of predicting IDAs, which is *more challenging* than traditional GDAs prediction, due to the lack of IDAs and the complex relationship between isoforms and genes. This task can provide a deeper understanding of the pathology of complex diseases. To combat this task, we introduce a computational solution (IDAPred) to predict IDAs in large scale by fusing genomic and transcriptome data and by distributing gene-disease associations to individual isoforms. IDAPred firstly introduces a dispatch and aggregation term to dispatch GDAs to individual isoforms and reversely aggregate these dispatched IDAs to affiliated genes based on the gene-isoform relations. To remedy incomplete GDAs, it fuses nucleic acid sequences and interactome of genes to further fulfil the to-be-dispatched GDAs. As well as that, it leverages multiple RNA-seq datasets to construct tissue-specific isoform co-expression networks and to induce a linear classifier to predict IDAs. In addition, it introduces an indicator matrix to differentiate the observed GDAs from the further fulfilled ones and thus to alleviate the bias toward observed ones. Finally, IDAPred merges these objectives into a unified objective function and predicts IDAs in a coherent way. Experimental results show that IDAPred achieves better results across various evaluation metrics than other competitive approaches that are introduced for predicting GDAs [32] or isoform functions [14,34,40].

2   Method

2.1   Materials and Pre-processing

Suppose there are n genes, the i-th gene produces $n_i \geq 1$ isoforms, and the total number of isoforms is $m = \sum_{i=1}^{n} n_i$. We adopt the widely-used Fragments Per Kilobase of exon per Million fragments mapped fragments (FPKM) values to quantify the expression of isoforms. Particularly, we downloaded 596 RNA-seq runs (of total 298 samples from different tissues and conditions) of Human from the ENCODE project [5] (access date: 2019-11-10). These datasets are heterogeneous in terms of library preparation procedures and sequencing platform. Following the pre-process done in [14,34], for each tissue, we control the quality of these RNA-seq datasets and quantify the expression value of isoforms as follows:

(i) We firstly align the short-reads of each RNA-seq dataset of the Human genome (build GRCh38.90) from Ensemble using HISAT2(v.2-2.1.0) [12], and A GTF annotation file of the same build with an option of no-novel-junction.

(ii) Then, we use Stringtie(v.1.3.3b) [21] to calculate the relative abundance of the transcript as Fragments Per Kilobase of exon per Million fragments mapped fragments (FPKM). We separately compute the FPKM values of a total of 57,964 genes with 219,288 isoforms for each sample.

(iii) The FPKM values of very short isoforms are exceptionally higher. Therefore, we discard the isoforms with less than 100 nucleotides.

(iv) To further control the quality of isoforms, we use known protein coding gene names to map those genes obtained in step (iii). Due to the prohibitive runtime on such a large number of isoforms and sufficient nonzero values in the expression vector are required to induce a predictor, we refilter the data. Particularly, we set all FPKM values lower than 0.3 as 0, and then remove isoform with all FPKM values of 0. To ensure data filtered at the gene level, we do a further filtering: if an isoform of a gene is filtered, this gene and its all spliced isoforms are removed also. Finally, we obtain 7,549 genes with 39,559 isoforms, whose values are stored in the corresponding data matrix $\mathbf{X} \in \mathbb{R}^{m \times d}$. We further normalize $\mathbf{X}$ by $\mathbf{X}_{nor} = \mathbf{X}./max(\mathbf{X})$. We use the normalized $\mathbf{X}$ for subsequent experiments.

We downloaded the gene-disease associations file and the mappings file UMLS CUI to Disease Ontology (DO) [25] vocabularies from DisGeNET [22]. Next, we directly use the available gene-disease associations and DO hierarchy to specify the gene-term association matrix $\mathbf{Y} \in \mathbb{R}^{n \times c}$ between n genes and c DO terms. Specifically, if a DO term s, or s's descendant terms are positively associated with gene i, then $\mathbf{Y}(i, s) = 1$. Otherwise, $\mathbf{Y}(i, s) = 0$.

We collected the gene interaction data from BioGrid (https://thebiogrid.org), which is a curated biological database of genetic interactions, chemical interactions, and post-translational modifications of gene products. Let $\mathbf{S}_{11}^{(v)} \in \mathbb{R}^{n \times n}$ encode the gene-level interaction, $\mathbf{S}_{11}^{(1)}(i, j) > 0$ if the gene i has a physical interaction with gene j, $\mathbf{S}_{11}^{(1)}(i, j) = 0$ otherwise, and the weight of $\mathbf{S}_{11}^{(1)}(i, j)$ is determined by the interaction strength. We collected the gene sequence data from NCBI (https://www.ncbi.nlm.nih.gov/). We adopted conjoint triad method [27] to represent nucleic acid sequences by numeric features and then adopted cosine similarity to construct another gene similarity network $\mathbf{S}_{11}^{(2)} \in \mathbb{R}^{n \times n}$.

2.2 Isoform-Disease Associations Prediction

Owing to the lack of DO annotations of isoforms, traditional supervised learning cannot be directly applied to predict IDAs. A bypass solution is to distribute the collected gene-level GDAs (stored in $\mathbf{Y}$) to individual isoforms spliced from the genes using the readily available gene-isoform relations (stored in $\mathbf{R}_{12} \in \mathbb{R}^{n \times m}$, $\mathbf{R}_{12}(i, j) = 1$ if isoform j is spliced from gene i, $\mathbf{R}_{12}(i, j) = 0$ otherwise). Suppose $\mathbf{Z} \in \mathbb{R}^{m \times c}$ stores the latent associations between m isoforms and c distinct DO terms. Following the MIL principle that the labels of a bag is responsible by at least one instance of this bag [2,17], a GDA should also be responsible by at least one isoform spliced from this gene. To concrete this principle, we define a dispatch and aggregation objective to push the gene-level associations to isoform-level and reversely aggregate the associations to gene-level in a compatible way as follows:

$$\mathbf{Y} = \mathbf{\Lambda} \mathbf{R}_{12} \mathbf{Z} \tag{1}$$

where $\mathbf{\Lambda} \in \mathbb{R}^{n \times n}$ is a diagonal matrix, $\mathbf{\Lambda}(i, i) = 1/n_i$, n_i represents the number of distinct isoforms spliced from the i-th gene. Given the known $\mathbf{Y}$, $\mathbf{\Lambda}$ and $\mathbf{R}_{12}$, we can optimize $\mathbf{Z}$ by minimizing $\|\mathbf{Y} - \mathbf{\Lambda} \mathbf{R}_{12} \mathbf{Z}\|_F^2$, and thus to predict the associations between m isoforms and c DO terms. Next, we can induce a linear predictor based on $\mathbf{Z}$ as follows:

$$\min \Omega(\mathbf{W}, \mathbf{Z}) = \|\mathbf{Z} - \mathbf{X} \mathbf{W}\|_F^2 + \|\mathbf{Y} - \mathbf{\Lambda} \mathbf{R}_{12} \mathbf{Z}\|_F^2 + \alpha \|\mathbf{W}\|_F^2 \tag{2}$$

where $\mathbf{W} \in \mathbb{R}^{d \times c}$ is the coefficient matrix for the linear predictor, which maps the numeric expression features $\mathbf{X}$ of isoforms onto c distinct DO terms. The Frobenius norm and scale parameter α are added to control the complexity of linear predictor.

The above equation can simultaneously distribute GDAs to individual isoforms and induce a classifier to predict IDAs. However, it ignores the important genomics data, which carry important information to boost the performance of isoform function prediction and to identify the genetic determinants of disease [3,37]. Similarly, the incorporation of genomic data can also improve the performance of predicting IDAs. Furthermore, the collected GDAs are still incomplete. As a consequence, the distributed IDAs are also not sufficient to induce a reliable predictor and the predictor may be mislead by the collected GDAs, which are imbalanced and biased by the research interests of the community [8,9]. To alleviate these issues, we replenish GDAs by fusing gene-gene interactions and nucleic acid sequence data, and update the above equation as follows:

$$
\begin{aligned}
\min \Omega(\mathbf{W}, \mathbf{Z}, \mathbf{F}) = {} & \|\mathbf{Z} - \mathbf{X}\mathbf{W}\|_F^2 + \alpha \|\mathbf{W}\|_F^2 + \|\mathbf{F} - \mathbf{\Lambda}\mathbf{R}_{12}\mathbf{Z}\|_F^2 + \|\mathbf{H} \odot (\mathbf{F} - \mathbf{Y})\|_F^2 \\
& + \frac{1}{2V_n} \sum_{v=1}^{V_n} \sum_{i,j=1}^{n} \|\mathbf{F}(i,\cdot) - \mathbf{F}(j,\cdot)\|_F^2 \mathbf{S}_n^{(v)}(i,j) \\
= {} & \|\mathbf{Z} - \mathbf{X}\mathbf{W}\|_F^2 + \alpha \|\mathbf{W}\|_F^2 + \|\mathbf{F} - \mathbf{\Lambda}\mathbf{R}_{12}\mathbf{Z}\|_F^2 + \|\mathbf{H} \odot (\mathbf{F} - \mathbf{Y})\|_F^2 \\
& + \frac{1}{V_n} \sum_{v=1}^{V_n} tr(\mathbf{F}^T \mathbf{L}_n^{(v)} \mathbf{F}))
\end{aligned}
\tag{3}
$$

where $\mathbf{F} \in \mathbb{R}^{n \times c}$ stores the latent IDAs between n genes and c DO terms. $\mathbf{H} = \mathbf{Y}$, $\odot$ means the element-wise multiplication. $\|\mathbf{H} \odot (\mathbf{F} - \mathbf{Y})\|_F^2$ is introduced to enforce latent IDAs being consistent with the collected ones and also to differentiate the observed ones from latent ones, and thus to reduce the bias toward observed ones. $\frac{1}{V_n} \sum_{v=1}^{V_n} tr(\mathbf{F}^T \mathbf{L}_n^{(v)} \mathbf{F}))$ is introduced to replenish IDAs by fusing diverse gene-level data, and V_n is the number of genomic data sources. Here, we specify the elements of $\mathbf{S}_n^{(v)}$ using the gene interaction network and nucleic acid sequences (as stated in the data preprocess subsection). $\mathbf{L}_n^{(v)} = \mathbf{D}_n^{(v)} - \mathbf{S}_n^{(v)}$, $\mathbf{D}_n^{(v)}$ is a diagonal matrix with $\mathbf{D}_n^{(v)}(i,i) = \sum_{j=1}^{n} \mathbf{S}_n^{(v)}(i,j)$.

The co-expression pattern of isoforms also carry important information about the functions of isoforms [3,40], whose usage also boosts the prediction of IDAs. In addition, the expression of isoforms has tissue specificity [7,38]. To make use of tissue-specific co-expression patterns, we update the objective function of IDAPred as follows:

$$
\begin{aligned}
\min \Omega(\mathbf{W}, \mathbf{Z}, \mathbf{F}) = {} & \|\mathbf{Z} - \mathbf{X}\mathbf{W}\|_F^2 + \frac{1}{V_m} \sum_{v=1}^{V_m} tr(\mathbf{Z}^T \mathbf{L}_m^{(v)} \mathbf{Z}) + \alpha \|\mathbf{W}\|_F^2 \\
& + \beta(\|\mathbf{F} - \mathbf{\Lambda}\mathbf{R}_{12}\mathbf{Z}\|_F^2 + \frac{1}{V_n} \sum_{v=1}^{V_n} tr(\mathbf{F}^T \mathbf{L}_n^{(v)} \mathbf{F}) + \|\mathbf{H} \odot (\mathbf{F} - \mathbf{Y})\|_F^2)
\end{aligned}
\tag{4}
$$

where V_m counts the number of tissues that are used to construct the expression profile feature vectors of m isoforms in $\mathbf{X}$, $\mathbf{L}_m^{(v)} = \mathbf{D}_m^{(v)} - \mathbf{S}_m^{(v)}$, and $\mathbf{S}_m^{(v)} \in \mathbb{R}^{m \times m}$ encodes the co-expression patterns of m isoforms from the v-th tissue. $\mathbf{D}_m^{(v)}$ is a diagonal matrix with $\mathbf{D}_m^{(v)}(i,i) = \sum_{j=1}^{m} \mathbf{S}_m^{(v)}(i,j)$. β is introduced to balance the information sources from the gene-level and isoform-level.

The optimization problem in Eq. (4) is non-convex with respect to $\mathbf{W}$, $\mathbf{Z}$ and $\mathbf{F}$ altogether. It is difficult to seek the global optimal solutions for them at the same time. We follow the idea of alternating direction method of multipliers (ADMM) [1] to alternatively optimize one variable by fixing the other two variables in an iterative way. IDAPred often converges in 60 iterations on our used datasets. The optimization detail is omitted here for page limit.

3 Experiment Results and Analysis

3.1 Experimental Setup

To assess the performance of IDAPred for predicting IDAs, we collect multiple RNA-Seq datasets from ENCODE project, gene-disease associations data from DisGeNET, gene interaction data from BioGrid, sequence data of genes from NCBI. We only consider the genes within all the four types of data for experiments. The pre-processed GDAs and isoforms of the genes are listed in Table 1.

Table 1. Statistics of isoforms and collected GDAs. 'associations' is the number of GDAs for experiment.

genes (n)	isoforms (m)	terms (c)	associations
2,482	14,484	2,949	73,515

To comparatively study the performance of IDAPred, we take the state-of-the-art isoform function prediction methods (iMILP [14], IsoFun [40], Disofun [34]) and gene-disease association prediction method (PRINCE [32]) as comparing methods. The input parameters of these comparing methods are fixed/optimized as the original papers or shared codes. For IDAPred, we choose α and β in $\{10^{-4}, 10^{-3}, \ldots, 10^3, 10^4\}$. Due to the lack of IDAs, we surrogate the evaluation by aggregating the predicted IDAs to affiliated genes, this approximate evaluation was also adopted in isoform function prediction [14,40]. In addition, we further compare IDAPred against its degenerated variants to further study the contribution components of IDAPred.

The task of predicting IDAs can be evaluated alike gene function prediction [11,39], and multi-instance multi-label learning by taking each gene as bag, the spliced isoforms as instances and associated diseases (DO terms) as distinct labels [36,41]. Given that, we adopt five evaluation metrics $MicroF1$, $MacroF1$,

$1 - RankLoss$, $Fmax$ and $AUPRC$, which are widely-used in gene function prediction and multi-label learning. $MicroF1$ computes the F1-score on the predictions of different DO terms as a whole; $MacroF1$ calculates the F1-score of each term, and then takes the average value across all DO terms; $RankLoss$ computes the average fraction of incorrectly predicted associations ranking ahead of the ground-truth associations. $Fmax$ is the global maximum harmonic mean of recall and precision across all possible thresholds. $AUPRC$ calculates the area under the precision-recall curve of each term, and then computes the average value of these areas as the overall performance. The higher the value of $MicroF1$, $MacroF1$, $1 - RankLoss$, $Fmax$ and $AUPRC$, the better the performance is. We want to remark that these five metrics quantify the prediction results from different aspects, and it is difficult for one method to always outperform another one across all these metrics.

3.2   Results Evaluation at Gene-Level

We adopt five-fold cross-validation at the gene-level for experiment. For each test fold, we randomly initialize the test part of $\mathbf{F}$ and $\mathbf{Y}$ in Eq. (4). We initialize the isoform-term association matrix $\mathbf{Z}$ by the gene-term association matrix $\mathbf{F}$, say all the diseases associated with a gene are also initialized as temporarily associated with its spliced isoforms. The GDAs in the validation set are considered as unknown during training and prediction, and only used for validation. Table 2 reports the results of IDAPred and of compared methods.

Table 2. Experimental results of five-fold cross-validation. ●/○ indicates IDAPred performing significantly better/worse than the other comparing method, with significance assessed by pairwise t-test at 95% level.

	PRINCE	iMILP	IsoFun	Disofun	IDAPred
MicroF1	0.3122 ± 0.0274●	0.2349 ± 0.0273●	0.2829 ± 0.0195●	0.3306 ± 0.0092●	0.8248 ± 0.0118
MacroF1	0.2863 ± 0.0341●	0.0645 ± 0.0269●	0.1232 ± 0.0254●	0.0398 ± 0.0046●	0.4250 ± 0.0241
1-RankLoss	0.8591 ± 0.0434●	0.0836 ± 0.0473●	0.6877 ± 0.0536●	0.8699 ± 0.0016●	0.9966 ± 0.0003
Fmax	0.3281 ± 0.0097●	0.1559 ± 0.0750●	0.2140 ± 0.0143●	0.2250 ± 0.0109●	0.6795 ± 0.0068
AUPRC	0.3596 ± 0.0025●	0.0092 ± 0.0051●	0.0413 ± 0.0031●	0.0430 ± 0.0049●	0.4782 ± 0.0067

IDAPred gives significantly better results than the compared methods across all the five evaluation metrics. $MicroF1$, $MacroF1$ and $AUPRC$ are disease term-centric metrics, while $1 - Rankloss$ and $Fmax$ are gene-centric metrics. The significant improvement shows that IDAPred can more reliably predict the GDAs (IDAs) from both the gene (isoforms) and DO term perspectives. Three factors contribute to this improvement. (i) IDAPred fuses the gene sequence and interaction data to complete GDAs, along with the isoform expression data, while these compared methods either use only the interaction data and/or the expression data. (ii) IDAPred accounts for tissue specificity and fuses co-expression networks of different tissues, while IsoFun and Disofun concatenate the expression profiles of different tissues into a single feature vector and then construct a single co-expression network; as a result, they do not make use of the important tissue specificity patterns of alternative splicing. (iii) IDAPred models the incompleteness of the gene-term associations and introduces the indicator matrix $\mathbf{H}$ to enforce latent IDAs being consistent with the collected ones, and to differentiate the observed ones from latent ones.

PRINCE directly predicts GDAs based on the topology of gene interaction networks, and it outperforms most comparing methods (except our proposed IDAPred). One explanation is that the evaluation is approximately made at the gene-level, not the targeted isoform-level, and these compared methods more focus on using the transcriptomics expression data. Last but not least, we want to remark that IDAPred is an inductive approach that can directly predict the associations between diseases and a new isoform, whereas these compared methods can only work in transductive setting, they have to include this isoform for retraining the model and then to make the prediction.

Overall, these comparisons indirectly prove the effectiveness of IDAPred in predicting the associations between isoforms and diseases.

3.3 Further Analysis

Ablation Study. To further study the contribution components, we introduce five variants of IDAPred, which are IDAPred(L), IDAPred(P), IDAPred(S), IDAPred(A) and IDAPred(H). IDAPred(L) removes the $\frac{1}{V_n}\sum_{v=1}^{V_n} tr(\mathbf{F}^T\mathbf{L}_n^{(v)}\mathbf{F})$ in Eq. (4), namely both the gene sequence and interaction data are excluded; IDAPred(P) only uses the gene interaction data; IDAPred(S) only utilizes the gene sequence data; IDAPred(A) concatenates the isoform expression profile feature vectors of different tissues into a single one, and then directly constructs a single isoform co-expression network using cosine similarity also. IDAPred(H) removes the indicator $\mathbf{H}$ in $\|\mathbf{H} \odot (\mathbf{F} - \mathbf{Y})\|_F^2$ in Eq. (4), say it does not consider the bias toward the observed GDAs. Figure 1 reports the performance results of IDAPred and of its variants. The experimental settings are the same as the evaluation at the gene-level.

It is easy to observe that IDAPred manifests the highest performance among its variants. IDAPred(L) has much lower performance values than IDAPred. This fact corroborates our assumption that the observed GDAs are incomplete, and also the contribution of fusing gene interaction and sequence data to complete the GDAs, which then improve the prediction of IDAs. IDAPred(P) and IDAPred(S) manifest better results than IDAPred(L), but they both are outperformed by IDAPred. This comparison not only shows that gene interaction network data and gene sequence data can help to replenish GDAs, but also expresses the joint benefit of fusing gene interaction and sequence data. IDAPred(P) has a lower performance than IDAPred(S), this facts the gene sequence data is more positively related with the isoform/gene-disease associations than the incomplete

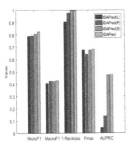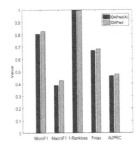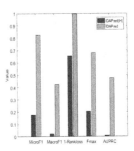

Fig. 1. Performance results of IDAPred and its variants, which fuse fewer data or do not alleviate the bias toward observed GDAs.

gene interaction data. IDAPred(A) also loses to IDAPred, which proves the necessity of combining isoform co-expression patterns from tissue-wise, instead from sample-wise. There is a big performance margin between IDAPred and IDAPred(H), which expresses the importance to explicitly account for the incompleteness of observed GDAs and to alleviate the bias toward observed GDAs, which is overlooked by most compared methods.

In summary, the ablation study also confirms the effectiveness of our unified objective function in fusing genomics and trascriptomics data, and in handling the difficulty of predicting IDAs.

Parameter Sensitivity Analysis. There are two input parameters (α and β) involved with IDAPred. α controls the complexity of linear predictor, and β balances the information sources from the gene-level and isoform-level. We vary α and β in the grid of $\{10^{-4}, 10^{-3}, \cdots, 10^3, 10^4\}$, and visualize the results of IDAPred under different combinations of α and β in Fig. 2.

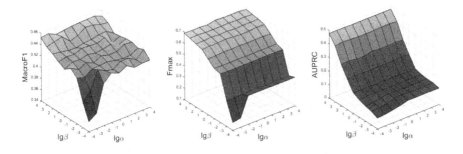

Fig. 2. Performance results vs. α and β.

We observe that IDAPred firstly has a clearly increased performance as α growing from 10^{-4} to 10^{-2}, and then holds a relatively stable performance as α further growing. As β growing from 10^{-4} to 10^{-1}, IDAPred also shows a sharply increased performance trend, and a slowly increased trend as β further growing from 10^{-1} to 10^4. This trend again confirms that the gene-level data should be leveraged for predicting IDAs. We also find β playing more important role than α. That is because α only controls the complexity of predictor, while the complexity is also inherently controlled by the simple linear classifier. When both α and β are fixed with too small values, IDAPred has the lowest performance. This observation again expresses the effectiveness of the unified objective function for handling the difficulty of predicting IDAs. Based on these results, we adopt $\alpha = 10^{-2}$ and $\beta = 10^4$ for experiments.

4   Conclusion

In this paper, we proposed an approach called IDAPred to computationally predict isoform-disease associations by data fusion. IDAPred makes use of multi-instance learning to bypass the lack of the ground-truth isoform-disease associations and to push gene-disease associations onto individual isoforms. It fuses nucleic acid sequences and interactome of genes to further fulfil the incomplete GDAs. In addition, it leverages multiple RNA-seq datasets to construct tissue-specific isoform co-expression networks and to induce a linear classifier to predict IDAs. Experimental results show that IDAPred significantly outperforms related comparing methods, which target to identify gene-disease associations or isoform functions.

Acknowledgements. This research is supported by NSFC (61872300), Fundamental Research Funds for the Central Universities (XDJK2019B024 and XDJK2020B028), Natural Science Foundation of CQ CSTC (cstc2018jcyjAX0228).

References

1. Boyd, S., Vandenberghe, L.: Convex Optimization. Cambridge University Press, Cambridge (2004)
2. Carbonneau, M.A., Cheplygina, V., Granger, E., Gagnon, G.: Multiple instance learning: a survey of problem characteristics and applications. Pattern Recogn. **77**, 329–353 (2018)
3. Chen, H., Shaw, D., Zeng, J., Bu, D., Jiang, T.: Diffuse: predicting isoform functions from sequences and expression profiles via deep learning. Bioinformatics **35**(14), i284–i294 (2019)
4. Claussnitzer, M., et al.: A brief history of human disease genetics. Nature **577**(7789), 179–189 (2020)
5. Consortium, E.P., et al.: An integrated encyclopedia of DNA elements in the human genome. Nature **489**(7414), 57 (2012)
6. Eksi, R., et al.: Systematically differentiating functions for alternatively spliced isoforms through integrating rna-seq data. PLoS Comput. Biol. **9**(11), e1003314 (2013)
7. Ellis, J.D., et al.: Tissue-specific alternative splicing remodels protein-protein interaction networks. Mol. Cell **46**(6), 884–892 (2012)

8. Gaudet, P., Dessimoz, C.: Gene ontology: pitfalls, biases, and remedies. In: The Gene Ontology Handbook, pp. 189–205. Humana Press, New York (2017)
9. Holman, L., Head, M.L., Lanfear, R., Jennions, M.D.: Evidence of experimental bias in the life sciences: why we need blind data recording. PLoS Biol. **13**(7), e1002190 (2015)
10. Holtzman, D.M., et al.: Apolipoprotein E isoform-dependent amyloid deposition and neuritic degeneration in a mouse model of Alzheimer's disease. Proc. Nat. Acad. Sci. **97**(6), 2892–2897 (2000)
11. Jiang, Y., et al.: An expanded evaluation of protein function prediction methods shows an improvement in accuracy. Genome Biol. **17**(1), 184 (2016)
12. Kim, D., Langmead, B., Salzberg, S.L.: HISAT: a fast spliced aligner with low memory requirements. Nat. Methods **12**(4), 357 (2015)
13. Li, H.D., Menon, R., Omenn, G.S., Guan, Y.: The emerging era of genomic data integration for analyzing splice isoform function. Trends Genet. **30**(8), 340–347 (2014)
14. Li, W., et al.: High-resolution functional annotation of human transcriptome: predicting isoform functions by a novel multiple instance-based label propagation method. Nucleic Acids Res. **42**(6), e39–e39 (2014)
15. Lundberg, A.K., Jonasson, L., Hansson, G.K., Mailer, R.K.: Activation-induced FOXP3 isoform profile in peripheral CD4+ T cells is associated with coronary artery disease. Atherosclerosis **267**, 27–33 (2017)
16. Luo, P., Li, Y., Tian, L.P., Wu, F.X.: Enhancing the prediction of disease-gene associations with multimodal deep learning. Bioinformatics **35**(19), 3735–3742 (2019)
17. Maron, O., Lozano-Pérez, T.: A framework for multiple-instance learning. In: NeurIPS, pp. 570–576 (1998)
18. Natarajan, N., Dhillon, I.S.: Inductive matrix completion for predicting gene-disease associations. Bioinformatics **30**(12), i60–i68 (2014)
19. Neagoe, C., et al.: Titin isoform switch in ischemic human heart disease. Circulation **106**(11), 1333–1341 (2002)
20. Pan, Q., Shai, O., Lee, L.J., Frey, B.J., Blencowe, B.J.: Deep surveying of alternative splicing complexity in the human transcriptome by high-throughput sequencing. Nat. Genet. **40**(12), 1413 (2008)
21. Pertea, M., Pertea, G.M., Antonescu, C.M., Chang, T.C., Mendell, J.T., Salzberg, S.L.: Stringtie enables improved reconstruction of a transcriptome from RNA-seq reads. Nat. Biotechnol. **33**(3), 290 (2015)
22. Piñero, J., et al.: The disgenet knowledge platform for disease genomics: 2019 update. Nucleic Acids Res. **48**(D1), D845–D855 (2020)
23. Pletscher-Frankild, S., Pallejà, A., Tsafou, K., Binder, J.X., Jensen, L.J.: Diseases: text mining and data integration of disease-gene associations. Methods **74**, 83–89 (2015)
24. Sanan, D.A., et al.: Apolipoprotein E associates with beta amyloid peptide of Alzheimer's disease to form novel monofibrils. isoform apoE4 associates more efficiently than apoE3. J. Clin. Invest. **94**(2), 860–869 (1994)
25. Schriml, L.M., et al.: Disease ontology: a backbone for disease semantic integration. Nucleic Acids Res. **40**(D1), D940–D946 (2012)
26. Shaw, D., Chen, H., Jiang, T.: Deepisofun: a deep domain adaptation approach to predict isoform functions. Bioinformatics **35**(15), 2535–2544 (2019)
27. Shen, J., et al.: Predicting protein-protein interactions based only on sequences information. Proc. Nat. Acad. Sci. **104**(11), 4337–4341 (2007)
28. Skotheim, R.I., Nees, M.: Alternative splicing in cancer: noise, functional, or systematic? Int. J. Biochem. Cell Biol. **39**(7–8), 1432–1449 (2007)

29. Smith, L.M., Kelleher, N.L.: Proteoforms as the next proteomics currency. Science **359**(6380), 1106–1107 (2018)
30. Strittmatter, W.J., et al.: Binding of human apolipoprotein E to synthetic amyloid beta peptide: isoform-specific effects and implications for late-onset Alzheimer disease. Proc. Nat. Acad. Sci. **90**(17), 8098–8102 (1993)
31. Sun, P.G., Gao, L., Han, S.: Prediction of human disease-related gene clusters by clustering analysis. Int. J. Biol. Sci. **7**(1), 61 (2011)
32. Vanunu, O., Magger, O., Ruppin, E., Shlomi, T., Sharan, R.: Associating genes and protein complexes with disease via network propagation. PLoS Comput. Biol. **6**(1), e1000641 (2010)
33. Wang, E.T., et al.: Alternative isoform regulation in human tissue transcriptomes. Nature **456**(7221), 470 (2008)
34. Wang, K., Wang, J., Domeniconi, C., Zhang, X., Yu, G.: Differentiating isoform functions with collaborative matrix factorization. Bioinformatics **36**(6), 1864–1871 (2020)
35. Wang, X., Gulbahce, N., Yu, H.: Network-based methods for human disease gene prediction. Brief. Funct. Genomics **10**(5), 280–293 (2011)
36. Xing, Y., Yu, G., Domeniconi, C., Wang, J., Zhang, Z., Guo, M.: Multi-view multi-instance multi-label learning based on collaborative matrix factorization. In: AAAI, pp. 5508–5515 (2019)
37. Xiong, H.Y., et al.: The human splicing code reveals new insights into the genetic determinants of disease. Science **347**(6218), 1254806 (2015)
38. Yeo, G., Holste, D., Kreiman, G., Burge, C.B.: Variation in alternative splicing across human tissues. Genome Biol. **5**(10), R74 (2004). https://doi.org/10.1186/gb-2004-5-10-r74
39. Yu, G., Rangwala, H., Domeniconi, C., Zhang, G., Yu, Z.: Protein function prediction using multilabel ensemble classification. IEEE/ACM Trans. Comput. Biol. Bioinf. **10**(4), 1045–1057 (2013)
40. Yu, G., Wang, K., Domeniconi, C., Guo, M., Wang, J.: Isoform function prediction based on bi-random walks on a heterogeneous network. Bioinformatics **36**(1), 303–310 (2020)
41. Zhou, Z.H., Zhang, M.L., Huang, S.J., Li, Y.F.: Multi-instance multi-label learning. Artif. Intell. **176**(1), 2291–2320 (2012)

EpIntMC: Detecting Epistatic Interactions Using Multiple Clusterings

Huiling Zhang[1], Guoxian Yu[1], Wei Ren[1], Maozu Guo[2], and Jun Wang[1(✉)]

[1] College of Computer and Information Science,
Southwest University,
Chongqing 400715, China
kingjun@swu.edu.cn
[2] College of Electronics and Information Engineering,
Beijing University of Civil Engineering and Architecture,
Beijing 100044, China

Abstract. Detecting epistatic interaction between multiple single nucleotide polymorphisms (SNPs) is crucial to identify susceptibility genes associated with complex human diseases. Stepwise search approaches have been extensively studied to greatly reduce the search space for follow-up SNP interactions detection. However, most of these stepwise methods are prone to filter out significant polymorphism combinations and thus have a low detection power. In this paper, we propose a two-stage approach called EpIntMC, which uses multiple clusterings to significantly shrink the search space and reduce the risk of filtering out significant combinations for the follow-up detection. EpIntMC firstly introduces a matrix factorization based approach to generate multiple diverse clusterings to group SNPs into different clusters from different aspects, which helps to more comprehensively explore the genotype data and reduce the chance of filtering out potential candidates overlooked by a single clustering. In the search stage, EpIntMC applies Entropy score to screen SNPs in each cluster, and uses Jaccard similarity to merge the most similar clusters into candidate sets. After that, EpIntMC uses exhaustive search on these candidate sets to precisely detect epsitatic interactions. Extensive simulation experiments show that EpIntMC has a higher (comparable) power than related competitive solutions, and results on Wellcome Trust Case Control Consortium (WTCCC) dataset also expresses its effectiveness.

Keywords: Genome-wide association study · Epistatic interactions · Stepwise search · Multiple clusterings · Matrix factorization

1 Introduction

Genome-wide association study (GWAS) measures and analyzes DNA sequence variations from genomes, with an effort to detect hundreds of single-nucleotide polymorphisms (SNPs) associated with complex diseases [30]. In genetics, SNPs

© Springer Nature Switzerland AG 2020
Z. Cai et al. (Eds.): ISBRA 2020, LNBI 12304, pp. 56–67, 2020.
https://doi.org/10.1007/978-3-030-57821-3_6

are notably a type of common genetic variation, and multiple SNPs are now believed to affect individual susceptibility to complex diseases. Various joint effects of genetic variants are often called as epistasis or epistatic interactions [18]. Nevertheless, many efforts focus on single factors, which cannot completely explain the pathology of complex diseases [20]. Single locus-based methods ignore these interactions. As a consequence, two (or multi)-locus SNPs of epistasis detection has been a critical demand.

There are twofold challenges in epistasis detection. The first is statistic. Traditional statistic tests used in univariate SNP-phenotype associations are inadequate to find epistasis. The second challenge is the heavy computational burden. The overall complexity is linear when detecting single loci, but it becomes exponential when the order increases [21]. An exhaustive search of epistatic interactions of order ≥ 3 would lead to the 'curse of dimensionality' or 'combinatorial explosion'. To combat with the first challenge, some statistic tests, such as likelihood test and chi-squared test [4], have been developed to quantify the association effect. To handle the second challenge, computationally efficient and/or memory-saving algorithms have been proposed [17,25].

The computational methods of detecting epistasis can be divided into three categories: exhaustive, heuristic and stepwise search-based ones. Exhaustive search-based approaches usually use the Chi-squared test, logistic regression and other traditional statistical analysis methods to evaluate all possible combinations of SNPs. Ritchie *et al.* [22] proposed a nonparametric and model-free method named Multifactor-Dimensionality Reduction (MDR). With MDR, all multi-locus genotypes are pooled into high-risk and low-risk groups to actualize 'Dimension Reduction'. Heuristic search methods are often occurred in ant colony optimization (ACO) based algorithms. Wang *et al.* proposed AntEpiseeker [28] which uses two-stage design on ant colony optimization to better enhance the power of ACO algorithms. Stepwise search methods carefully pre-select SNP subsets or pairwise interaction candidates before the time-consuming detection. HiSeeker [16] employs Chi-squared test and logistic regression model to obtain candidate pairwise combinations in the screening stage, and then detects high-order interactions on candidate combinations. ClusterMI (Clustering combined with Mutual Information) [7] utilizes mutual information and conditional mutual information to group and screen significant pairwise SNP combinations in each cluster, and then use the exhaustive (or improved heuristic) search to obtain a high detection accuracy. DualWMDR [8] combines a dual screening strategy with a MDR based weighted classification evaluation to detect epistasis.

However, these aforementioned methods still have several limitations. Most of methods are designed to detect only pairwise interactions and cannot afford the higher-order interaction search space. For the large genome-wide data, screening the candidate set from all pairwise SNP combinations is computationally overwhelming. In addition, most single-clustering based epistasis detection approaches usually just utilize a single clustering result to obtain significant SNP combinations, which may reduce the coverage of SNP combinations and suffer the risk of filtering out too many significant combinations.

In this paper, we propose a two-stage approach named EpIntMC (Epistatic Interactions detection based on Multiple Clusterings) to more precisely detect SNP combinations associated with disease risk. In the first stage, EpIntMC introduces matrix factorization based multiple clusterings algorithm to generate diverse clusterings, each of which divide SNPs into different clusters. This stage greatly reduces the search space and groups associated SNPs together from different perspectives. In the search stage, EpIntMC applies Entropy score to select high-quality SNPs in each cluster, then uses Jaccard similarity to merge the most similar clusters into candidate sets, which have strongly associated SNPs but with much smaller sizes. After that, EpIntMC uses exhaustive search on candidate sets and report SNP combinations with the highest entropy scores as the detected interactions. Extensive experiments on simulated datasets show that EpIntMC is more powerful in detecting epistatic interactions than competitive methods (EDCF [31], DCHE [12], DECMDR [32], MOMDR [33], and ClusterMI [7]). Experiments on real Wellcome Trust Case Control Consortium (WTCCC) dataset also show that EpIntMC is feasible for identifying high-order SNP interactions on large GWAS data.

2   Materials and Methods

Suppose a genome-wide case-control dateset $\mathbf{X} \in \mathbb{R}^{d \times n}$ with d SNPs and n samples. Let n_0 and n_1 denote the number of controls (i.e. normal individuals) and of cases (i.e. disease individuals), respectively. We use uppercase letters (A, B) to denote major alleles and lowercase letters (a, b) to denote minor alleles. For three genotypes, we use 0, 1, 2 to represent the homozygous reference genotype (AA), heterozygous genotype (Aa), homozygous variant genotype (aa), respectively. EpIntMC is a stepwise approach. In the first stage, multiple clusterings algorithm is introduced to divide SNPs into different clusters. In the search stage, EpIntMC applies Entropy score and Jaccard similarity to obtain candidate sets. After that, EpIntMC utilizes the exhaustive search on candidate sets to detect epistatic interactions. The next two subsections elaborate on these two stages.

2.1   Stage 1: Select SNPs by Multiple Clusterings

Clustering is one of the popular techniques in detecting epistasis [7,8,12,31], it attempts to divide SNPs into different groups based on the similarity between them and filter out non-significant ones. In this way, it reduces the search space for the follow-up epistasis detection. However, existing clustering-based epistasis detection approaches all only use a single clustering result to filter out 'non-significant' SNP combinations, which may be significant ones from another perspective [7,8,12,31]. In addition, they group SNPs using all SNPs and may lead a too overwhelming computational load. Thus, to boost the power of detecting epistasis interactions, we resort to the multiple clusterings technique to generate different groupings of SNPs, and to minimize the chance of wrongly filtering out significant SNP combinations. Compared with typical clustering [13], multiple clusterings is much less studied, it aims generate a set of diverse clusterings and to explore different aspects of the same data [3].

Semi-nonnegative matrix factorization based clustering [11] has been extended for generating multiple clusterings [26, 27, 29, 34, 35]. Semi-NMF aims to factorize the data matrix $\mathbf{X} \approx \mathbf{GZ}$. By taking $\mathbf{Z} \in \mathbb{R}^{n \times k}$ as k cluster centroids in the n-dimensional sample space, and $\mathbf{G} \in \mathbb{R}^{d \times k}$ as the soft membership indicators of d SNPs to these centroids. To obtain the diversity between generated alternative clusterings and to reduce the computational burden of factorizing a data matrix with massive SNPs, we project d SNPs into different subspaces by a series of projective matrices $\{\mathbf{P}_h \in \mathbb{R}^{d \times d_h}\}_{h=1}^m$ to seek m diverse clusterings in these subspaces. And to enforce the diversity between m alternative clusterings, we reduce the overlap between two projective matrices using a Frobenius norm. Besides, to reduce the impact of noisy SNPs, we further add an l_1-norm on each $\mathbf{P}_h$. Thus, for concreting these, we introduce a matrix factorization based multiple clustering objective as follows:

$$\min J(\mathbf{G}_h, \mathbf{Z}_h, \mathbf{P}_h) = \frac{1}{m} \sum_{h=1}^m \| \mathbf{P}_h^T \mathbf{X} - \mathbf{G}_h \mathbf{Z}_h^T \|_F^2 + \lambda_1 tr(\mathbf{G}_h^T \mathbf{L} \mathbf{G}_h)$$

$$+ \frac{\lambda_2}{m} \sum_{h=1}^m \| \mathbf{P}_h \|_1 + \frac{\lambda_3}{m^2} \sum_{h=1, h \neq h_2}^m \| \mathbf{P}_h^T \mathbf{P}_{h_2} \|_F^2 \qquad (1)$$

$$s.t. \ \mathbf{G}_h \geq 0$$

where the scale parameter λ_1 balances these four terms, $\lambda_2 > 0$ controls the sparsity of projective matrices, and $\lambda_3 > 0$ balances the quality and diversity of m clusterings in the projected subspaces, two normalization factors $1/m^2$ and $1/m$ are introduced to reduce the scale impact. $\mathbf{L} = \mathbf{D} - \mathbf{W}$, where $\mathbf{W}$ reflects the association strength between two SNPs and $\mathbf{D}$ is a diagonal matrix with diagonal entry equal to the row sum of $\mathbf{W}$ of respective row. Here, we adopt Entropy score (ES)[9] to quantify the non-linear direct associations between two SNPs as follows:

$$\mathbf{W}(S_1, S_2) = \frac{\min\{S_1, S_2\} - S_{1,2}}{\min\{S_1, S_2\}} \qquad (2)$$

where S_1 and S_2 are the single loci entropy for SNP1 and SNP2, respectively. $S_1 = \sum_{i=0}^2 P_{1i} h_{1i}$, where $h_{1i} = -p_i \log p_i - (1 - p_i) \log(1 - p_i)$, $p_i = n_{case}(i)/n_{total}(i)$, $n_{total}(i)$, $n_{case}(i)$ are the number of samples with genotype i in all samples and case samples, respectively. $P_{1i} = n_{total}(i)/n$ is the weighted co-efficient for h_{1i} and $\sum_{i=0}^2 P_{1i} = 1$, n is the number of all samples. Performing the same calculation on S_2 and $S_{1,2}$. For single loci entropy, $i = 0, 1, 2$ denotes the three genotypes (AA, Aa, aa) of SNP. For two-way interactions, $i = 1, 2, \cdots, 9$ denotes SNP1$\star$SNP2 = AABB, AABb,..., aabb, respectively. Based on Eq. (2), we can quantify the association between interacting SNPs and disease status. A higher ES value indicates a stronger association between two SNPs and they will be more probably grouped into the same cluster.

Note, the Eq. (1) explore and exploit the intrinsic local geometrical structure of SNPs and thus to boost the grouping effect of SNPs, also has the flavor of Elastic net [37], which encourages a grouping effect where strongly correlated predictors tend to be in or out of the model together. Due to page limit, the detailed optimization is not presented here.

2.2 Stage 2: Significant SNP Interactions Detection

Since SNPs in the same cluster have higher genetic association than those in different clusters, we can select high quality SNPs that have strong association with other SNPs in the same cluster and reduce the interference of noisy SNPs. Given that, for each cluster, we also use Entropy score to calculate the interaction effect and adopt the maximum entropy score to approximate the effect of SNP_i:

$$S_i = \max_{i \neq j, j \in \mathcal{C}(i)} \{ES_{ij}\}; \tag{3}$$

where ES_{ij} is the association score (quantified by Eq. (2)) between SNP_i and SNP_j, and $j \in \mathcal{C}(i)$ means SNP_j is within the same cluster as SNP_i. Next, top T SNPs with the highest scores are selected from each cluster.

To obtain the useful dominant SNPs and search the SNPs combinations strongly associated with disease, we use Jaccard similarity coefficient [2] to calculate the similarity of different clusters from different clusterings. The larger the Jaccard coefficient value, the greater the similarity is. Here we employ the Jaccard similarity to measure the similarity between two clusters of two clusterings as follows:

$$J(\mathcal{C}_h^i, \mathcal{C}_{h'}^j) = \frac{|\mathcal{C}_h^i \bigcap \mathcal{C}_{h'}^j|}{|\mathcal{C}_h^i \bigcup \mathcal{C}_{h'}^j|} \tag{4}$$

where $\mathcal{C}_h^i$ ($\mathcal{C}_{h'}^j$) is the $i(j)$-th cluster in the $h(h')$-th clustering, $h, h' \leq m, h' \neq h$, and $i, j \leq k$. Note, each cluster only retains the T high quality SNP locus. We firstly calculate the similarity of different clusters from different clusterings, then merge the clusters among m clusterings with high similarity and discard other clusters with low similarity. In this way, we obtain k candidate sets which contains dominant SNPs.

Till now, we obtain k much smaller candidate sets, each set contains accurate and strong association SNPs. Since exhaustive analysis is feasible to exhaustively analyze all high-order interactions when the number of candidates is very small, EpIntMC uses exhaustive search strategy on candidate sets to detect epistasis. Each K-SNP combination is evaluated by the entropy score within k candidate sets. Then, all combinations from individual sets are merged into a set $\mathcal{S}$ and sorted in descending order. EpIntMC reports top N combinations with the highest entropy score as the detected K-SNP interactions.

3 Experimental Result and Analysis

In this section, we evaluate the performance of EpIntMC on both simulated and real datasets on detecting epistasis interactions. In the simulation study, we compare EpIntMC with EDCF [31], DCHE [12], DECMDR [32], MOMDR [33], and ClusterMI [7] on two-locus and three-locus epistatic models. We use a canonically used evaluation metric *Power* [25] to quantify the performance of EpIntMC and of other comparing methods. *Power* is defined as the proportion of the total generated datasets in which the specific disease-associated combination is detected. For the real study, we test EpIntMC on breast cancer (BC) dataset collected from WTCCC.

3.1 Parameter Analysing and Setting

EpIntMC has several parameters which may affect its performance. For better detecting epistasis interactions, we firstly study the parameter sensitivity. Power was used as the evaluate metric on 50 simulation datasets with 1600 samples and 1000 SNPs. Figure 1 reveals the results under different parameter settings of m and k, and different settings of λ_2 and λ_3.

Given the multiplicity of genetic data, we study the number of clusterings m in $[2, 5]$ and the number of clusters in each clustering k in $[2, 9]$ to seek the suitable input values. We observe Fig. 1a that $k = 3$ and $m = 3$ gives the highest power, and the further increase of k reduces the power. That is because two SNPs with high association are more likely grouped into different clusters as the growth of k, which compromises the power. As to m, we find $m > 5$ gives a greatly reduced power, since strongly associated SNPs are easily separated when merging the selected SNPs from different clusters of different clusterings as the growth of alternative clusterings, which also reduces the quality of individual clusterings. We want to remark that the increase of m and k requires more time to converge and merge the SNPs. Give these results, we adopt $m = 3$ and $k = 3$ for the following simulation experiments.

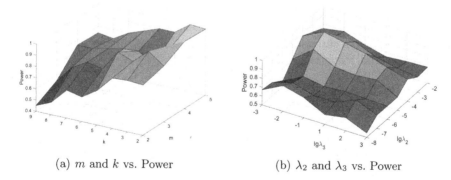

(a) m and k vs. Power   (b) λ_2 and λ_3 vs. Power

Fig. 1. Parameter analysis of EpIntMC in the screening stage. (a) EpIntMC under different values of m and k; (b) EpIntMC under different values of λ_2 and λ_3.

λ_1 balances the multiple clusterings and the grouping effect of SNPs, we find that a larger λ_1 contributes to a larger power. The power can be even higher when

$\lambda_1 > 100$. However, such a large λ_1 can lead most SNPs grouped into one cluster, it increases the coverage but brings a very big candidate SNP combinations for follow-up detection. Given that, we adopt $\lambda_1 = 100$ for experiment. With $m = 3$, $k = 3$ and $\lambda_1 = 100$, we also study λ_2 and λ_3 in $[10^{-8}, 10^{-2}]$ and $[10^{-3}, 10^2]$, respectively. Figure 1b shows the power under combinations of λ_2 and λ_3. When $\lambda_2 = 10^{-4}$ the power of EpIntMC achieves the highest value. A too large λ_2 will exclude the significant SNPs, while a too small λ_2 cannot exclude noisy ones. λ_3 is used to adjust the diversity among m clusterings, the power is higher when $\lambda_3 = 10^{-2}$. We also observe that a too small λ_3 can not ensure the diversity among alternative clusterings but reduce power, while a too large λ_3 increases the diversity but decrease the quality of individual clusterings. We also test the power when $\lambda_2 = 0$ and $\lambda_3 = 0$, which is lower than any combination in Fig. 1b. These observations support the necessity to control the sparsity of projective matrices and to balance the quality and diversity of multiple clusterings. Based on these results, we adopted $\lambda_1 = 100$, $\lambda_2 = 10^{-4}$ and $\lambda_3 = 10^{-2}$ for experiments. For simplicity, we fix $d_h = d$.

3.2 Experiments on Simulation Datasets

In the simulation data experiments, we use two two-locus epistatic models: Model 1 involves two-locus multiplicative effects [19], Model 2 is the well known XOR (exclusive OR) model [15]. Two three-locus models: Model 3 obtained from [36] is a three-locus model with marginal effect. Model 4 is a three-locus model without marginal effect proposed by Culverhouse et al. [10]. Marginal effect size λ of a disease locus in each model is defined as [36]:

$$\lambda = \frac{P_{Aa}/P_{AA}}{(1 - P_{Aa}/(1 - P_{AA}))} - 1 \tag{5}$$

where P_{AA} and P_{Aa} denote the penetrance of genotype AA and Aa, respectively. For two-locus models, $\lambda = 0.2$, minor allele frequencies (MAFs) of the disease loci are the same at three levels: 0.1, 0.2 and 0.4, linkage disequilibrium (LD) between loci and associated markers (measured by r^2) is set to 0.7 and 1.0. We simulate 100 datasets under each parameter setting for each disease model and each dataset contains $d = 1000$ SNPs. For two-locus models, each dataset contains $n = 1600$ (or $n = 4000$) samples with balanced samples design. For Model 3, $\lambda = 0.3$, MAFs is at three levels: MAF $= 0.1, 0.2, 0.5$, LD is also at two levels: $r^2 = 0.7$ and 1. And each dataset contains $n = 2000$ (or $n = 4000$) samples with the balanced design. Model 4 which yields maximum genetic heritability h^2 with the population penetrance $p \in (0, 1/16]$ and MAF $= 0.5$, heritability h^2 ranges from 0.01 to 0.4. The sample size of each dataset varies from 400 to 800.

Figure 2a–2b reveals the performance of compared methods on two-locus epistatic models. The performance of all methods increases with the growth of the sample size. The power of these models improves significantly when r^2 changing from 0.7 to 1. But in Model 1, the performance of most methods decreases when MAFs varies from 0.2 to 0.4, the trend is consistent with the

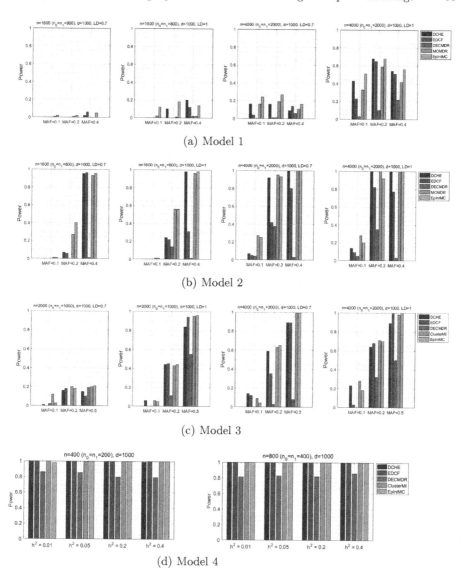

Fig. 2. Powers of DCHE, DECMDR, EDCF, MOMDR, ClusterMI and EpIntMC on four disease models under different allele frequencies (MAF), sample sizes (n) and Linkage disequilibrium (LD). n_0 is the number of controls, n_1 is the number of cases, and d is the number of SNPs. The absence of a bar indicates no power.

results in Marchini *et al.* [19] and Wan *et al.* [25]. For Model 2, the power of most methods increases as the MAFs growing from 0.1 to 0.4. EpIntMC uses multiple clusterings to generate diverse clusterings and minimize the chance of wrongly filtering out significant SNP combinations. As a result, EpIntMC has a better performance than DCHE, and they both outperform EDCF. In some case, when MAF = 0.1 and MAF = 0.2, EpIntMC loses to DCHE and MOMDR. That is because DCHE measures significance via the chi-squared test with low degrees of freedom and MOMDR uses the correct classification and likelihood rates to simultaneously evaluate interactions. EpIntMC only uses the entropy score to evaluate the significance of candidate combinations. In practice, our EpIntMC has a higher coverage of significant SNP combinations in the screening stage than other comparing methods, which prove the availability of multiple clusterings. These comparisons show the effectiveness of EpIntMC in detecting two-locus epistasis.

Figure 2c–2d shows the performance of comparing methods on three-locus models for detecting high-order interactions. Here, we do not use MOMDR since the extensive computational burden of MOMDR in high-order epistasis detection. Instead, we include another high-order epistasis detection approach (ClusterMI [7]) for experiments. For these two models, the power of all methods increases as the growth of heritability and sample size in most cases. EpIntMC has a higher (or comparable) power than other compared methods in most cases. This is consistent with the results on the above two-locus datasets. When $h^2 = 0.01$ and $n = 400$ in Model 4, EpIntMC loses little performance than some methods, which may due to the low heritability and small sample size affects the effect of multiple clustering together. On the whole, all methods shows the highest performance in Model 4, except DECMDR with a relatively poor performance in selecting epistatic combinations. In summary, the results on detecting high-order epistasis interactions again expresses the effectiveness of EpIntMC.

3.3   Experiments on Real Dataset

The Wellcome Trust Case Control Consortium (WTCCC) is a collaborative effort between 50 British research groups established in 2005. A real breast cancer dataset (BC) collected from WTCCC project [6] is used to evaluate EpIntMC. This dataset contains genotypes of 15,347 SNPs from 1,045 affected individuals and 2,073 controls. Quality control is performed to exclude very low call rate samples and SNPs. After the quality control, the BC dataset contains 1,045 case samples and 2,070 control samples with 5,607 SNPs.

Some significant two-locus and three-locus combinations on BC dataset identified by EpIntMC and several significant and representative combinations are listed in Table 1. rs3742303 is located at gene $USPL1$ which is evidenced that $USPL1$ is associated with breast cancer [5]. rs3739523 belongs to gene $PRUNE2$ on chromosome 9. The protein encoded by this gene plays roles in many cellular processes including apotosis, cell transformation, and synaptic function. rs2301572 is located at gene $PLXND1$ which is characterized as a receptor for semaphorins and is known to be essential for axonal guidance and vascular patterning. Emerging data show that expression of $PLXND1$ is deregulated in several cancers [24]. rs4987117 is located at gene $BRCA2$ in chromosome 13. $BRCA2$ is considered a tumor suppressor gene, as tumors with $BRCA2$ mutations generally exhibit loss of heterozygosity (LOH) of the wild-type allele.

It is evidenced that inherited mutations in $BRCA1$ and $BRCA2$ confer increased lifetime risk of developing breast cancer [1].

Table 1. Significant two-locus and three-locus combinations identified by EpIntMC on WTCCC Breast cancer data.

Significant combinations	Chromosome and related genes	Combination entropy score
(rs2297089, rs3739523)	(chr9:CEMIP2,chr9:PRUNE2)	0.6375
(rs3742303, rs3749671)	(chr13:USPL1,chr5:MRPL22)	0.5866
(rs2301572, rs3749671)	(chr3:PLXND1,chr5:MRPL22)	0.5818
(rs4987117, rs1800961)	(chr13:BRCA2,chr20:HNF4A)	0.5835
(rs3749671, rs13362036)	(chr5:MRPL22,chr5:CCNJL)	0.5866
(rs2297089, rs1800058, rs3784635)	(chr9:CEMIP2,chr11:ATM,chr15:VPS13C)	0.8544

In the three-locus combination, (rs2297089, rs1800058, rs3784635) as the representative one is reported in the last row of Table 1. rs2297089 is located at gene $CEMIP2$ in chromosome 9. It is reported that the activation of $CEMIP2$ by transcription factor SOX4 in breast cancer cells mediates the pathological effects of SOX4 on cancer progression [14]. rs1800058 is located at gene ATM. Independent ATM (oxidized ATM) can enhance the glycolysis and aberrant metabolism-associated gene expressions in breast Cancer-associated fibroblasts (CAFs) [23]. These significant two-locus combinations and three-locus combinations associated with breast cancer demonstrate the effectiveness of EpIntMC in detecting SNP interactions on genome-wide data.

4   Conclusion

In this paper, we proposed a multi-clustering based approach EpIntMC to effectively detect high-order SNP interactions from genome-wide case-control data. In the first stage, considering the multiplicity of genome-wide data, EpIntMC firstly introduces a matrix factorization based approach to generate multiple diverse clusterings to group SNPs into different clusters from different aspects. This stage greatly reduces the search space and also the risk of filtering out potential candidates overlooked by a single clustering. In the search stage, EpIntMC applies Entropy score to select high-quality SNPs in each cluster, then uses Jaccard similarity to merge the most similar clusters into candidate sets, which have strongly associated SNPs but with much smaller sizes. Next, EpIntMC uses exhaustive search on these candidate sets to precisely detect epistatic interactions. A series of simulation experiments on two-locus and three-locus disease models show that EpIntMC has a better performance that state-of-art methods. Experiments on real WTCCC dataset corroborate that EpIntMC is feasible for identifying high-order SNP interactions from genome-wide data.

Acknowledgements. This research is supported by NSFC (61872300), Fundamental Research Funds for the Central Universities (XDJK2020B028 and XDJK2019B024), Natural Science Foundation of CQ CSTC (cstc2018jcyjAX0228).

References

1. Abdulrashid, K., AlHussaini, N., Ahmed, W., Thalib, L.: Prevalence of BRCA mutations among hereditary breast and/or ovarian cancer patients in Arab countries: systematic review and meta-analysis. BMC Cancer **19**(1), 256 (2019). https://doi.org/10.1186/s12885-019-5463-1
2. Albatineh, A.N., Niewiadomska-Bugaj, M.: Correcting Jaccard and other similarity indices for chance agreement in cluster analysis. Adv. Data Anal. Classif. **5**(3), 179–200 (2011). https://doi.org/10.1007/s11634-011-0090-y
3. Bailey, J.: Alternative clustering analysis: a review. In: Data Clustering, pp. 535–550. Chapman and Hall/CRC (2018)
4. Balding, D.J.: A tutorial on statistical methods for population association studies. Nat. Rev. Genet. **7**(10), 781 (2006)
5. Bermejo, J.L., et al.: Exploring the association between genetic variation in the SUMO isopeptidase gene *USPL1* and breast cancer through integration of data from the population-based genica study and external genetic databases. Int. J. Cancer **133**(2), 362–372 (2013)
6. Burton, P.R., et al.: Association scan of 14,500 nonsynonymous SNPs in four diseases identifies autoimmunity variants. Nat. Genet. **39**(11), 1329 (2007)
7. Cao, X., Yu, G., Liu, J., Jia, L., Wang, J.: ClusterMI: detecting high-order SNP interactions based on clustering and mutual information. Int. J. Mol. Sci. **19**(8), 2267 (2018)
8. Cao, X., Yu, G., Ren, W., Guo, M., Wang, J.: DualWMDR: detecting epistatic interaction with dual screening and multifactor dimensionality reduction. Hum. Mutat. **40**, 719–734 (2020)
9. Chattopadhyay, A.S., Hsiao, C.L., Chang, C.C., Lian, I.B., Fann, C.S.: Summarizing techniques that combine three non-parametric scores to detect disease-associated 2-way SNP-SNP interactions. Gene **533**(1), 304–312 (2014)
10. Culverhouse, R., Suarez, B.K., Lin, J., Reich, T.: A perspective on epistasis: limits of models displaying no main effect. Am. J. Hum. Genet. **70**(2), 461–471 (2002)
11. Ding, C.H., Li, T., Jordan, M.I.: Convex and semi-nonnegative matrix factorizations. TPAMI **32**(1), 45–55 (2010)
12. Guo, X., Meng, Y., Yu, N., Pan, Y.: Cloud computing for detecting high-order genome-wide epistatic interaction via dynamic clustering. BMC Bioinform. **15**(1), 102 (2014)
13. Jain, A.K., Murty, M.N., Flynn, P.J.: Data clustering: a review. ACM Comput. Surv. **31**(3), 264–323 (1999)
14. Lee, H., Goodarzi, H., Tavazoie, S.F., Alarcón, C.R.: TMEM2 is a SOX4-regulated gene that mediates metastatic migration and invasion in breast cancer. Cancer Res. **76**(17), 4994–5005 (2016)
15. Li, W., Reich, J.: A complete enumeration and classification of two-locus disease models. Hum. Hered. **50**(6), 334–349 (2000)
16. Liu, J., Yu, G., Jiang, Y., Wang, J.: HiSeeker: detecting high-order SNP interactions based on pairwise SNP combinations. Genes **8**(6), 153 (2017)

17. Ma, L., Runesha, H.B., Dvorkin, D., Garbe, J.R., Da, Y.: Parallel and serial computing tools for testing single-locus and epistatic SNP effects of quantitative traits in genome-wide association studies. BMC Bioinform. **9**(1), 315 (2008). https://doi.org/10.1186/1471-2105-9-315

18. Mackay, T.F., Moore, J.H.: Why epistasis is important for tackling complex human disease genetics. Genome Med. **6**(6), 42 (2014). https://doi.org/10.1186/gm561

19. Marchini, J., Donnelly, P., Cardon, L.R.: Genome-wide strategies for detecting multiple loci that influence complex diseases. Nat. Genet. **37**(4), 413 (2005)

20. Moore, J.H., Asselbergs, F.W., Williams, S.M.: Bioinformatics challenges for genome-wide association studies. Bioinformatics **26**(4), 445–455 (2010)

21. Niel, C., Sinoquet, C., Dina, C., Rocheleau, G.: A survey about methods dedicated to epistasis detection. Front. Genet. **6**, 285 (2015)

22. Ritchie, M.D., et al.: Multifactor-dimensionality reduction reveals high-order interactions among estrogen-metabolism genes in sporadic breast cancer. Am. J. Hum. Genet. **69**(1), 138–147 (2001)

23. Sun, K., et al.: Oxidized ATM-mediated glycolysis enhancement in breast cancer-associated fibroblasts contributes to tumor invasion through lactate as metabolic coupling. EBioMedicine **41**, 370–383 (2019)

24. Vivekanandhan, S., Mukhopadhyay, D.: Divergent roles of Plexin D1 in cancer. Biochimica et Biophysica Acta (BBA)-Rev. Cancer **1872**(1), 103–110 (2019)

25. Wan, X., et al.: BOOST: a fast approach to detecting gene-gene interactions in genome-wide case-control studies. Am. J. Hum. Genet. **87**(3), 325–340 (2010)

26. Wang, J., Wang, X., Yu, G., Domeniconi, C., Yu, Z., Zhang, Z.: Discovering multiple co-clusterings with matrix factorization. IEEE Trans. Cybern. **99**(1), 1–14 (2020)

27. Wang, X., Wang, J., Domeniconi, C., Yu, G., Xiao, G., Guo, M.: Multiple independent subspace clusterings. In: AAAI, pp. 5353–5360 (2019)

28. Wang, Y., Liu, X., Robbins, K., Rekaya, R.: AntEpiSeeker: detecting epistatic interactions for case-control studies using a two-stage ant colony optimization algorithm. BMC Res. Notes **3**(1), 117 (2010). https://doi.org/10.1186/1756-0500-3-117

29. Wei, S., Wang, J., Yu, G., Zhang, X., et al.: Multi-view multiple clusterings using deep matrix factorization. In: AAAI, pp. 1–8 (2020)

30. Welter, D., et al.: The NHGRI GWAS catalog, a curated resource of SNP-trait associations. Nucleic Acids Res. **42**(D1), D1001–D1006 (2013)

31. Xie, M., Li, J., Jiang, T.: Detecting genome-wide epistases based on the clustering of relatively frequent items. Bioinformatics **28**(1), 5–12 (2011)

32. Yang, C.H., Chuang, L.Y., Lin, Y.D.: CMDR based differential evolution identifies the epistatic interaction in genome-wide association studies. Bioinformatics **33**(15), 2354–2362 (2017)

33. Yang, C.H., Chuang, L.Y., Lin, Y.D.: Multiobjective multifactor dimensionality reduction to detect SNP-SNP interactions. Bioinformatics **34**(13), 2228–2236 (2018)

34. Yao, S., Yu, G., Wang, J., Domeniconi, C., Zhang, X.: Multi-view multiple clustering. In: IJCAI, pp. 4121–4127 (2019)

35. Yao, S., Yu, G., Wang, X., Wang, J., Domeniconi, C., Guo, M.: Discovering multiple co-clusterings in subspaces. In: SDM, pp. 423–431 (2019)

36. Zhang, Y., Liu, J.S.: Bayesian inference of epistatic interactions in case-control studies. Nat. Genet. **39**(9), 1167 (2007)

37. Zou, H., Hastie, T.: Regularization and variable selection via the elastic net. J. Roy. Stat. Soc. B **67**(2), 301–320 (2005)

Improving Metagenomic Classification Using Discriminative k-mers from Sequencing Data

Davide Storato and Matteo Comin[✉]

Department of Information Engineering, University of Padua, 35100 Padua, Italy
{storatod,comin}@dei.unipd.it

Abstract. The major problem when analyzing a metagenomic sample is to taxonomically annotate its reads to identify the species they contain. Most of the methods currently available focus on the classification of reads using a set of reference genomes and their k-mers. While in terms of precision these methods have reached percentages of correctness close to perfection, in terms of recall (the actual number of classified reads) the performances fall at around 50%. One of the reasons is the fact that the sequences in a sample can be very different from the corresponding reference genome, e.g. viral genomes are highly mutated. To address this issue, in this paper we study the problem of metagenomic reads classification by improving the reference k-mers library with novel discriminative k-mers from the input sequencing reads. We evaluated the performance in different conditions against several other tools and the results showed an improved F-measure, especially when close reference genomes are not available.

Availability: https://github.com/davide92/K2Mem.git

Keywords: Metagenomic reads classification · Discriminative k-mers · Minimizers

1 Introduction

Metagenomics is the study of the heterogeneous microbes samples (e.g. soil, water, human microbiome) directly extracted from the natural environment with the primary goal of determining the taxonomical identity of the microorganisms residing in the samples. It is an evolutionary revise, shifting focuses from the individual microbe study to a complex microbial community. The classical genomic-based approaches require the prior clone and culturing for further investigation [21]. However, not all bacteria can be cultured. The advent of metagenomics succeeded to bypass this difficulty. Microbial communities can be analyzed and compared through the detection and quantification of the species they contain [26]. In this paper, we will focus on the detection of species in a sample using a set of reference genomes, e.g. bacteria and virus. The reference-based metagenomics

© Springer Nature Switzerland AG 2020
Z. Cai et al. (Eds.): ISBRA 2020, LNBI 12304, pp. 68–81, 2020.
https://doi.org/10.1007/978-3-030-57821-3_7

classification methods can be broadly divided into two categories: (1) alignment-based methods, (2) sequence-composition-based methods, which are based on the nucleotide composition (e.g. k-mers usage). Traditionally, the first strategy was to use BLAST [1] to align each read with all sequences in GenBank. Later, faster methods have been deployed for this task, popular examples are MegaBlast [36] and Megan [17]. However, as the reference databases and the size of sequencing data sets have grown, alignment has become computationally infeasible, leading to the development of metagenomics classifiers that provide much faster results.

The fastest and most promising approaches belong to the composition-based one [20]. The basic principles can be summarized as follows: each genome of reference organisms is represented by its k-mers, and the associated taxonomic label of the organisms, then the reads are searched and classified throughout this k-mers database. For example, Kraken [33] constructs a data structure that is an augmented taxonomic tree in which a list of significant k-mers are associated to each node, leaves and internal nodes. Given a node on this taxonomic tree, its list of k-mers is considered representative for the taxonomic label and it will be used for the classification of metagenomic reads. Clark [24] uses a similar approach, building databases of species- or genus-level specific k-mers, and discarding any k-mers mapping to higher levels. The precision of these methods is as good as MegaBlast [36], nevertheless, the processing speed is much faster [20]. Several other composition-based methods have been proposed over the years. In [14] the number of unassigned reads is decreased through reads overlap detection and species imputation. Centrifuge and Kraken 2 [19,34] try to reduce the size of the k-mer database with the use respectively of FM-index and minimizers. The sensitivity can be improved by filtering uninformative k-mers [22,27] or by using spaced seeds instead of k-mers [7].

The major problem with these reference-based metagenomics classifiers is the fact that most bacteria found in environmental samples are unknown and cannot be cultured and separated in the laboratory [13]. As a consequence, the genomes of most microbes in a metegenomic sample are taxonomically distant from those present in existing reference databases. This fact is even more important in the case of viral genomes, where the mutation and recombination rate is very high and as a consequence, the viral reference genomes are usually very different from the other viral genomes of the same species.

For these reasons, most of the reference-based metagenomics classification methods do not perform well when the sample under examination contains strains that are different from the genomes used as references. Indeed, e.g. CLARK [24] and Kraken [33] report precisions above 95% on many datasets. On the other hand, in terms of recall, i.e. the percentage of reads classified, both Clark and Kraken usually show performances between 50% and 60%, and sometimes on real metagenomes, just 20% of reads can be assigned to some taxa. In this paper we address this problem and we propose a metagenomics classification tool, named *K2Mem*, that is based, not only on a set of reference genomes but also it uses discriminative k-mers from the input metagenomics reads to improve the classification. The basic idea is to add memory to a classification pipeline, so that previously analyzed reads can be of help for the classification.

2   Methods

To improve the metagenomics classification our idea is based on the following considerations. All reference-based metagenomics classification methods need to index a set of reference genomes. The construction of this reference database is based on a set of genomes, represented by its k-mers (a piece of the genome with length k), and the taxonomic tree. For example, Kraken [33] constructs a data structure that is an augmented taxonomic tree in which a list of discriminative k-mers is associated with each node, leaves and internal nodes. Given a node on this taxonomic tree, its list of k-mers is considered representative for the taxonomic label and it is used for the classification of metagenomic reads. However, for a given genome only a few of its k-mers will be considered discriminative. As a consequence, only the reads that contains these discriminative k-mers can be classified to this species.

Given a read with length n, each of its $n - k + 1$ k-mers have the first $k - 1$ bases in common with the previous k-mer, except the first k-mer. Furthermore, it is possible that reads belonging to the input sequencing data can have many k-mers in common.

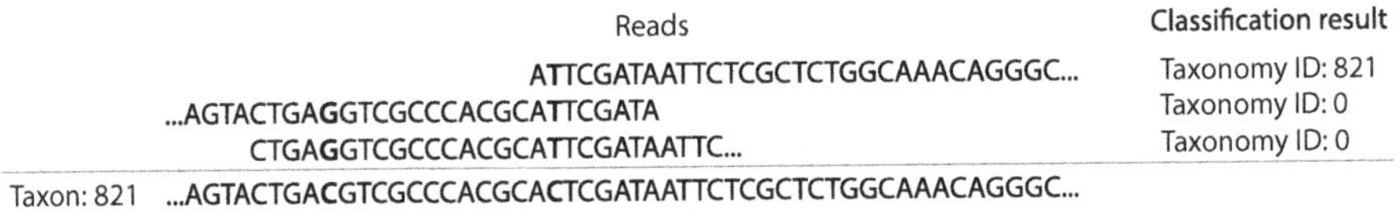

Reads	Classification result
ATTCGATAATTCTCGCTCTGGCAAACAGGGC...	Taxonomy ID: 821
...AGTACTGAGGTCGCCCACGCATTCGATA	Taxonomy ID: 0
CTGAGGTCGCCCACGCATTCGATAATTC...	Taxonomy ID: 0

Taxon: 821   ...AGTACTGACGTCGCCCACGCACTCGATAATTCTCGCTCTGGCAAACAGGGC...

Fig. 1. Example of a reference-based metagenomics classifier behaviour. In red the k-mer in common between the reads, in green the k-mer associated to a species' taxonomy ID (in this case 821) present in the classifier's database, and in bold the mutations' positions in the reads. A taxonomy ID of zero indicates that the classifier wasn't able to classify the read. (Color figure online)

As can be seen in Fig. 1, in this example we have three reads containing the same k-mer (in red) but only one is classified thanks to the presence in the read of a discriminative k-mer (in green), with a taxonomy ID associated, contained in the classifier's database. The second read could not be classified because none of its k-mers are in the k-mer reference library, as there is a mutation (in bold) respect to the reference genome. However, the k-mers of the first read, that are not present in the classifier's database, can belong to the same species to which the read classified belongs to. With reference to Fig. 1, if we associate to the shared k-mer (the red one) the taxonomy ID of the first read then, we can classify the other two reads. Thus, using the above considerations, one can try to extend the taxonomy ID of a classified read to all its k-mers.

The idea is to equip the classifier with a memory from previous classifications, thus adding novel discriminative k-mers found in the input sequencing data. To obtain this memory effect, one needs to modify a given classifier with additional data structures and a new classification pipeline. Note that, this idea can be applied to any reference-based metagenomics classifiers that are based on a database of k-mers. In this paper we choose to use Kraken 2 [34], that was recently introduced, and that is reported to be the current state of the art. Before to describe our classification tool, for completeness here we give a brief introduction of Kraken 2 to better understand our contribution.

2.1 Background on Kraken 2

Kraken 2 is an improved version of the classifier Kraken [33] regarding memory usage and classification time. These improvements were obtained thanks to the use of the minimizers and a probabilistic compact data structure, instead of the k-mers and a sorted list used in Kraken.

Instead of utilizing the complete genome as reference, Kraken considers only its k-mers, as well as many other tools [2,24], thus a genome sequence is alternatively represented by a set of k-mers, which plays a role of efficiently indexing a large volume of target-genomes database, e.g., all the genomes in RefSeq. This idea is borrowed from alignment-free methods [32] and some researchers have verified its availability in different applications. For instance, the construction of phylogenetic trees, traditionally is performed based on multiple-sequence alignment, whereas with alignment-free methods it can be carried out on the whole genomes [8,31]. Recently some variations of k-mers-based methods have been devised for the detection of enhancers in ChIP-Seq data [5,9,10,15,18], entropic profiles [3,4], and NGS data compression [29,30,35]. Also, the assembly-free comparison of genomes and metagenomes based on NGS reads and k-mers counts has been investigated in [11,12,23,28]. For a comprehensive review of alignment-free measures and applications we refer the reader to [32].

At first, Kraken 2 needs to build a database starting from a set of reference genomes. To build the database Kraken 2 downloads from the NCBI the taxonomy and reference sequences libraries required. With the taxonomy data, Kraken 2 builds a tree within each node a taxonomy ID. In each tree node, a list of minimizer is stored that is useful for the classification. Precisely, for each minimizer ($k = 35, l = 31$) if it is unique to a reference sequence then, it is associated with the node with the sequence's taxonomy ID. Instead, if the minimizer belongs to more than one reference sequence then, the minimizer is moved to the node with taxonomy ID equals to the Lowest Common Ancestor (LCA) of the two sequences it belongs. All the minimizer-taxonomy ID pairs are saved in a probabilistic Compact Hash Table (CHT) which allows a reduction of memory with respect to Kraken.

Once the database is built, Kraken 2 classifies the input reads in a more efficient manner than Kraken due to the smaller number of accesses to the CHT map. This is due to the fact that only distinct minimizers from the read trigger the research in the map. When the minimizer is queried in the compact table the node ID of the augmented tree is returned, and the counter in the node is increased. Once all the minimizers have been analyzed, the read is classified by choosing the deepest node from the root with the highest-weight path in the taxonomy tree.

2.2 K2Mem

Here we present K2Mem (Kraken 2 with Memory) a classifier based on Kraken 2. In order to implement the idea explained above, K2Mem needs to detect new discriminative minimizers, to store them in memory and to use this additional information in the classification of reads. The new classification pipeline of K2Mem discovers novel discriminative minimizers from the input sequencing data and it saves them in a map of additional minimizers.

The data structure used to store these additional minimizers is an unordered map that stores pairs composed of the novel discriminative minimizer, not present in the classifier's database, and the taxonomy ID associated to the read that contains the minimizer. An unordered map is an associative container that contains key-value pairs with a unique key. The choice of this structure is due to the fact that search, insertion and removal of elements have average constant-time complexity. Internally, the elements are not ordered in any particular order but are organized in buckets. Which bucket an element is placed into depends entirely on the hash of its key. This allows fast access to the single element, once the hash is computed, it refers to the exact bucket where the element has been inserted. The key and value are both 64-bit unsigned integer. This choice was made to keep the complete minimizers ($l = 31$) on the map without loss of information due to the CHT hash function and to contains the taxonomy ID in case the number of taxonomy tree nodes increases in future.

K2Mem has two main steps, in the first phase all reads are processed and novel discriminative minimizers are discovered and stored in the additional minimizers map. In the second phase, the same input reads are re-classified using the Compact Hash Table and also the additional minimizers obtained in the first phase.

An overview of the first phase, the discovery of novel discriminative minimizers, is reported in Fig. 2. The population of the additional minimizers map works as follow: for each read, its minimizers ($k = 35$, $l = 31$) are computed one at a time and, for each of them, the Compact Hash Table (CHT) is queried (1). If it returns a taxonomy value equal to zero, then the additional map is queried if it is not empty (2). If the minimizer is not found in the additional minimizers map, this means that the minimizer is not in the Kraken 2's database and no taxonomy ID has been assigned to it or is the first time the minimizer is found. In that case, the minimizer is added to a temporary list of not taxonomically assigned minimizers (3). Instead, if the CHT or the additional minimizers map query returns a taxonomy ID not equal to zero, then the taxonomy ID count is updated (4). Then, the read is classified (5), based on the highest-weight path in the taxonomy tree, and the resulting taxonomy ID is checked if it is at the

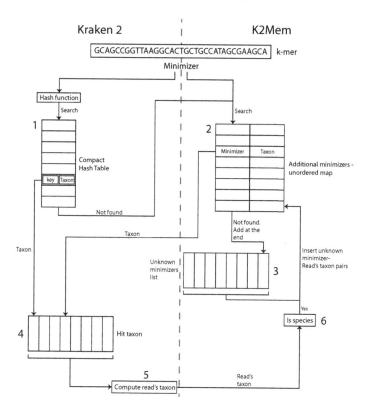

Fig. 2. An overview of K2Mem and the discovery of additional minimizers.

species level or below (6). If it is, then the minimizers in the unknown minimizers list are added to the additional minimizers map with key the minimizer and value the taxonomy ID obtained by the read classification. If the minimizer is already in the additional map the LCA of the input and stored taxonomy IDs value is saved. Instead, if the taxonomy ID obtained after the read classification is at a level above the species then, the minimizers are not added and the list is emptied. In the first phase, this procedure is repeated for all the reads in the dataset. Then, once the population of the additional minimizers map is completed, in the second phase all reads are re-classified. In this phase the classification only uses the CHT and the additional minimizers map to classify the reads in the same dataset, but without to update the additional map content. With these additional minimizers, the dataset is processed to obtain a better classification. K2Mem starts querying the CHT for each minimizer in the read (1). If the minimizer is not found the additional map is queried (2). If the minimizer is not present in the latest map a taxon of zero is assigned to the minimizer. Instead if in the CHT or additional map the minimizer is found the associated taxon is returned and the taxon's number of hit is updated (4). Once all the read's minimizer is analyzed the read's taxon is computed (5), classifying the read. When K2Mem ends, it generates the same output files as Kraken 2.

3   Results

To analyze the performance of K2Mem we compare it with five of the most popular classifiers: Centrifuge [19], CLARK [24], Kraken [33], Kraken 2 [34], and KrakenUniq [6]. We test all these tools on several simulated datasets with different bacteria and viral genomes from NCBI. The experimental setup is described in the next section.

3.1   Experimental Setup

To assess the performance of K2Mem we follow the experimental setup of Kraken 2 [34], that is the strain exclusion experiment. This experiment consists of downloading the archaea, bacteria, and viral reference genomes and the taxonomy from NCBI. From this set of reference genomes 50 are removed (40 bacteria and 10 viruses), called the origin strains, that are used for testing. All the remaining reference genomes are used to build the classifiers' databases, one for each tool in this study. This setup tries to mimic a realistic scenario in which a reference genome is available, but the sequenced genome is a different strain from the same species.

With the origin strains chosen we build 10 datasets using Mason 2 [16] to simulate 100 bps paired-end Illumina sequence data. Of these datasets, 7 are built by varying the number of reads, from 50k to 100M, using the default Illumina error profile. These datasets are used to test the impact of sequencing coverage on the performance of the tools under examination. We also constructed other 3 datasets, all with the same number of reads 100M, but with different mutation rates from the original strains: 2%, 5%, and 10%. With these datasets, we evaluate another scenario in which a close reference genome is not available.

To compare K2Mem with the other tools we use the same evaluation metrics as in [27,34]; precisely we use Sensitivity, Positive Predicted Value (PPV), F-measure, and Pearson Correlation Coefficient (PCC) of the abundance profile.

3.2   Performance Evaluation

In this section, we analyze the performance results at the genus level of K2Mem respect to the other classifiers. All tools are used with the default parameters and run in multithreading using 16 threads. Their databases are built using the same set of reference genomes obtained from the strain exclusion experiment.

The obtained results are reported below in different figures to better understand the performance and the impact of the different configurations.

In Fig. 3 and 4 are shown the full results obtained with the 100M reads dataset for bacteria and viruses respectively. We analyze the 100M dataset as the behaviour of K2Mem with the other datasets is similar. As it can be seen

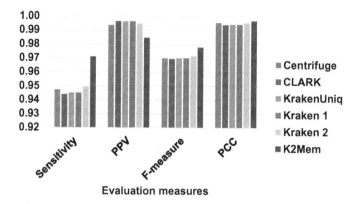

Fig. 3. Bacteria evaluation at genus level on the 100M reads dataset.

in Fig. 3, with bacteria K2Mem obtains an F-measure improvement of at least 0.5 percentage points (pps) respect to the closest competitor, Kraken 2, and the other classifiers. This improvement is due to an increase of sensitivity of at least 2 pps despite a worsening of the PPV of about 1 pps respect to the other tools. Moreover, K2Mem obtains the best PCC value with an improvement of at least 0.1 pps respect to Centrifuge and Kraken 2.

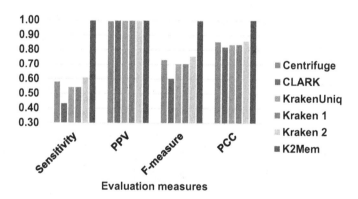

Fig. 4. Viruses evaluation at genus level on the 100M reads dataset.

On the viral dataset, as it can be observed in Fig. 4, K2Mem obtains an F-measure improvement of almost 40 pps respect to Kraken 2 and the other tools. This improvement is due to an increase of sensitivity, whereas the PPV of all tools is to close to 1. Moreover, K2Mem shows the best PCC with an improvement of about 13 pps respect to Centrifuge and Kraken 2.

In summary, with the results reported above, we can observe that thanks to the additional information provided by the new discriminative minimizers

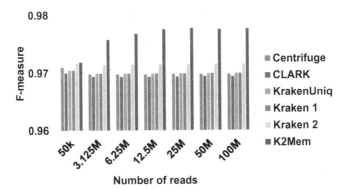

Fig. 5. F-measure values for bacteria at genus level varying the number of reads.

K2Mem obtains a moderate improvement in bacteria's classification and a significant improvement in viruses' classification. Similar results are observed also for species-level classification (data not shown for space limitation).

In Fig. 5 and 6 are shown the F-measure values obtained varying the number reads in the dataset for bacteria and viruses respectively. For bacteria, as it can be seen in Fig. 5, K2Mem gets better F-measure values than the other tools as the number of reads increases; obtaining improvements up to almost 1 pps. This improvement is given mainly from the increase of sensitivity.

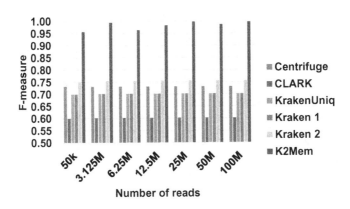

Fig. 6. F-measure values for viruses at genus level varying the number of reads.

For viruses, as it can be seen in Fig. 6, K2Mem achieves greater F-measure improvement than bacteria, always thanks to an increase of the sensitivity. It is interesting to note that the performances of all other tools are independent of the size of the dataset. However, we can observe that for K2Mem the greater the amount of data the better the classification results, due to a greater possibility of finding new discriminative minimizers.

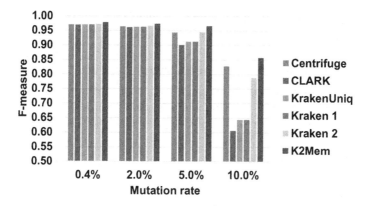

Fig. 7. F-measure values for bacteria at genus level varying the mutation rate.

In Fig. 7 and 8 are shown the F-measure values obtained for the 100M reads dataset while varying the mutation rate, for bacteria and viruses respectively. The mutation rate respect to the origin strains rages between 0.4% to 10%.

As it can be seen in Fig. 7, the first observation is that the performance of all tools decreases as the mutation rate increases. This is expected because the reference genomes are no longer similar to the genomes in the sample data. However, K2Mem obtains the best F-measure values in all cases. Moreover, the F-measure improvement increases up to 2.5 pps w.r.t. Centrifuge and 7 pps with Kraken 2, as the mutation rate increases.

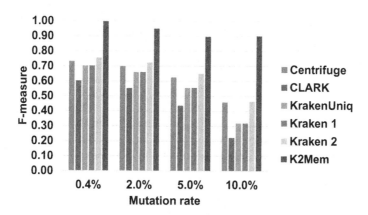

Fig. 8. F-measure values for viruses at genus level varying the mutation rate.

With viruses, as reported in Fig. 8, K2Mem has the same behaviour described for the bacteria, but with a bigger performance gap up to 45 pps respect to Kraken 2. From the results above, the increase in the mutation rate leads to worse performance for all tools as expected. However, K2Mem is the classifier that has suffered less from the presence of mutations in the sample.

3.3 Execution Time and Memory Usage

The execution time and memory usage of each tool during the datasets classification are shown in Fig. 9. For this analysis, the execution time and memory usage values are reported for the largest dataset with 100M reads.

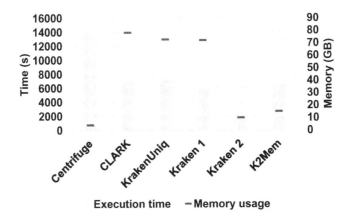

Fig. 9. Execution time and memory usage for a dataset with 100M reads.

The execution time of K2Mem, as expected, it is larger than Kraken 2 due to the new discriminative minimizers search phase. However, the classification time is in line with other tools. As for the memory usage K2Mem needs more memory than Kraken 2, due to the new map. However, K2Mem requires only 15 GB of RAM in line with the best tools.

4 Conclusions

We have presented K2Mem, a classifier based on Kraken 2 with a new classification pipeline and an additional map to store new discriminative k-mers from the input sequencing reads. The experimental results have demonstrated that K2Mem obtains higher values of F-measure, mainly by an improved sensitivity, and PCC respect to the most popular classifiers thanks to the greater number of reference k-mers available during the classification. We showed that the performance improvement of K2Mem increases as the size of the input sequencing data grows, or when reads originate from strains that are genetically distinct from those in the reference database. As possible future developments it could be interesting to increase the PPV, e.g. using unique k-mers, to speed up the classification algorithm with a better implementation and to test other data structures, e.g. counting quotient filter [25], to decrease the memory requirements.

References

1. Altschul, S.F., Gish, W., Miller, W., Myers, E.W., Lipman, D.J.: Basic local alignment search tool. J. Mol. Biol. **215**(3), 403–410 (1990)
2. Ames, S.K., Hysom, D.A., Gardner, S.N., Lloyd, G.S., Gokhale, M.B., Allen, J.E.: Scalable metagenomic taxonomy classification using a reference genome database. Bioinformatics **29** (2013). https://doi.org/10.1093/bioinformatics/btt389
3. Antonello, M., Comin, M.: Fast entropic profiler: an information theoretic approach for the discovery of patterns in genomes. IEEE/ACM Trans. Comput. Biol. Bioinform. **11**(3), 500–509 (2014). https://doi.org/10.1109/TCBB.2013.2297924
4. Comin, M., Antonello, M.: Fast computation of entropic profiles for the detection of conservation in genomes. In: Ngom, A., Formenti, E., Hao, J.-K., Zhao, X.-M., van Laarhoven, T. (eds.) PRIB 2013. LNCS, vol. 7986, pp. 277–288. Springer, Heidelberg (2013). https://doi.org/10.1007/978-3-642-39159-0_25
5. Antonello, M., Comin, M.: Fast alignment-free comparison for regulatory sequences using multiple resolution entropic profiles. In: Proceedings of the International Conference on Bioinformatics Models, Methods and Algorithms (BIOSTEC 2015), pp. 171–177 (2015). https://doi.org/10.5220/0005251001710177
6. Breitwieser, F., Baker, D., Salzberg, S.L.: KrakenUniq: confident and fast metagenomics classification using unique k-mer counts. Genome Biol. **19**(1), 198 (2018)
7. Břinda, K., Sykulski, M., Kucherov, G.: Spaced seeds improve k-mer-based metagenomic classification. Bioinformatics **31**(22), 3584 (2015). https://doi.org/10.1093/bioinformatics/btv419
8. Comin, M., Verzotto, D.: Whole-genome phylogeny by virtue of unic subwords. In: 2012 23rd International Workshop on Database and Expert Systems Applications (DEXA), pp. 190–194, September 2012. https://doi.org/10.1109/DEXA.2012.10
9. Comin, M., Verzotto, D.: Beyond fixed-resolution alignment-free measures for mammalian enhancers sequence comparison. IEEE/ACM Trans. Comput. Biol. Bioinform. **11**(4), 628–637 (2014). https://doi.org/10.1109/TCBB.2014.2306830
10. Comin, M., Antonello, M.: On the comparison of regulatory sequences with multiple resolution entropic profiles. BMC Bioinform. **17**(1), 130 (2016). https://doi.org/10.1186/s12859-016-0980-2
11. Comin, M., Leoni, A., Schimd, M.: Clustering of reads with alignment-free measures and quality values. Algorithms Mol. Biol. **10**(1), 1–10 (2015). https://doi.org/10.1186/s13015-014-0029-x
12. Comin, M., Schimd, M.: Assembly-free genome comparison based on next-generation sequencing reads and variable length patterns. BMC Bioinform. **15**(9), 1–10 (2014). https://doi.org/10.1186/1471-2105-15-S9-S1
13. Eisen, J.A.: Environmental shotgun sequencing: its potential and challenges for studying the hidden world of microbes. PLoS Biol. **5**, e82 (2007)
14. Girotto, S., Comin, M., Pizzi, C.: Higher recall in metagenomic sequence classification exploiting overlapping reads. BMC Genomics **18**(10), 917 (2017)
15. Goke, J., Schulz, M.H., Lasserre, J., Vingron, M.: Estimation of pairwise sequence similarity of mammalian enhancers with word neighbourhood counts. Bioinformatics **28**(5), 656–663 (2012). https://doi.org/10.1093/bioinformatics/bts028
16. Holtgrewe, M.: Mason: a read simulator for second generation sequencing data (2010)
17. Huson, D.H., Auch, A.F., Qi, J., Schuster, S.C.: MEGAN analysis of metagenomic data. Genome Res. **17**, 377–386 (2007)

18. Kantorovitz, M.R., Robinson, G.E., Sinha, S.: A statistical method for alignment-free comparison of regulatory sequences. Bioinformatics **23** (2007). https://doi.org/10.1093/bioinformatics/btm211

19. Kim, D., Song, L., Breitwieser, F., Salzberg, S.: Centrifuge: rapid and sensitive classification of metagenomic sequences. Genome Res. **26**, 1721–1729 (2016). https://doi.org/10.1101/gr.210641.116

20. Lindgreen, S., Adair, K., Gardner, P.: An Evaluation of the Accuracy and Speed of Metagenome Analysis Tools. Cold Spring Harbor Laboratory Press (2015)

21. Mande, S.S., Mohammed, M.H., Ghosh, T.S.: Classification of metagenomic sequences: methods and challenges. Brief. Bioinform. **13**(6), 669–681 (2012). https://doi.org/10.1093/bib/bbs054

22. Marchiori, D., Comin, M.: SKraken: fast and sensitive classification of short metagenomic reads based on filtering uninformative k-mers. In: BIOINFORMATICS 2017–8th International Conference on Bioinformatics Models, Methods and Algorithms, Proceedings; Part of 10th International Joint Conference on Biomedical Engineering Systems and Technologies, BIOSTEC 2017, vol. 3, pp. 59–67 (2017)

23. Ondov, B.D., et al.: Mash: fast genome and metagenome distance estimation using MinHash. Genome Biol. **17**, 132 (2016). https://doi.org/10.1186/s13059-016-0997-x

24. Ounit, R., Wanamaker, S., Close, T.J., Lonardi, S.: CLARK: fast and accurate classification of metagenomic and genomic sequences using discriminative k-mers. BMC Genomics **16**(1), 1–13 (2015). https://doi.org/10.1186/s12864-015-1419-2

25. Pandey, P., Bender, M.A., Johnson, R., Patro, R.: A general-purpose counting filter: making every bit count. In: Proceedings of the 2017 ACM International Conference on Management of Data, pp. 775–787. ACM (2017)

26. Qian, J., Comin, M.: MetaCon: unsupervised clustering of metagenomic contigs with probabilistic k-mers statistics and coverage. BMC Bioinform. **20**(367) (2019). https://doi.org/10.1186/s12859-019-2904-4

27. Qian, J., Marchiori, D., Comin, M.: Fast and sensitive classification of short metagenomic reads with SKraken. In: Peixoto, N., Silveira, M., Ali, H.H., Maciel, C., van den Broek, E.L. (eds.) BIOSTEC 2017. CCIS, vol. 881, pp. 212–226. Springer, Cham (2018). https://doi.org/10.1007/978-3-319-94806-5_12

28. Schimd, M., Comin, M.: Fast comparison of genomic and meta-genomic reads with alignment-free measures based on quality values. BMC Med. Genomics **9**(1), 41–50 (2016). https://doi.org/10.1186/s12920-016-0193-6

29. Shibuya, Y., Comin, M.: Better quality score compression through sequence-based quality smoothing. BMC Bioinform. **20**(302) (2019)

30. Shibuya, Y., Comin, M.: Indexing k-mers in linear-space for quality value compression. J. Bioinform. Comput. Biol. **7**(5), 21–29 (2019)

31. Sims, G.E., Jun, S.R., Wu, G.A., Kim, S.H.: Alignment-free genome comparison with feature frequency profiles (FFP) and optimal resolutions. Proc. Nat. Acad. Sci. **106** (2009). https://doi.org/10.1073/pnas.0813249106

32. Vinga, S., Almeida, J.: Alignment-free sequence comparison-a review. Bioinformatics **19** (2003). https://doi.org/10.1093/bioinformatics/btg005

33. Wood, D., Salzberg, S.: Kraken: ultrafast metagenomic sequence classification using exact alignments. Genome Biol. **15**, 1–12 (2014)

34. Wood, D.E., Lu, J., Langmead, B.: Improved metagenomic analysis with Kraken 2. Genome Biol. **20**(1), 257 (2019). https://doi.org/10.1186/s13059-019-1891-0
35. Yu, Y.W., Yorukoglu, D., Peng, J., Berger, B.: Quality score compression improves genotyping accuracy. Nat. Biotechnol. **33**(3), 240–243 (2015)
36. Zhang, Z., Schwartz, S., Wagner, L., Miller, W.: A greedy algorithm for aligning DNA sequences. J. Comput. Biol. **7**(1–2), 203–214 (2004)

Dilated-DenseNet for Macromolecule Classification in Cryo-electron Tomography

Shan Gao[1,2,3], Renmin Han[4], Xiangrui Zeng[3], Xuefeng Cui[5], Zhiyong Liu[1], Min Xu[3(✉)], and Fa Zhang[1(✉)]

[1] High Performance Computer Research Center, Institute of Computing Technology, Chinese Academy of Sciences, Beijing 100190, China
{gaoshan,zyliu,zhangfa}@ict.ac.cn
[2] University of Chinese Academy of Sciences, Beijing, China
[3] Computational Biology Department, School of Computer Science, Carnegie Mellon University, Pittsburgh, PA, USA
{xiangruz,mxu1}@cs.cmu.edu
[4] Research Center for Mathematics and Interdisciplinary Sciences, Shandong University, Qingdao 266237, People's Republic of China
hanrenmin@gmail.com
[5] School of Computer Science and Technology, Shandong University, Qingdao 266237, People's Republic of China
xfcui@email.sdu.edu.cn

Abstract. Cryo-electron tomography (cryo-ET) combined with subtomogram averaging (STA) is a unique technique in revealing macromolecule structures in their near-native state. However, due to the macromolecular structural heterogeneity, low signal-to-noise-ratio (SNR) and anisotropic resolution in the tomogram, macromolecule classification, a critical step of STA, remains a great challenge.

In this paper, we propose a novel convolution neural network, named 3D-Dilated-DenseNet, to improve the performance of macromolecule classification in STA. The proposed 3D-Dilated-DenseNet is challenged by the synthetic dataset in the SHREC contest and the experimental dataset, and compared with the SHREC-CNN (the state-of-the-art CNN model in the SHREC contest) and the baseline 3D-DenseNet. The results showed that 3D-Dilated-DenseNet significantly outperformed 3D-DenseNet but 3D-DenseNet is well above SHREC-CNN. Moreover, in order to further demonstrate the validity of dilated convolution in the classification task, we visualized the feature map of 3D-Dilated-DenseNet and 3D-DenseNet. Dilated convolution extracts a much more representative feature map.

Keywords: Cryo-electron Tomography · Subtomogram averaging · Object classification · Convolutional neural network

© Springer Nature Switzerland AG 2020
Z. Cai et al. (Eds.): ISBRA 2020, LNBI 12304, pp. 82–94, 2020.
https://doi.org/10.1007/978-3-030-57821-3_8

1   Introduction

The cellular process is dominated by the interaction of groups of macromolecules. Understanding the native structures and spatial organizations of macromolecule inside single cells can help provide better insight into biological processes. To address this issue, cryo-Electron Tomography (cryo-ET), with the ability to visualize macromolecular complexes in their native state at sub-molecular resolution, has become increasingly essential for structural biology [1]. In cryo-ET, a series of two-dimensional (2D) projection images of a frozen-hydrated biological sample is collected under the electron microscopy with different tilted angles. From such a series of tilted images, a 3D cellular tomogram with sub-molecular resolution can be reconstructed [2] with a large number of macromolecules in the crowded cellular environment. To further obtain macromolecular 3D view with higher resolution, multiple copies (subtomograms) of the macromolecule of interest need to be extracted, classified, aligned [3] and averaged, which is named as subtomogram averaging (STA) [4]. However, due to the macromolecular structural heterogeneity, the anisotropic resolution caused by the missing wedge effect and the particularly poor signal-to-noise-ratio (SNR), macromolecule classification is still a great challenge in STA.

One pioneering classification method is template matching [5], where subtomograms are classified by comparing with established template images. However, the accuracy of template matching can be severely affected by the template image. Because the template image can misfit its targets when the template image and the targets are from different organisms or have different conformation. To overcome the limitations of using template images, a few template-free classification methods have been developed [6,7]. Most template-free methods use iterative clustering methods to group similar structures. Because the clustering of a large number of 3D volumes is very time-consuming and computationally intensive, template-free method is only suitable to small datasets with few structural classes.

Recently, with the blowout of deep learning, convolution neural network (CNN) has been applied to the macromolecule classification task [8,9]. CNN classification methods recognize objects by extracting macromolecular visual shape information. Extracting discriminative features is the key to guaranteeing model classification performance. However, due to the high level of noise and complex cellular environment, it is challenging for CNN models to extract accurate visual shape information. Moreover, in traditional CNN architecture, with each convolution layer directly connected, the current convolution layer only feed in features from its adjacent previous layer. Because different depth convolution layer extracts image feature of different level, the lack of reusing features from other preceding convolution layer further limits the accuracy in macromolecule classification.

In this article, we focus on improving classification performance by designing a CNN model (Dilated-DenseNet) that highly utilizes the image multi-level features. We enhance the utilization of image multi-level features by following two ways: 1) Use dense connection to enhance feature reuse during the forward propagation. 2) Adapt dilated convolution in dense connection block to enrich feature map multi-level information. For the convenience of further discussion, here we denote this adapted block as *dilated-dense block*. In our dilated-dense block, with dense connection, each convolution layer accepts features extracted from all preceding convolution layers. And by gradually increasing the dilated ration of dilated convolution layers, the dilated convolution layer performs convolution with an increasingly larger gap on the image to get multi-level information.

In order to verify the effectiveness of the above two ways for classification task, we designed a 3D-DenseNet [10], a 3D-Dilated-DenseNet and compared these two models with the state-of-the-art CNN method (SHREC-CNN) of SHREC, a public macromolecule classification contest [9], on synthetic data [9] and experimental data [11,12]. Our synthetic data is SHREC dataset [9] which contains twelve macromolecular classes and is classified by SHREC into four sizes: tiny, small, medium and large. Our experimental data is extracted from EMPIAR [11,12] with seven categories of macromolecules. The results on both synthetic data and experimental data show that 3D-Dilated-DenseNet outperforms the 3D-DenseNet but 3D-DenseNet is well above SHREC-CNN. On synthetic data, 3D-Dilated-Dense network can improve the classification accuracy by an average of 2.3% for all the categories of the macromolecules. On experimental data, the 3D-Dilated-Dense network can improve the classification accuracy by an average of 2.1%. Moreover, in order to further demonstrate the validity of 3D-Dilated-DenseNet, we visualized the feature map of 3D-Dilated-DenseNet and the result shows that our model can extract more representative features.

The remaining of the paper is organized as follows. Section 2 presents the theory and implementation of our new CNN model 3D-Dilated-DenseNet. Section 3 shows dataset description, experiment details and classification performance of 3D-Dilated-DenseNet by comparing with widely used methods. Section 4 presents the conclusions.

2   Method

2.1   3D-Dilated-DenseNet Architecture

Figure 1A shows the architecture of our 3D-Dilated-DenseNet, the network mainly consists of three parts: dilated dense block (Sect. 2.2), transition block, and the final global average pooling (GAP) layer. Each block comprises several layers which is a composite operations such as convolution (Conv), average pooling (AvgPooling), batch normalization (BN), or rectified linear units (ReLU).

For a given input subtomogram, represented as a 3D array of $\mathbb{R}^{n \times n \times n}$, after the first shallow Conv, the extracted features are used as input for the following dilated dense block. In dilated dense block, we denote the input of block as x_0, the composite function and output of layer l ($l = 1,...,4$) as $H_l(\cdot)$ and x_l. With dense shortcuts interconnect between each layer, the layer l receives the feature maps from its all preceding layers (x_1, ..., x_{l-1}), and we denote the input of layer l as: $x_l = H_l(x_0, x_1, ..., x_{l-1})$. Let each layer outputs k feature maps, so the input feature map of the current l layer is $k_0 + k \times (l - 1)$ where k_0 is the number of block input feature map. Thus the whole block contains $L \times k_0 + k \times L(L - 1)/2$ feature maps. Such large number of feature maps can cause enormous memory consumption. In order to reduce model memory requirement, the layer is designed with a feature map compress module. So the composite function of layer (Fig. 1B) includes two consecutive convolution operations: 1) a $1 \times 1 \times 1$ convolution operation which is used to compress the number of input feature map, and 2) a $3 \times 3 \times 3$ dilated convolution which is used to extract image multi-level information. The detailed information of dilated dense block which focuses on dilated convolution is shown in the next section.

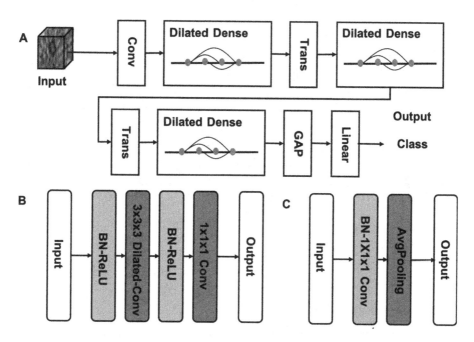

Fig. 1. The architecture of 3D-Dilated-DenseNet. (A) The model framework of 3D-Dilated-DenseNet. (B) The composite function of each layer in dilated dense block. (C) The composite function of transition block in 3D-Dilated-DenseNet.

Because dense connection is not available between size changed feature maps, all feature maps in dilated dense block maintain the same spatial resolution. If these high spatial resolution feature maps go through the entire network without down-sampling, the computation consumption in following block is huge. So we design a transition block between two dilated dense blocks to reduce the size of feature maps. Due to the number of input feature map of transition block is $L \times k_0 + k \times L(L - 1)/2$, the transition block is also defined with a feature map compress module. Therefore, transition block includes following operations:

batch normalization (BN) followed by a $1 \times 1 \times 1$ convolution (Conv) and average pooling layer.

After a series of convolution and down-sampling block, the given input subtomogram is represented as a patch of highly abstract feature maps for final classification. Usually, fully connection (FC) is used to map the final feature maps to a categorical vector which shows the probabilities assigned to each class. In order to increase the non-linearly, traditional CNN always contains multiple FC layers. However, FC covers most of the parameters of the network which can easily cause model overfitting. To reduce model parameter and avoid overfitting, GAP is introduced to replace the first FC layer [13]. The GAP does average pooling to the whole feature map, so all feature maps become a 1D vector. Then the last FC layer with fewer parameters maps these 1D vectors to get the category vector.

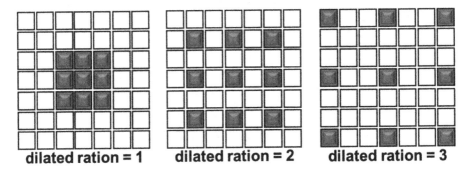

dilated ration = 1   **dilated ration = 2**   **dilated ration = 3**

Fig. 2. A 2D example of dilated convolution layers with 3×3 kernel, and the dilated ration is 1, 2, 3.

2.2   Dilated Dense Block

In order to obtain feature map with representative shape information from the object of interest, we introduce dilated convolution [14] in the dilated dense block. Fig. 2 shows a 2D dilated convolution example. By enlarging a small $k \times k$ kernel filter to $k + (k-1)(r-1)$ where r is dilation ratio, the size of receptive field is increase to the same size. Thus, with enlarged receptive field of the convolution layer, the model can extract multi-level information of subtomogram. And with the stack of convolution layers, the multi-level features can be integrated to present macromolecular shape with less noise.

However, when stack dilated convolution layer with same dilation rate, adjacent pixels in the output feature map are computed from completely separate sets of units in the input, which can easily cause grid artifacts [15]. To solve this problem, we design our dilated convolution layers by following hybrid dilated convolution rule (HDC) [15]. First, the dilated ration of stacked dilated convolution cannot have a common divisor greater than 1. In each dilated dense block, we choose 2 and 3 as dilated ration. Second, the dilated ration should be designed as a zigzag structure such as [1, 2, 5, 1, 2, 5]. We put the dilated convolution layer at the mid of block.

2.3   Visualization of Image Regions Responsible for Classification

To prove that the dilated convolution layer can extract feature map with clearly respond regions from our interest object, we visualize the class activation mapping (CAM) [16] image by global average pooling (GAP). For GAP, the input is the feature map extracted from last convolution layer, and the output is a 1D average vector of each feature map. In a trained CNN model with GAP module followed by FC layer and softmax classification layer, the FC layer has learned a weight matrix that maps the 1D average vector to a category vector. With the computation of softmax classifier, the category vector can show the probability of the input image assigned to each class. For a predicted class that has the highest probability in the category vector, it is easy to get its corresponding weight vector from weight matrix. In the weight vector, each value represents the contribution of its corresponding feature map to classification. Therefore, we can get the class activation mapping image by using a weighted summation of feature map extracted from the last convolution layer and the weight value learned from FC layer.

Here, we denote the kth feature map of the last convolution layer as $f_k(x, y, z)$. After $f_k(x, y, z)$ goes through the GAP block, each $f_k(x, y, z)$ are computed as $\sum_{x,y,z} f_k(x, y, z)$ that is denoted as F_k. From the linear layer followed by GAP, we can get a weight matrix w which shows the contribution of each feature to every category. Inputting an image, predicted with c class, we can get w^c. Each item of w^c records the contribution of last convolution feature map to c class. Then we can compute the class active mapping CAM_c by

$$CAM_c(x, y, z) = \sum_k w_k^c f_k(x, y, z) \tag{1}$$

Due to the fact that CAM_c is the weighted sum of feature maps extracted from last convolution. The size of $CAM_c(x, y, z)$ is generally smaller than origin input image. In order to conveniently observe the extracted features with the input image as a reference, we then up-sample the $CAM_c(x, y, z)$ to get an image with same size as input data.

3   Experiments and Results

3.1   Data

The synthetic subtomogram data is extracted from SHREC dataset [9], consisting of ten 512 * 512 * 512 tomograms and the corresponding ground truth tables. Each tomogram is with 1 nm/1 voxel resolution and contains ∼2500 macromolecules of 12 categories. All macromolecules are uniformly distributed in tomograms with random rotation. And the ground truth table records the location, Euler angle and category for each macromolecule. These 12 macromolecules have various size and have been classified by SHREC to tiny, small, medium and large size. Table 1 shows the protein data bank (PDB) identification of the 12 macromolecules and their size category.

Table 1. The PDB ID and corresponding size of each macromolecule in synthetic data

Macromolecule size	PDB ID
Tiny	1s3x, 3qm1, 3gl1
Small	3d2f, 1u6g, 2cg9, 3h84
Medium	1qvr, 1bxn, 3cf3
Large	4b4t, 4d8q

According to the ground truth table, we extract subtomograms of size $32 \times 32 \times 32$ with the macromolecules located in center. From Fig. 3A we can see the SNR of these subtomograms is low. In order to provide a noise-free subtomogram as a reference for CAM [16] images, we generated the corresponding ground truth using their PDB information. We first download each macromolecules structures from PDB, then generate a corresponding density map by IMOD [17]. Finally, we create an empty volume of $512 \times 512 \times 512$ and put each macromolecule density map into the volume according to the location and Euler angle recorded in the ground truth table (Fig. 3B).

The experimental data are extracted from EMPIAR [11,12], which is a public resource for electron microscopy images. Seven cryo-ET single particle datasets are downloaded as the experiment data[1]. Each dataset on EMPIAR is an aligned 2D tilt series and only contains purified macromolecule of one category. The categories of these macromolecule are rabbit muscle aldolase, glutamate dehydrogenase, DNAB helicase-helicase, T20S proteasome, apoferritin, hemagglutinin, and insulin-bound insulin receptor. To obtain subtomograms, we first reconstruct the tilt series by IMOD and get the 3D tomogram. Then we manually picked up 400 macromolecules for each category.

3.2   Training Details

In this work, our 3D-DenseNet and 3D-Dilated-DenseNet is implemented with Pytorch. During training, the weights of convolution layer and fully connected layer in both networks are initialed by the Xavier algorithm [18]. In particular, we set the parameters to random values uniformly drawn form $[-a, a]$, where $a = \sqrt{\frac{6}{n_{in} + n_{out}}}$, n_{in} and n_{out} denotes size of input and output channel. For batch

---

[1] The EMPIAR indexes of these datasets are 10131, 10133, 10135, 10143, 10169, 10172 and 10173.

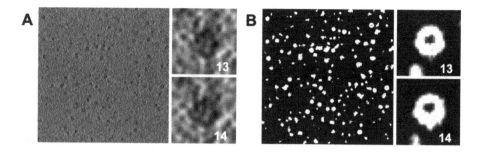

Fig. 3. The example of synthetic data. (A) The middle slice of one $512 \times 512 \times 512$ tomogram. The right 32×32 slices are the consecutive slices of a subtomogram with PDB ID 4d8q. The number of right corner are their slice index. (B) Ground truth corresponding to Fig. (A).

normalization layer, y is set to 1, b is set to 0, all biases are set to 0. The number of feature map output from each convolution layer in dense or dilated dense block (growth rate) is set to 12.

All our network is trained for 30 epochs on 2 GTX 1080TI GPUs with batch size of 64. With the limit memory of GPUs, our network only contains three dilated dense blocks. In fact, users can add more dilated dense blocks according to their GPU memory. According to the training experience, we used Adam [19] as the optimizer and the learning rate is set at 0.1 and scaled down by a factor of 0.1 after every 10 epochs. In order to get efficient training, we adapted various techniques mentioned in the [20] including learning rate warmup strategy, and linear scaling learning rate strategy.

3.3   The Performance of 3D-Dilated-DenseNet on Synthetic Data

In order to compare the classification performance of 3D-DenseNet and 3D-Dilated-DenseNet with the state-of-the-art method on SHREC contest (SHREC-CNN), we chose the same test data and F1 metric as SHREC contest. The computation of F1 metric is given by Eq. 2 which shows the balance of recall and precision.

$$F1 = \frac{2 * \text{precision} * \text{recall}}{\text{precision} + \text{recall}} = \frac{2\text{TP}}{2\text{TP} + \text{FN} + \text{FP}} \tag{2}$$

In Eq. 2. TP means true positive, FN means false negative and FP means false positive.

Table 2 shows the classification performance of above three models on each macromolecule. We counted number of TP (true positive), FN (false negative) and FP (false positive) of each macromolecule and got corresponding F1 score of each category. Judging from the result, we find that the 3D-Dilated-DenseNet performs better than 3D-DenseNet, but 3D-DenseNet performs better than SHREC-CNN. Second, we find the classification performance has high relationship to macromolecule size. Here, we analyze the average F1 value of each model

Table 2. Each macromolecule classification F1 score on SHREC

Model	PDB ID											
	Tiny			Small				Medium			Large	
	1s3x	3qm1	3gl1	3d2f	1u6g	2cg9	3h84	1qvr	1bxn	3cf3	4b4t	4d8q
SHREC	0.154	0.193	0.318	0.584	0.522	0.343	0.332	0.8	0.904	0.784	0.907	0.951
3D-DenseNet	0.522	**0.519**	0.644	0.712	0.580	0.504	0.563	0.795	0.958	0.807	1	**0.997**
3D-Dilated-DenseNet	**0.684**	0.485	**0.675**	**0.778**	**0.652**	**0.565**	**0.635**	**0.855**	**0.971**	**0.846**	1	**0.997**

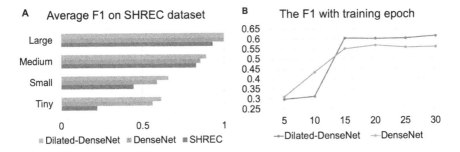

Fig. 4. Dilated-DenseNet performance on synthetic data. (A) Average F1 value on macromolecules according to different size of 3D-DenseNet and 3D-Dilated-DenseNet. (B) The relationship between F1 value and training epoch of 3D-DenseNet and 3D-Dilated-DenseNet.

on macromolecules according to tiny, small, medium and large size (Fig. 4A). The F1 value in Fig. 4A and B is the average of macromolecules F1 scores from same size category. According to Fig. 4A, for all networks, the classification of large size macromolecules has the best performance. Especially, for 3D-DenseNet and 3D-Dilated-DenseNet, the F1 value is close to 1. As the size of macromolecule becomes smaller, the model gets poorer performance. This result is actually valid since that compared to smaller macromolecules larger ones can preserve more shape (or structure) information during pooling operations and get better classification results. Furthermore, with the decreasing of macromolecule size, the performance gap between 3D-DenseNet and 3D-Dilated-DenseNet becomes larger. In Table 2, compared with 3D-DenseNet, 3D-Dilated-DenseNet averagely increased macromolecule classification by 3.7% on medium size, 5.3% on small size, and 6.8% on tiny size respectively.

Also, we test the convergence speed of 3D-DenseNet and 3D-Dilated-DenseNet, we find that dilated convolution does not affect 3D-Dilated-DenseNet convergence. Here, we analyze the performance of a series of 3D-DenseNet and 3D-Dilated-DenseNet which are trained up 30 epochs at intervals of 5. Figure 4B shows the relationship of epoch number and network performance on tiny size macromolecules. According to Fig. 4B, although in the first 13 epoch, the convergence speed of 3D-Dilated-DenseNet is slow, at epoch 15, both models reaches stability and the performance of 3D-Dilated-DenseNet is better than 3D-DenseNet.

3.4   Visualization the Class Active Mapping of 3D-Dilated-DenseNet

In order to demonstrate the effectiveness of dilated convolution in improving classification performance, we visualize the feature map extracted from 3D-DenseNet and 3D-Dilated-DenseNet. Generally, there are two ways to assess feature map validity: 1) showing correct spatial information, in particular, the area which contains macromolecule in the tomogram, and 2) presenting object distinguishable shape information. Because the raw input image has high level noise, we further compare the CAM image of 3D-DenseNet and 3D-Dilated-DenseNet with the ground-truth. In Fig. 5, each row shows one macromolecule with the input image, ground truth, CAM image of 3D-DenseNet and 3D-Dilated-DenseNet. Because the subtomogram data is 3D, we only show the center slice. Here, we explain the image content of each data that is presented in Fig. 5. In the input image, the cluster black regions present macromolecule, and this region is located generally in the center. Oppositely, in ground truth data, the black regions represent background and white regions represent macromolecule. In the CAM image of 3D-DenseNet and 3D-Dilated-DenseNet, the response region is presented with bright pixel and the pixel value reveals the contribution of the

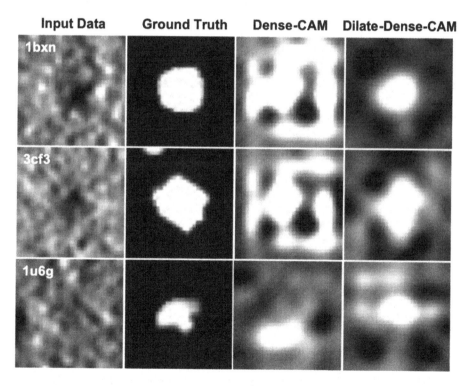

Fig. 5. Class active mapping image of 3D-DenseNet and 3D-Dilated-DenseNet. Each row represents one macromolecule. And the column images are raw input data, ground truth, CAM image of 3D-DenseNet and CAM image of 3D-Dilated-DenseNet

corresponding region of input data to classification. The higher the pixel value, the more contribution to classification.

Judging from Fig. 5, we can find that compared with the CAM image of 3D-DenseNet, the CAM image of 3D-Dilated-DenseNet shows more representative shape information of macromolecule. First, 3D-Dilated-DenseNet CAM shows less response to subtomogram background region. Second, the high response region of 3D-Dilated-DenseNet CAM is more consistent with the macromolecule region in input data. Moreover, the high response region of 3D-Dilated-DenseNet contains clear boundaries that can help network easily distinguish macromolecule region and background region which also arouse slight response.

3.5   The Performance of 3D-Dilated-DenseNet on Experimental Data

We also test the classification performance of 3D-DenseNet and 3D-Dilated-DenseNet on experimental data with F1 metric (Table 3). Compared with synthetic data, the experimental data has higher SNR. Therefore, the classification performance on experimental data is better than that on synthetic data. Because we do not know the PDB id of each macromolecule in experimental data, we cannot compute the relationship of particle size to model performance.

Judging from the Table 3, we can find that the F1 score of category DNAB helicase-helicase, apoferritin is the same, both equal to 1, which means that for these two category macromolecules, the balance between precision and recall is the same. However, for macromolecule of other categories, 3D-Dilated-DenseNet outperforms 3D-DenseNet. Overall, 3D-Dilated-DenseNet improved by 2.1% compared with 3D-DenseNet. Thus, dilated convolution do have a promotion for macromolecule classification task.

Table 3. Macromolecule classification F1 score on experimental data

Model	Particle class						
	Rabbit muscle aldolase	Glutamate dehydrogenase	DNAB helicase - helicase	T20S proteasome	Apoferritin	Hemagglutinin	Insulin-bound insulin receptor
3D-DenseNet	0.9231	0.9558	1.0	0.9339	1.0	0.9569	0.9958
3D-Dilated-DenseNet	0.9915	0.9655	1.0	0.9917	1.0	0.9677	1.0

4   Conclusion

As a significant step in STA procedure, macromolecule classification is important for obtaining macromolecular structure view with sub-molecular resolution. In this work, we focus on improving classification performance of the CNN-based method (3D-Dilated-DenseNet). By adapting dense connection and dilated convolution, we enhance the ability of the network to utilize image multi-level features. In order to verify the effectiveness of dense connection and dilated convolution in improving classification, we implement 3D-DenseNet, 3D-Dilated-DenseNet and compared these two models with the SHREC-CNN (the state-of-the-art model on SHREC contest) on the SHREC dataset and the experimental dataset. The results show that 3D-Dilated-DenseNet significantly outperforms 3D-DenseNet but 3D-DenseNet is still well above the SHREC-CNN. To further demonstrate the validity of dilated convolution in the classification task, we visualized the feature map of 3D-DenseNet and 3D-Dilated-DenseNet. The results show that the dilated convolution can help network extract a much more representative feature map. Although our model has significant improvements in the macromolecule classification task. The small-sized macromolecule is still a bottleneck for our method. And due to the lack of suitable labeled experimental data, we have not fully explored the 3D-Dilated-DenseNet performance on experimental data according to macromolecule sizes. In future works, we will focus on improving classification performance on small size macromolecule and explore the method performance with abundant cryo-ET tomogram experimental data.

Acknowledgments. This research is supported by the Strategic Priority Research Program of the Chinese Academy of Sciences Grant (No. XDA19020400), the National Key Research and Development Program of China (No. 2017YFE0103900 and 2017YFA0504702), Beijing Municipal Natural Science Foundation Grant (No. L182053), the NSFC projects Grant (No. U1611263, U1611261 and 61672493), Special Program for Applied Research on Super Computation of the NSFC-Guangdong Joint Fund (the second phase). This work is supported in part by U.S. National Institutes of Health (NIH) grant P41 GM103712. This work is supported by U.S. National Science Foundation (NSF) grant DBI-1949629. XZ is supported by a fellowship from Carnegie Mellon University's Center for Machine Learning and Health. And SG is supported by Postgraduate Study Abroad Program of National Construction on High-level Universities funded by China Scholarship Council.

References

1. Grünewald, K., Medalia, O., Gross, A., Steven, A.C., Baumeister, W.: Prospects of electron cryotomography to visualize macromolecular complexes inside cellular compartments: implications of crowding. Biophys. Chem. **100**(1–3), 577–591 (2002)
2. Han, R., et al.: AuTom: a novel automatic platform for electron tomography reconstruction. J. Struct. Biol. **199**(3), 196–208 (2017)
3. Han, R., Wang, L., Liu, Z., Sun, F., Zhang, F.: A novel fully automatic scheme for fiducial marker-based alignment in electron tomography. J. Struct. Biol. **192**(3), 403–417 (2015)
4. Wan, W., Briggs, J.: Cryo-electron tomography and subtomogram averaging. In: Methods in Enzymology, vol. 579, pp. 329–367. Elsevier (2016)
5. Ortiz, J.O., Förster, F., Kürner, J., Linaroudis, A.A., Baumeister, W.: Mapping 70s ribosomes in intact cells by cryoelectron tomography and pattern recognition. J. Struct. Biol. **156**(2), 334–341 (2006)

6. Bartesaghi, A., Sprechmann, P., Liu, J., Randall, G., Sapiro, G., Subramaniam, S.: Classification and 3D averaging with missing wedge correction in biological electron tomography. J. Struct. Biol. **162**(3), 436–450 (2008)

7. Xu, M., Beck, M., Alber, F.: High-throughput subtomogram alignment and classification by Fourier space constrained fast volumetric matching. J. Struct. Biol. **178**(2), 152–164 (2012)

8. Che, C., Lin, R., Zeng, X., Elmaaroufi, K., Galeotti, J., Xu, M.: Improved deep learning-based macromolecules structure classification from electron cryo-tomograms. Mach. Vis. Appl. **29**(8), 1227–1236 (2018). https://doi.org/10.1007/s00138-018-0949-4

9. Gubins, I., et al.: Classification in cryo-electron tomograms. In: SHREC 2019 Track (2019)

10. Huang, G., Liu, Z., Van Der Maaten, L., Weinberger, K.Q.: Densely connected convolutional networks. In: Proceedings of the IEEE Conference on Computer Vision and Pattern Recognition, pp. 4700–4708 (2017)

11. Noble, A.J., et al.: Routine single particle CryoEM sample and grid characterization by tomography. Elife **7**, e34257 (2018)

12. Noble, A.J., et al.: Reducing effects of particle adsorption to the air-water interface in cryo-EM. Nat. Methods **15**(10), 793–795 (2018)

13. Lin, M., Chen, Q., Yan, S.: Network in network. arXiv preprint arXiv:1312.4400 (2013)

14. Yu, F., Koltun, V.: Multi-scale context aggregation by dilated convolutions. arXiv preprint arXiv:1511.07122 (2015)

15. Wang, P., et al.: Understanding convolution for semantic segmentation. In: 2018 IEEE Winter Conference on Applications of Computer Vision (WACV), pp. 1451–1460. IEEE (2018)

16. Zhou, B., Khosla, A., Lapedriza, A., Oliva, A., Torralba, A.: Learning deep features for discriminative localization. In: Proceedings of the IEEE Conference on Computer Vision and Pattern Recognition, 2921–2929 (2016)

17. Kremer, J.R., Mastronarde, D.N., McIntosh, J.R.: Computer visualization of three-dimensional image data using IMOD. J. Struct. Biol. **116**(1), 71–76 (1996)

18. Glorot, X., Bengio, Y.: Understanding the difficulty of training deep feedforward neural networks. In: Proceedings of the Thirteenth International Conference on Artificial Intelligence and Statistics, pp. 249–256 (2010)

19. Kingma, D.P., Ba, J.: Adam: a method for stochastic optimization. arXiv preprint arXiv:1412.6980 (2014)

20. He, T., Zhang, Z., Zhang, H., Zhang, Z., Xie, J., Li, M.: Bag of tricks for image classification with convolutional neural networks. In: Proceedings of the IEEE Conference on Computer Vision and Pattern Recognition, pp. 558–567 (2019)

Ess-NEXG: Predict Essential Proteins by Constructing a Weighted Protein Interaction Network Based on Node Embedding and XGBoost

Nian Wang, Min Zeng, Jiashuai Zhang, Yiming Li, and Min Li[(✉)]

School of Computer Science and Engineering, Central South University, Changsha 410083, People's Republic of China
{wang_nian,zengmin,jiashuaizhang,lynncsu}@csu.edu.cn,
limin@mail.csu.edu.cn

Abstract. Essential proteins are indispensable in the development of organisms and cells. Identification of essential proteins lays the foundation for the discovery of drug targets and understanding of protein functions. Traditional biological experiments are expensive and time-consuming. Considering the limitations of biological experiments, many computational methods have been proposed to identify essential proteins. However, lots of noises in the protein-protein interaction (PPI) networks hamper the task of essential protein prediction. To reduce the effects of these noises, constructing a reliable PPI network by introducing other useful biological information to improve the performance of the prediction task is necessary. In this paper, we propose a model called Ess-NEXG which integrates RNA-Seq data, subcellular localization information, and orthologous information, for the prediction of essential proteins. In Ess-NEXG, we construct a reliable weighted network by using these data. Then we use the node2vec technique to capture the topological features of proteins in the constructed weighted PPI network. Last, the extracted features of proteins are put into a machine learning classifier to perform the prediction task. The experimental results show that Ess-NEXG outperforms other computational methods.

Keywords: Essential proteins · RNA-Seq data · Subcellular localization · Weighted protein-protein interaction network · Node embedding · XGBoost

1 Introduction

Essential proteins are very important in organisms and play a crucial role in the life process [1]. If the absence of a certain protein would lead to organisms to become disability or death, it can be said that this protein is essential [2]. Identification of essential proteins not only helps us to deepen the understanding of the life activities of cells

---

The authors wish it to be known that, in their opinion, the first two authors should be regarded as Joint First Authors.

© Springer Nature Switzerland AG 2020
Z. Cai et al. (Eds.): ISBRA 2020, LNBI 12304, pp. 95–104, 2020.
https://doi.org/10.1007/978-3-030-57821-3_9

but also provides a theoretical basis for the study of the pathogenesis of complex diseases and the discovery of drug targets [3, 4]. Thus, it is important for biologists to identify essential proteins. Conventional methods for the identification of essential proteins are biological experiments including RNA interference [5], conditional knockout [6], and single-gene knockout [7]. However, these experimental methods are expensive and time-consuming. Therefore, it is necessary to identify essential proteins by using computational approaches.

The rule of Centrality-Lethality, which indicates that nodes with high connectivity in the networks tend to be essential proteins, has been proposed in 2001 [8]. After that, several computational methods have been developed to identify essential proteins. These computational methods can be roughly divided into two classes: topology-based and machine learning-based methods. There are many topology-based methods, such as Degree Centrality (DC) [9], Betweenness Centrality (BC) [10], Closeness Centrality (CC) [11], Subgraph Centrality (SC) [12], Eigenvector Centrality (EC) [13], Information Centrality (IC) [14], and Local Average Connectivity (LAC) [15]. These methods focus on node centrality and provide a decent performance to identify essential proteins.

With the development of high-throughput sequencing technology, an increasing number of protein data are available to obtain. These protein data lay the foundation for the identification of the essential proteins. Many researchers integrated PPI network and biological information to improve the performance of the essential protein identification. The representative methods are PeC [16], UDoNC [17], ION [18], and CoTB [19]. Besides, many traditional machine learning algorithms are applied to this task. These machine learning algorithms include support vector machine (SVM) [20], Naïve Bayes [21], genetic algorithm [22], and decision tree [23]. Recently, deep learning techniques also have been applied to essential protein prediction and achieve good performance. Zeng et al. [24] proposed a novel computational framework to predict essential proteins based on deep learning techniques which can automatically learn features from three kinds of biological data. Zeng et al. also proposed a method named DeepEP [25, 26] which integrates PPI network and gene expression profiles.

Both in topology-based and machine learning-based methods, PPI networks play an important role. Studies showed that there are many false positive and false negative edges in PPI networks [27, 28], which can influence the performance of essential protein prediction [29]. Thus, to reduce the effects of these noises, it is imperative to construct a reliable weighted network to improve the performance of essential protein prediction by using other biological information. In this study, we used three kinds of biological data: RNA-Seq data, the subcellular localization information, and the orthologous information.

In this paper, we propose a novel computational framework named Ess-NEXG to identify essential proteins. First, to eliminate the noises in the PPI network, the PPI network is weighted by integrating RNA-Seq data, subcellular localization information, and orthologous information. Different from using score function in traditional computational methods, the weights of edges are calculated by dimension reduction from these data. Second, the network representation learning technique is used to learn the topological features of each protein in the weighted PPI network. Finally, the extracted features are used as the input of XGBoost model to identify potential essential proteins.

The effectiveness of Ess-NEXG is validated on the PPI network of saccharomyces cerevisiae (S. cerevisiae) [30]. Compared with the current topology-based methods including BC, CC, EC, IC, LAC, NC, SC, PeC, SPP [31], WDC [32], RSG [33] and NIE [34], Ess-NEXG achieves a better performance. Besides, Ess-NEXG also outperforms other machine learning-based methods.

2 Materials and Methods

2.1 Data Source and Preprocessing

In this study, we used multiple biological data to identify essential proteins: PPI network, RNA-Seq data, subcellular localization information, and orthologous information. These biological data are widely used in the prediction of essential proteins. The PPI network dataset is downloaded from BioGRID database. After the removal of self-cycle interactions and discrete nodes, there are 5,501 proteins and 52,271 interactions in the dataset. Proteins and interactions represent nodes and edges in the PPI network, respectively.

The essential proteins are downloaded from Four databases: MIPS [35], SGD [36], DEG [37], OGEE [38]. After integrating information of essential proteins in the four databases, the dataset contains 1285 essential proteins. The RNA-Seq data is collected from the NCBI SRA database by Lei et al. [39]. This dataset contains gene expression data of 7108 proteins. The subcellular localization information is downloaded from the knowledge channel of COMPARTMENTS database [40]. The orthologous information is gathered from InParanoid database [41].

2.2 Constructing Weighted PPI Network

Formally, the PPI network is described as an undirected graph G (V, E) consisting of a set of nodes $V = \{v_1, v_2, \ldots, v_n\}$ and edges $E = \{e(v_i, v_j)\}$. A node $v_i \in V$ represents a protein and an edge $e(v_i, v_j) \in E$ represents the interaction between protein v_i and v_j. As mentioned above, the PPI network plays an indispensable role in essential protein prediction. However, recent studies showed that there are some noises in the current PPI network, which can affect the identification performance. In order to improve the performance of the essential protein prediction, it is necessary to construct a reliable PPI network.

In this paper, we used RNA-Seq data, subcellular localization information, and orthologous information to weigh the PPI network to reduce the effects of noises. The three types of biological data represent the co-expression of two interacting proteins, the spatiality of proteins, and the conservatism of proteins, respectively. Thus, they can be used to filter the noises and calculate the weight of interacting proteins.

2.2.1 Obtain Better Representation with Principal Component Analysis

We have three different types of biological data. If we combine them directly, each protein has a 24-dimensional feature vector. However, the three kinds of biological data are from different sources and have the following properties:

1. The ranges of values in the three types of biological data vary a lot. The range of values in RNA-Seq data is from zero to tens of thousands; the values in subcellular localization information are binary (0 and 1); the range of values in orthologous information is from 0 to 99.
2. The subcellular localization information is very sparse; the RNA-Seq data and orthologous information are dense.
3. The dimensionality of three kinds of data is different. To order to extract useful features, we used principal component analysis (PCA) to reduce the dimensionality and obtain better representations of proteins.

It is important to find a good way that combines three kinds of biological data for calculating how strong two proteins interact. In consideration of the differences in those three kinds of biological data, we use PCA to reduce the dimension of the 24-dimensional vector to get a better protein representation vector. PCA is a useful tool for feature extraction. The samples are projected from high-dimensional space into low-dimensional space by PCA through linear transformation, which can obtain a dense protein vector and be more suitable for calculating the weight of edges. After the steps of PCA, we can obtain a dense vector which is a better representation.

2.2.2 Calculate the Strength of Interacting Proteins by Pearson's Correlation Coefficient

Pearson's correlation coefficient (PCC) is used to calculate how strong two proteins interact in the raw PPI network. After PCA, each protein has a dense representation vector $W_i = (\omega_1, \omega_2, \ldots \omega_{n'})$. So the strength of two interacting proteins $v_i = (x_1, x_2, \ldots x_{n'})$ and protein $v_j = (y_1, y_2, \ldots y_{n'})$ is calculated by PCC. The value of PCC ranges from -1 to 1, if PCCv_i, v_j is a positive value, it means that the relationship between protein v_i and v_j is positive. On the contrary, if PCCv_i, v_j is a negative value, it means that the relationship between protein and v_j is negative.

Finally, the weight of edges in the network is Weight$(v_i, v_j) = \text{PCC}(v_i, v_j)$. So far, the raw PPI network has been weighted by integrating three types of biological data. Figure 1 plots the workflow of the weighting process.

2.3 Identification of Essential Proteins Based on Network Representation Learning and XGboost

In order to identify essential proteins more correctly, it is necessary to learn better topological features for proteins. In this study, we use node2vec [42] to learn the topological features. Node2vec technique was developed in 2016, it is inspired by word2vec [43] and DeepWalk [44]. It projects every node in the network to a low-dimensional space vector based on unsupervised learning. Node2vec defines two parameters p and q that are used to balance the depth-first search (DFS) and the breadth-first search (BFS), which can preserve the local neighbor node relations and global structure information.

After getting topological features of proteins, the next step is choosing a suitable classifier for essential protein prediction. XGBoost (eXtreme Gradient Boosting) [45] is one of the best available machine learning methods. XGBoost algorithm uses a simple

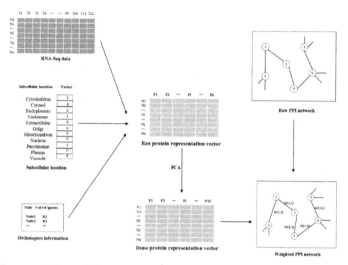

Fig. 1. A diagram of the weighted PPI network construction. The raw protein representation vector is constituted by using RNA-Seq data, subcellular location, and orthologous information. In order to obtain a better representation, we use the PCA technique to reduce the dimension of the raw protein representation vector. Finally, we weight the PPI network by PCC based on the dense representation vector and the raw PPI network.

model to fit the data that can get a general performance. Then, simple models are added to the whole XGBoost model constantly. Until the whole model approaches the complexity of the sample data, the performance of this model is best to identify essential proteins. Figure 2 plots the whole workflow.

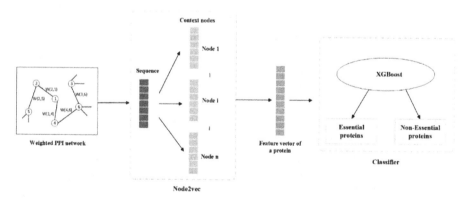

Fig. 2. A diagram of the essential proteins identifying. We use node2vec to extract the protein features, and then use the features we extract as the input of XGBoost to classify proteins.

3   Results

3.1   Comparisons with Current Topology-Based Methods

To validate the performance of Ess-NEXG, we compared Ess-NEXG with some current topology-based methods (BC, CC, EC, IC, LAC, NC, SC, PeC, SPP, WDC, RSG, NIE). In these methods, every node has a score according to corresponding score function. Because there are 1285 essential proteins in the PPI network, we select the top 1285 proteins as candidate essential proteins, and the rest 4216 proteins are candidate non-essential proteins. According to the true label, we calculated the accuracy, precision, recall, and F-score of the 12 computation-based methods. The results of Ess-NEXG and other topology-based methods are shown in Table 1.

Table 1. Comparison of the values of accuracy, precision, recall, and F-score of Ess-NEXG and other topology-based methods.

Methods	Accuracy	Precision	Recall	F-score
BC	0.728	0.411	0.383	0.396
CC	0.670	0.278	0.259	0.268
EC	0.732	0.420	0.391	0.405
IC	0.746	0.454	0.423	0.438
LAC	0.763	0.492	0.458	0.475
NC	0.762	0.490	0.457	0.473
SC	0.732	0.420	0.391	0.405
PeC	0.758	0.480	0.447	0.463
SPP	0.706	0.561	0.479	0.516
WDC	0.758	0.481	0.448	0.464
RSG	0.758	0.475	0.518	0.495
NIE	0.757	0.473	0.528	0.499
Ess-NEXG	0.819	0.600	0.580	0.590

From Table 1, we can see that all assessment metrics obtained by Ess-NEXG are higher than other topology-based methods. According to the results of these topology-based methods, we find that the accuracy of LAC, the precision of SPP, the recall of NIE, and the F-score of SPP are the highest values in these four assessment metrics among these topology-based methods. Compare with the four assessment metrics, Ess-NEXG improves the performance by 7.3%, 7.0%, 9.8%, and 14.3% respectively. In summary, the results indicate that Ess-NEXG outperforms other topology-based methods.

3.2   Comparisons with Other Machine Learning Algorithms

In Ess-NEXG, we choose the XGBoost classifier to identify essential proteins. In order to validate the performance of Ess-NEXG, we compared Ess-NEXG with other machine learning algorithms including support vector machine (SVM), Naïve Bayes, and decision tree, random forest [46], AdaBoost [47]. To ensure equitably, we also use the same input features of proteins and assessment metrics. The results are shown in Table 2. From Table 2, we can see that Ess-NEXG has the best performance. Figure 3 plots the ROC curve of Ess-NEXG and other machine learning algorithms. We can see that the ROC curve of Ess-NEXG is significantly higher than other machine learning algorithms. Table 2 and Fig. 3 show that Ess-NEXG is better than other machine learning algorithms.

Table 2. Comparison of the values of accuracy, precision, recall, F-score, and AUC of Ess-NEXG and other machine learning algorithms.

Model	Accuracy	Precision	Recall	F-score	AUC
SVM	0.70	0.38	0.62	0.47	0.73
Naïve Bayes	0.79	0.50	0.38	0.43	0.72
Decision tree	0.71	0.35	0.40	0.37	0.62
Random forest	0.80	0.58	0.26	0.36	0.71
AdaBoost	0.79	0.51	0.29	0.37	0.71
Ess-NEXG	0.82	0.60	0.58	0.59	0.82

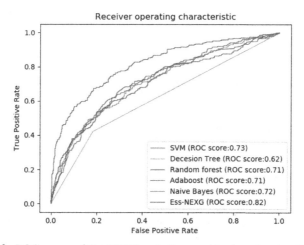

Fig. 3. ROC curves of Ess-NEXG and other machine learning algorithms.

4 Conclusions

Essential proteins are very important proteins in the life process, which can help us understanding life activities in living organisms. The identification of essential proteins is helpful in drug design and disease prediction. In this paper, we propose a novel computational framework to identify essential proteins in the PPI network. Previous studies have shown that the noises in the PPI network affect the performance of essential protein identification. In order to reduce the effects of noises in the PPI network, we propose a weighted method that integrates RNA-Seq data, subcellular localization information, and orthologous information. After obtaining the protein representation vector by using PCA technique, PCC is used to calculate the edge weights in the PPI network. Then node2vec is applied to extract topological features from the weighted PPI network. Finally, the topological features are fed into XGboost model to identify essential proteins. In order to evaluate the performance of Ess-NEXG, we compared it with current topology-based methods. The results show that Ess-NEXG outperforms them. In addition, we also compared Ess-NEXG with machine learning algorithms to show effectiveness. While Ess-NEXG outperforms other computational models, it still has some limitations. The biggest limitation is that we have to collect biological data for each new species, which is expensive and cumbersome. In the future, we would further improve the performance of essential protein prediction by using powerful deep learning techniques [48] and useful biological information [49].

Acknowledgement. This work was supported in part by the National Natural Science Foundation of China under Grants (No. 61832019), the 111 Project (No. B18059), Hunan Provincial Science and Technology Program (2018WK4001).

References

1. Winzeler, E.A., et al.: Functional characterization of the S. cerevisiae genome by gene deletion and parallel analysis. Science **285**, 901–906 (1999)
2. Clatworthy, A.E., Pierson, E., Hung, D.T.: Targeting virulence: a new paradigm for antimicrobial therapy. Nat. Chem. Biol. **3**, 541 (2007)
3. Furney, S.J., Albà, M.M., López-Bigas, N.: Differences in the evolutionary history of disease genes affected by dominant or recessive mutations. BMC Genom. **7**, 165 (2006). https://doi.org/10.1186/1471-2164-7-165
4. Zhao, J., Lei, X.: Detecting overlapping protein complexes in weighted PPI network based on overlay network chain in quotient space. BMC Bioinform. **20**, 1–12 (2019)
5. Roemer, T., et al.: Large-scale essential gene identification in Candida albicans and applications to antifungal drug discovery. Mol. Microbiol. **50**, 167–181 (2003)
6. Cullen, L.M., Arndt, G.M.: Genome-wide screening for gene function using RNAi in mammalian cells. Immunol. Cell Biol. **83**, 217–223 (2005)
7. Giaever, G., et al.: Functional profiling of the Saccharomyces cerevisiae genome. Nature **418**, 387 (2002)
8. Jeong, H., Mason, S.P., Barabási, A.-L., Oltvai, Z.N.: Lethality and centrality in protein networks. Nature **411**, 41 (2001)
9. Hahn, M.W., Kern, A.D.: Comparative genomics of centrality and essentiality in three eukaryotic protein-interaction networks. Mol. Biol. Evol. **22**, 803–806 (2004)

10. Joy, M.P., Brock, A., Ingber, D.E., Huang, S.: High-betweenness proteins in the yeast protein interaction network. Biomed. Res. Int. **2005**, 96–103 (2005)
11. Wuchty, S., Stadler, P.F.: Centers of complex networks. J. Theor. Biol. **223**, 45–53 (2003)
12. Estrada, E., Rodriguez-Velazquez, J.A.: Subgraph centrality in complex networks. Phys. Rev. E **71**, 056103 (2005)
13. Bonacich, P.: Power and centrality: a family of measures. Am. J. Sociol. **92**, 1170–1182 (1987)
14. Stephenson, K., Zelen, M.: Rethinking centrality: methods and examples. Soc. Netw. **11**, 1–37 (1989)
15. Li, M., Wang, J., Chen, X., Wang, H., Pan, Y.: A local average connectivity-based method for identifying essential proteins from the network level. Comput. Biol. Chem. **35**, 143–150 (2011)
16. Li, M., Zhang, H., Wang, J.-X., Pan, Y.: A new essential protein discovery method based on the integration of protein-protein interaction and gene expression data. BMC Syst. Biol. **6**, 15 (2012). https://doi.org/10.1186/1752-0509-6-15
17. Peng, W., Wang, J., Cheng, Y., Lu, Y., Wu, F., Pan, Y.: UDoNC: an algorithm for identifying essential proteins based on protein domains and protein-protein interaction networks. IEEE/ACM Trans. Comput. Biol. Bioinform. (TCBB) **12**, 276–288 (2015)
18. Peng, W., Wang, J., Wang, W., Liu, Q., Wu, F.-X., Pan, Y.: Iteration method for predicting essential proteins based on orthology and protein-protein interaction networks. BMC Syst. Biol. **6**, 87 (2012). https://doi.org/10.1186/1752-0509-6-87
19. Qin, C., Sun, Y., Dong, Y.: A new computational strategy for identifying essential proteins based on network topological properties and biological information. PLoS ONE **12**, e0182031 (2017)
20. Hwang, Y.-C., Lin, C.-C., Chang, J.-Y., Mori, H., Juan, H.-F., Huang, H.-C.: Predicting essential genes based on network and sequence analysis. Mol. BioSyst. **5**, 1672–1678 (2009)
21. Cheng, J., et al.: Training set selection for the prediction of essential genes. PLoS ONE **9**, e86805 (2014)
22. Zhong, J., Wang, J., Peng, W., Zhang, Z., Pan, Y.: Prediction of essential proteins based on gene expression programming. BMC Genom. **14**, S7 (2013). https://doi.org/10.1186/1471-2164-14-S4-S7
23. Acencio, M.L., Lemke, N.: Towards the prediction of essential genes by integration of network topology, cellular localization and biological process information. BMC Bioinform. **10**, 290 (2009). https://doi.org/10.1186/1471-2105-10-290
24. Zeng, M., et al.: A deep learning framework for identifying essential proteins by integrating multiple types of biological information. IEEE/ACM Trans. Comput. Biol. Bioinform. (2019). https://doi.org/10.1109/TCBB.2019.2897679
25. Zeng, M., Li, M., Wu, F.-X., Li, Y., Pan, Y.: DeepEP: a deep learning framework for identifying essential proteins. BMC Bioinform. **20**, 506 (2019). https://doi.org/10.1186/s12859-019-3076-y
26. Zeng, M., Li, M., Fei, Z., Wu, F.-X., Li, Y., Pan, Y.: A deep learning framework for identifying essential proteins based on protein-protein interaction network and gene expression data. In: 2018 IEEE International Conference on Bioinformatics and Biomedicine (BIBM), pp. 583–588. IEEE (2018)
27. Zhang, F., et al.: A deep learning framework for gene ontology annotations with sequence-and network-based information. IEEE/ACM Trans. Comput. Biol. Bioinform. (2020). https://doi.org/10.1109/TCBB.2020.2968882
28. Zhang, F., Song, H., Zeng, M., Li, Y., Kurgan, L., Li, M.: DeepFunc: a deep learning framework for accurate prediction of protein functions from protein sequences and interactions. Proteomics **19**, 1900019 (2019)
29. Von Mering, C., et al.: Comparative assessment of large-scale data sets of protein–protein interactions. Nature **417**, 399 (2002)

30. Stark, C., Breitkreutz, B.-J., Reguly, T., Boucher, L., Breitkreutz, A., Tyers, M.: BioGRID: a general repository for interaction datasets. Nucleic Acids Res. **34**, D535–D539 (2006)

31. Li, M., Li, W., Wu, F.-X., Pan, Y., Wang, J.: Identifying essential proteins based on sub-network partition and prioritization by integrating subcellular localization information. J. Theor. Biol. **447**, 65–73 (2018)

32. Tang, X., Wang, J., Zhong, J., Pan, Y.: Predicting essential proteins based on weighted degree centrality. IEEE/ACM Trans. Comput. Biol. Bioinform. (TCBB) **11**, 407–418 (2014)

33. Lei, X., Zhao, J., Fujita, H., Zhang, A.: Predicting essential proteins based on RNA-Seq, subcellular localization and GO annotation datasets. Knowl.-Based Syst. **151**, 136–148 (2018)

34. Zhao, J., Lei, X.: Predicting essential proteins based on second-order neighborhood information and information entropy. IEEE Access **7**, 136012–136022 (2019)

35. Mewes, H.-W., et al.: MIPS: a database for genomes and protein sequences. Nucleic Acids Res. **30**, 31–34 (2002)

36. Cherry, J.M., et al.: SGD: saccharomyces genome database. Nucleic Acids Res. **26**, 73–79 (1998)

37. Zhang, R., Lin, Y.: DEG 5.0, a database of essential genes in both prokaryotes and eukaryotes. Nucleic Acids Res. **37**, D455–D458 (2008)

38. Chen, W.-H., Minguez, P., Lercher, M.J., Bork, P.: OGEE: an online gene essentiality database. Nucleic Acids Res. **40**, D901–D906 (2011)

39. Zhao, J., Lei, X., Wu, F.-X.: Predicting protein complexes in weighted dynamic PPI networks based on ICSC. Complexity **2017**, 1–11 (2017)

40. Binder, J.X., et al.: COMPARTMENTS: unification and visualization of protein subcellular localization evidence. Database **2014** (2014)

41. Östlund, G., et al.: InParanoid 7: new algorithms and tools for eukaryotic orthology analysis. Nucleic Acids Res. **38**, D196–D203 (2009)

42. Grover, A., Leskovec, J.: node2vec: scalable feature learning for networks. In: Proceedings of the 22nd ACM SIGKDD International Conference on Knowledge Discovery and Data Mining, pp. 855–864. ACM (2016)

43. Goldberg, Y., Levy, O.: word2vec explained: deriving Mikolov et al.'s negative-sampling word-embedding method. arXiv preprint arXiv:1402.3722 (2014)

44. Perozzi, B., Al-Rfou, R., Skiena, S.: Deepwalk: online learning of social representations. In: Proceedings of the 20th ACM SIGKDD International Conference on Knowledge Discovery and Data Mining, pp. 701–710. ACM (2014)

45. Chen, W., Fu, K., Zuo, J., Zheng, X., Huang, T., Ren, W.: Radar emitter classification for large data set based on weighted-xgboost. IET Radar Sonar Navig. **11**, 1203–1207 (2017)

46. Breiman, L.: Random forests. Mach. Learn. **45**, 5–32 (2001). https://doi.org/10.1023/A:1010933404324

47. Freund, Y., Schapire, R., Abe, N.: A short introduction to boosting. J.-Japn. Soc. Artif. Intell. **14**, 1612 (1999)

48. Zeng, M., Li, M., Fei, Z., Yu, Y., Pan, Y., Wang, J.: Automatic ICD-9 coding via deep transfer learning. Neurocomputing **324**, 43–50 (2019)

49. Zeng, M., Zhang, F., Wu, F.-X., Li, Y., Wang, J., Li, M.: Protein–protein interaction site prediction through combining local and global features with deep neural networks. Bioinformatics **36**, 1114–1120 (2020)

mapAlign: An Efficient Approach for Mapping and Aligning Long Reads to Reference Genomes

Wen Yang🄳 and Lusheng Wang$^{(\boxtimes)}$🄳

City University of Hong Kong, 83 Tat Chee Avenue, Kowloon, Hong Kong
cswangl@cityu.edu.hk

Abstract. Long reads play an important role for the identification of structural variants, sequencing repetitive regions, phasing of alleles, etc. In this paper, we propose a new approach for mapping long reads to reference genomes. We also propose a new method to generate accurate alignments of the long reads and the corresponding segments of reference genome. The new mapping algorithm is based on the longest common sub-sequence with distance constraints. The new (local) alignment algorithms is based on the idea of recursive alignment of variable size k-mers. Experiments show that our new method can generate better alignments in terms of both identity and alignment scores for both Nanopore and SMRT data sets. In particular, our method can align 91.53% and 85.36% of letters on reads to identical letters on reference genomes for human individuals of Nanopore and SMRT data sets, respectively. The state-of-the-art method can only align 88.44% and 79.08% letters of reads for Nanopore and SMRT data sets, respectively. Our method is also faster than the state-of-the-art method.

Availability: https://github.com/yw575/mapAlign

Keywords: Long read mapping · Local alignment of long reads · LCS with distance constraints · Variable length k-mer alignment

1 Introduction

The next-generation sequencing (NGS) technologies have changed biological studies in many fields. However, short length of reads poses a limitation for the identification of structural variants, sequencing repetitive regions, phasing of alleles, etc. The long-read sequencing technologies may offer improvements in solving those problems. Nevertheless, the current long read technology suffers from high error rates. The first two steps for DNA analyses are read mapping and aligning reads with the corresponding segments of reference genomes. Most

Supported by GRF grants [Project Number CityU 11256116 and CityU 11210119] from Hong Kong SAR government.

© Springer Nature Switzerland AG 2020
Z. Cai et al. (Eds.): ISBRA 2020, LNBI 12304, pp. 105–118, 2020.
https://doi.org/10.1007/978-3-030-57821-3_10

of the existing methods for read mapping do not work properly for new types of long reads that have high error rates. A few new methods are available now for read mapping, but they are still very slow. The accuracy of existing alignment methods also needs to be improved for long reads. Efficient algorithms are required before any further analysis such as SNP calling and haplotype assembly can be done.

For read mapping, the problem has been solved for short reads. Lots of tools have been developed for short reads that can map large size data sets, e.g., data sets for human individuals, in reasonable time. BWA might be the best tool for short reads and works well for data sets of human individuals [1]. However, tools for short reads cannot work well for large size long read data sets. Some tools for long read mapping have been developed. BLASR [2] is a tool developed by PacBio Company which is designed for SMRT data. GraphMap [3], which is proposed in 2016, is the first tool ever that was reported to design for Nanopore data. Mashmap [6] is a method to use minimizer technique. Minimap [4] and Minimap2 [5] also give solutions for long read mapping. The most updated version Minimap2 can do read mapping and produce alignments of long reads against the corresponding segments in reference genomes. Minimap2 uses the idea of minimizer to reduce the size of reference genomes and speed up the mapping algorithm. A dynamic programming approach is used to find the location of each read by calculating a score that represents number of matched minimizers/k-mers and others facts. SIMD technique is also a key for the success of Minimap2. The speed of Minimap2 is very fast.

Many tools have been developed for aligning reads with reference genomes. In [7], a seed DNA sequence is found based on a "hash table" containing all k-mers present in the first DNA sequence. The hash table is then used to locate the occurrences of the $k\text{-}mer$ sequence in the other DNA sequence. Subsequently, this seed is extended on both sides to complete the alignment. Tools as BLAT [8], SOAP [9], SeqMap [10], mrsFAST [11] and PASS [12] also applied the idea of seed. This implementation is simple and quick for shorter sequences, but is more memory-intensive for long sequences. An improvement is PatternHunter [13], which uses "spaced seeds". This approach is similar to the "seed and extend" approach, but requires only some positions of the seed to match. Many tools were developed based on this approach, including the Corona Lite mapping tool [14], BFAST [15] and MAQ [16]. Newer tools like SHRiMP [17] and RazerS [18] improve on this approach by using multiple seed hits and allowing indels. Other "retrieval-based" approaches was aimed at reducing the memory requirements for alignment and use "Burrows-Wheeler Transform" (BWT), an technique that was first used for data compression [19]. Several very fast tools like SSAHA2 [20], BWA-SW [1], YOABS [21] and BowTie [22] have been created based on this approach, being useful for mapping longer reads. Other alignment tools like SOAP3 [23], BarraCUDA [24] and CUSHAW [25] combine "retrieval-based" approaches with GPGPU computing, taking advantage of parallel GPU cores to accelerate the process. However, the above approaches are still very slow when handling Nanopore and SMRT data sets.

In this paper, we propose a new approach for mapping long reads to reference genomes. We also propose a new method to generate accurate alignments of the long reads and the corresponding segments of reference genome. The new mapping algorithm is based on the longest common sub-sequence with distance constraints. The new (local) alignment algorithms is based on the idea of recursive alignment of variable size k-mers. Experiments show that our new method can generate better alignments in terms of both identity and alignment scores for both Nanopore and SMRT data sets. In particular, our method can align 91.53% and 85.36% of letters on reads to identical letters on reference genomes for human individuals of Nanopore and SMRT data sets, respectively. The state-of-the-art method can only align 88.44% and 79.08% letters of reads for Nanopore and SMRT data sets, respectively. Our method is also faster than the state-of-the-art method. Here we did not use the SIMD technique as in the state-of-the-art method. Our new mapping method is based on the longest common subsequence with approximate distance constraint model ($LCSDC_\delta$). We designed an $O(m \log n)$ running time algorithm for $LCSDC_\delta$. The alignment algorithm is based on the idea of recursively aligning k-mers of different values of k. We expect that the more than 3–5% identically aligned letters will make some differences in the upcoming analysis such as SNP calling and haplotype assembly.

2   Methods

Our method contains two parts. The first part tries to identify the location of each read in the reference genome. The second part does accurate local alignment for the read and the corresponding segment in the reference genome.

2.1   Identifying the Location of a Read in the Reference Genome

To identify the location of a *long* read in the reference genome, a common technique for *all* the existing methods is to start with a *set* of k-mers from the read and based on the set of k-mers positions in the reference genome to find the correct location of the read in the reference genome. The only reason that multiple k-mers should be used is due to the fact that the length of long read could be more than 100k base pairs. Different methods use different strategies to select multiple k-mers and different approaches to find the location of the read in reference genome based on the positions in reference genomes of the set of selected k-mers.

Reference Genome Index. We use homopolymer compressed k-mers for reference genomes. A homopolymer compressed k-mer is obtained from a sequence by compressing every subsequence of identical letters into one letter. It contains k letters where any two consecutive letters in the k-mer are different. Homopolymer compressed (HPC) k-mers was first proposed by SmartDenovo (https://github. com/ruanjue/smartdenovo; J. Ruan) and it can improve overlap sensitivity for

SMRT reads [5]. To increase the speed of the algorithm, we directly use an array of size $4 \times 3^{k-1} - 1$ as a hash table. Each k-mer p corresponds to an integer $i(p)$ between 0 and $4 \times 3^{k-1} - 1$. We use $i(p)$ to indicate the location of k-mer p in the indexing array of size $4 \times 3^{k-1} - 1$. When k is very large, $i(p)$ can serve as the key in a normal hash table. Each cell in the hash table (array) corresponds to a list of positions in the reference genome where the k-mer appears.

Identifying the Location of a Read. The main difficulty to find the location of a read in reference genome is that each k-mer on the read may appear at many positions in the reference genome. To identify the true location of a read on reference genome, we need to use multiple k-mers and require efficient and effective algorithms to identify the location of the read.

Sample n k-mers from a Read: GraphMap decomposes the whole genome into a set of overlapping buckets and looks at the number of k-mers in each bucket [3]. The method is slow since the number of buckets is very large (proportional to the total length of the genome), and the number of k-mers in each bucket is an inaccurate measure without considering the order among the k-mers and the distance between two consecutive k-mers. As a result, they have to use all the k-mers of the read in the process and the running time is very slow comparing to the state-of-the-art tool such as Minimap2.

Minimap2 uses the idea of minimizer to reduce the number of k-mers used for each read and uses a dynamic programming approach to optimize a score which considers the number of matched k-mers and other facts.

Here we propose to use a relatively small number n of k-mers that are approximately evenly distributed over the read, where the default value is $n = 128$. Assume that the average length of each read is 6k to 18k bps, n = 128 is very small comparing to 18k (if using all the k-mers of a read) and 18k/10 (if using minimizers with window size 10). Since n is small, we need a more accurate measure to handle the n samples of k-mers. Therefore, we model the problem as the *longest common subsequence with distance constraints* problem.

The New Measure: For a read r, we select n k-mers that are evenly distributed over the read. Let $r = r_1 r_2 \ldots r_n$ be a sequence of k-mers, where the n k-mers r_i are evenly distributed over the read and the distance between two consecutive k-mers is denoted as d. We treat each r_i as a letter in the sequence r.

Each k-mer r_i may appear at many places over the genome. Let $g = g_1 g_2 \ldots g_m$ be a sequence, where each g_j is an occurrence of a k-mer r_i for some $1 \leq i \leq n$. (Here a genome is considered as a sequence. For genomes with multiple chromosomes, we can handle the chromosomes one by one.)

The Longest Common Subsequence with Distance Constraint Problem (LCSDC): Let $g = g_1 g_2 \ldots g_m$ and $r = r_1 r_2 \ldots r_n$ be sequences obtained by sampling n k-mers from the read, where m is the total number of occurrences of the n k-mers on the genome. When both r and g are viewed as sequences, we assume that each g_j is a r_i for some $1 \leq i \leq n$. Let d be the distance between two consecutive k-mers in r. Our task is to compute a longest common subsequence $s = s_1 s_2 \ldots s_t$ for g and r such that for any two consecutive letters s_i and s_{i+1} in s, the distance between s_i and s_{i+1} is exactly d.

Since there are indels in both reads and genomes, we have the approximate version, the *longest common subsequence with approximate distance constraint problem* ($LCSDC_\delta$), where we want to find a longest common subsequence $s = s_1 s_2 \ldots s_t$ for g and r such that for any two consecutive letters s_i and s_{i+1} in s, the distance between s_i and s_{i+1} is between $d - \delta$ and $d + \delta$.

An $O(m)$ Exact Algorithm for LCSDC: Let $r = r_1 r_2 \ldots r_n$ be the sequence of n k-mers for the read. For each r_i, there is a list of all the occurrences of r_i in the genome. Thus, g can be represented as n lists $L_1, L_2, \ldots, L_n$, where each list L_i stores the positions of the occurrences of r_i over the genome. Every L_i is sorted based on the positions of r_i over the genome.

For two consecutive lists, say, L_1 and L_2, we can merge them into a new list $L_{1,2}$ as follows:

Algorithm 1. Merge two sorted lists

Input: two sorted lists L_1 and L_2
$y \leftarrow$ first item in L_2
for each item x in L_1 **do**
    //x should always be before y on the genome
    while $d(x, y) < d$ **do**
        add y to the new list $L_{1,2}$
        $y = y.next$
    end while
    if d(x,y)== d **then**
        merge two items x and y as a new item (x,y)
        add (x,y) to the new list $L_{1,2}$
    else
        add x to the new list $L_{1,2}$
    end if
end for
add the rest of items in L_2 to the new list $L_{1,2}$

Obviously, the running time of Algorithm 1 is linear in terms of the total length of L_1 and L_2. Based on the merge process, we have the following algorithm:

We merge L_i and L_{i+1} for i = 1, 3, 5, ..., n−1 (assuming that n is even) for the first round. After the first round, we have $0.5n$ lists. We repeatedly merge two consecutive lists rounds by round. After $\log n$ rounds, there is only one list left. We go through the list once more to get the longest (merged) items. The longest item is actually the longest common subsequence with distance constraint. This will lead to an $O(m \log n)$ time algorithm if everything is correct.

Nevertheless, the merge process does not work properly for the second and the rest of rounds. The merge process heavily depends on the two facts: (1) the two lists are sorted based on the items position in the genome and (2) the "distance" between the two lists is a fixed value d. For item (1) the newly merged list can still be kept as sorted. The real problem is that the newly created lists contain items from different lists. When we merge two newly created lists, e.g., $L_{1,2}$ (obtained from merging L_1 and L_2) and $L_{3,4}$ (obtained from merging L_3 and L_4), the distance is not unique. There are four pairs, (L_1 and L_3), (L_1 and L_4), (L_2 and L_3) and (L_2 and L_4.).

To solve the problem, we need to move each item y in L_2 that cannot be merged to any item in L_1 forward by the distance d when creating the new list $L_{1,2}$ from L_1 and L_2. In this way, when we merge $L_{1,2}$ and $L_{3,4}$, we can always use the distance between L_1 and L_3 as the d value to merge. (Note that when obtaining $L_{3,4}$, we also modify the items in L_4 in the same way.) Now the $m \log n$ algorithm works.

Theorem 1. *LCSDC admits an $O(m)$ running time algorithm to get an optimal solution.*

Proof. To get the $O(m)$ running time algorithm, we just move every item in list L_i for $i = 2, 3, \ldots, n$ forward by the distance between L_1 and L_i, where such a distance is always known. After that at each newly obtained position, we have a counter to record the number of times such a position is visited during the "moving forward process". Finally, we output the maximum counter over all the (at most m) positions as the length of the LCSDC. A common subsequence can also be built for each of the (at most m) positions when the value of counter is updated.   □

An $O(m \log n)$ Running Time Heuristic Algorithm for $LCSDC_\delta$: The approximate version $LCSDC_\delta$ is more complicated to deal with. We do not know any efficient algorithm to obtain an optimal solution. Here we propose an $O(m \log n)$ time heuristic algorithm for $LCSDC_\delta$. The main difficulty for $LCSDC_\delta$ is that when many identical k-mers occur at nearby locations in the reference genome, those nearby identical k-mers compete for matching the k-mer on the read. It is hard to decide which nearby k-mer can best match the k-mer on the read. Here we propose a heuristic that works well in practice for $LCSDC_\delta$.

We add a pre-process to remove identical k-mers that a within distance 0.25δ from both g and r. After that, we use an algorithm that is similar to the $O(m \log n)$ exact algorithm for LCSDC. The only difference is that (1) We change the condition $d(x, y) == d$ to a new condition $d - \delta \leq d(x, y) \leq d + \delta$ and (2) When the first y in L_2 is found to satisfy $d - \delta \leq d(x, y) \leq d + \delta$, we look at the next y in L_2 and select the one with the smallest error. We do not use the $O(m)$ algorithm here because the δ is proportional to the distance d and d could be very large if the two consecutive letters are from two far away lists.

Shifting: The running time heavily depends on the total length of lists for k-mers. To further reduce the running time for finding $LCSDC_\delta$, for each of the

n k-mers of the read, we look at the next 6 consecutive base pairs and choose the one with the shortest list.

Implementation of the Algorithms: The algorithm is implemented in C++. For each k-mer, the list containing all occurrences of the k-mer in genome are actually stored in an array (instead of a linked list) and this is one of the keys to speed up the program. It is perhaps worth to emphasize that accessing a list for a k-mer in a huge size array/hash table is time consuming and we access the huge size array/hash table only once for each k-mer on the read. When the final list is obtained after $\log n$ rounds of the merge processes, we will select the item in the unique list that corresponds to the longest subsequence. When the length of $LCSDC$ is less than 4, we assume that we cannot find the location of the read over the genome.

The number of k-mers n is set to be 128 when the length of the read >640. If the length of the read is between 225 and 640, n is set to be 64. If the length of the read is between 80 and 225, $n = 32$. Any read with length <80 are ignored.

We use small size Nanopore and SMRT data sets (see Sect. 3: dataset 3 and dataset 4) to test the tool for the $LCSDC_\delta$ model. The histograms for the length of $LCSDC$ are given in Figs. 1. We can see that for the Nanopore and SMRT data sets, the average lengths of $LCSDC$ are 21.57 and 35.80, respectively.

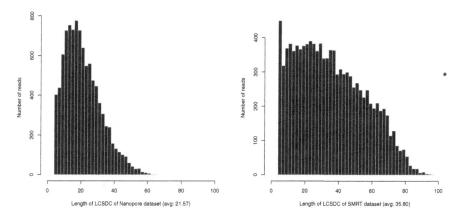

Fig. 1. The distributions of the LCSDC lengths for the dataset 3 (Nanopore) and dataset 4 (SMRT) (GRCh38 as the reference).

K-mer Size vs the Algorithm Speed: The running time of our read mapping algorithm is $O(m \log n)$, where n is the number of k-mers selected from the read and m is the total length of the n lists for k-mers. The running time is heavily depends on the k values of k-mers when indexing the reference genome. Let $l(k)$ be the expected length of a list for a k-mer. The expected total length of n list for k-mers is $m(k) = n \times l(k)$. The expected length of a list for $(k+1)$-mer is

$l(k + 1) = l(k)/4$ since the list of a k-mer will be decomposed into 4 lists when there is one more letter. Thus, the running time of our algorithm will be faster when the value of k increases. On the other hand, increasing the value of k will decreases the chance of finding a matched pair of k-mer.

To illustrate the speed of our algorithm for different k values, we did experiments on a Nanopore dataset and a SMRT dataset (See Sect. 3, dataset 3 and dataset 4). The results are given in Table 1. We can see that when k increases, the running time decreases and the average length of $LCSDC$ also decreases slightly. For $k > 16$, we need to use an efficient way to handle the hash table and will be implemented later.

Table 1. The running time of our algorithm for dataset 3 and dataset 4 with different k-values (GRCh38 as the reference).

Dataset 3	k-mer size $= 14$	k-mer size $= 15$	k-mer size $= 16$
Average $LCSDC$ length	24.38	22.69	21.57
Identity	75.41%	75.66%	75.80%
Align. score	89.42%	89.61%	89.70%
Failure cases	1.55%	1.97%	2.39%
CPU time	262 s	189 s	128 s
Peak Memory	19.6 G	22.1 G	25.7 G
Dataset 4	k-mer size $= 14$	k-mer size $= 15$	k-mer size $= 16$
Average $LCSDC$ length	35.32	35.92	35.80
Identity	81.10%	81.23%	81.30%
Align. score	86.34%	86.42%	86.53%
Failure cases	2.00%	2.29%	2.64%
CPU time	185 s	141 s	105 s
Peak memory	19.0 G	20.7 G	25.2 G

2.2   Aligning a Read Against the Reference Genome

Once the read location on reference genome is found, we will align the read with the corresponding segment on the reference genome. It is well known that a dynamic programming algorithm with quadratic running time can give the optimum solution. However, the algorithm is too slow to handle huge size data sets such as Nanopore or SMRT data for human individuals. Here will propose a heuristic algorithms for local alignment which is much more accurate than all the existing heuristic methods.

Our key idea is to view the DNA sequences (with 4 letters) as a sequence of k-mers. When the DNA sequence is long, we use large size k-mer to find an alignment of the k-mers between the two sequences. Such an alignment of k-mers between the two sequences decomposes the whole sequence region into many smaller size regions and for each pair of smaller size regions we can recursively use smaller size k-mers to do alignment. When the size of the pair of regions is small enough, e.g., <15, we use the dynamic programming algorithm to give the alignment. We refer to our algorithm as the *recursive variable length k-mer alignment algorithm*.

After the read mapping procedure, our program has found an LCSDC for k-mers with $k = 16$. Starting with a matched k-mer (with $k = 16$), we try to extend the matching to the two ends. The LCSDC for k-mers with $k = 16$ decomposes the whole read into many smaller size regions and the lengths of those regions could be from a few hundreds to a few thousands (if the read has 18k bps and the length of LCSDC is from 10 to 30). Each time, we look at a pair of segments with 256 DNA letters from both sequences. We then view the two segments as sequences of k-mers with $k = 9$ and use the heuristic Algorithm 2 to align the two sequences. The right ends of the two sequences stop at the positions where the last pair of matched k-mer with $k = 9$ ends. Our algorithm can still handle large size >256 indels since the extension from the other direction can handle this.

The Heuristic Algorithm for Alignment of K-mer Sequences. Let $L = L_1 L_2 \ldots L_n$ be a sequence of k-mers from the read. Each L_i is a list of positions on the reference genome that the k-mer appears. We use $|L_i|$ to represent the length of the list. Let p_0 and l_0 be integers representing the position in the reference genome and the location in the read, where the previous pair of k-mers match. Our algorithm looks at each L_i for $i = 1, 2, \ldots, n$. If the list L_i is empty, i.e., $|L_i| = 0$ then the corresponding k-mer does not have a match on the reference genome. If $|L_i| > 3$, then there are to many occurrences of the k-mer and a repeated region is found. In this case, we do not try to identify the match. When $|L_i| > 0$ and $|L_i| \leq 3$, our algorithm will try to find a match that satisfies the condition $|(l - l_0) - (p - p_0)| \leq \frac{\min\{l - l_0, p - p_0\}}{2}$. Such a condition is used to ensure that the lengths of the newly created pair of regions (by determining the pair of matched k-mer) on the reference genome and the read are roughly the same. Once a match of a pair of k-mers is found, the algorithm tries to extend the length of the match by looking at the next pair of DNA letters whenever possible.

Obviously, the running time of Algorithm 2 is linear in term of the input sequences and is much simpler and faster than the heuristic algorithm for $LCSDC_\delta$. The key point here is that once a pair of k-mers is matched, the region is decomposed into two parts of smaller sizes. Missing some matches of k-mers does not affect the final alignment as long as the matched pairs are correct.

After that, we re-start with the last pair of matched k-mers (with $k = 9$) and look at the next pair of segments with 256 DNA letters and repeat the process. The process stops if we meet the next matched k-mer with $k = 16$ (obtained in read mapping process) or there is no matched pair of k-mer for $k = 9$.

Algorithm 2. A heuristic algorithm to align two sequences of k-mers

---

Input: a sequence $L = L_1 L_2 \ldots L_n$ and integers p_0 and l_0.
STATE hit=0
for i=1 to n **do**
  if $|L_i| > 0$ and $|L_i| \leq 3$ **then**
    for each position p in L_i **do**
      Let l be the location of the i-th k-mer on the read
      if $|(l - l_0) - (p - p_0)| \leq \frac{\min\{l-l_0, p-p_0\}}{2}$ **then**
        match the current pair of k-mers and hit=1
        while The next pair of letters are identical **do**
          extend the match
        end while
      end if
    end for
  end if
end for
return hit

If the returned $hit == 0$, we will reduce the size of k by 1 and repeat the process. The process can be repeated by at most 6 times when k goes down from 9 to 4.

3 Results

In this section, we use some datasets to illustrate the running time and alignment quality of our method.

3.1 Data Sets

We downloaded two large size (human) datasets (dataset 1 and dataset 2) for full comparison purpose. Also, We generated two small size datasets (dataset 3 and dataset 4) for quick testing purpose. The human genome GRCh38 is used as the reference.

Dataset 1: We use a human dataset (NA12878) sequenced on Oxford Nanopore Technology. This dataset is downloaded at http://s3.amazonaws.com/nanopore-human-wgs/rel6/rel_6.fastq.gz [26]. There are 15666888 reads in this dataset.

Dataset 2: To do comparison with SMRT data, we use a raw sequence data resulting from PacBio SMRT Sequencing. This dataset is downloaded at http://bit.ly/chm1p5c3 [27]. This dataset has 22565609 reads.

Dataset 3: We randomly select 10000 reads from dataset 1.

Dataset 4: We randomly select 12000 reads from Ashkenazim trio-Child NA24385 dataset sequenced by PacBio SMRT.

3.2   Alignment Quality Comparison with BLAST

To compare the quality of the generated alignments, we use two different scores. The *alignment score* is defined as the number of identically matched pairs of letters over the total length of read. The *identity* score is define to be the number of identically matched pairs of letters over the total length of the alignment. Note that in many cases, the two ends of a read are excluded in the resulting alignments. Different programs may excluded different segments of the read ends. Thus identify score cannot fully represent the quality of the alignment. On the other hand, the alignment score defined here can only represent the percentage of read letters to be matched. There may be the case that all the letters in a read of length 100 can be perfectly matched to a region of length 10000. This is also not reasonable. Here we use both measures.

We did a preliminary comparison with BLAST (version: 2.9.0) which suppose to be good at doing alignment. We use dataset 3 against GRCh38. The results are as follows: The identity score and alignment score for BLAST are 77.77% and 70.40%, while the identity score and alignment score for our method are 75.80% and 89.70%, respectively. We can see that our method can align 89.70% of letters from the read to identical letters on the reference genome while BLAST can only align 70.40% of letters from the read. The identify score of our method is 75.80 which is slightly worse than that of BLAST. This means that the alignments generated by our method have slightly more number of letters from the reference genome that are aligned with spaces into order to get $89.70 - 70.40 = 19.3\%$ more letters from the read to be matched to identical letters. BLAST often cannot align well at the two ends of the read. The 77.77% identity score for BLAST is obtained by using the alignment length, where the two ends of read are not included in the alignment. BLAST is at least 30 times slower than our method. Thus, we do not give the details of running time comparison.

3.3   Comparison with Other Methods

GraphMap is at least 50 times slower than Minimap2. According to [5], Minimap2 also has better alignment quality. Here we did a comparison with Minimap2 using two large size datasets, i.e., dataset 1 and dataset 2. The results are shown in Table 2.

For the large Nanopore dataset (dataset 1). The failure cases for Minimap2 and mapAlign are 16.83% and 17.72%, respectively. We have a slightly higher failure case than Minimap2. However, the alignment quality (in terms of both the identify and alignment scores) of our method is better. Also, we use only 53% CPU time of Minimap2.

For the large SMRT dataset (dataset 2). The failure cases for Minimap2 and mapAlign are 4.38% and 6.26%, respectively. Both our identify and alignment scores are about 5% higher than Minimap2. Also, our method is faster than Minimap2.

Table 2. Comparison between Minimap2 and mapAlign for dataset 1 and dataset 2(GRCh38 as the reference).

Dataset 1	Minimap2	mapAlign
Identity	81.71%	83.45%
align. score	88.44%	91.53%
CPU time	266375 sec	141559 sec
Failure cases	16.83%	17.72%
Peak Memory	13.5 G.	27.1 G.
Dataset 2	Minimap2	mapAlign
Identity	74.17%	79.12%
align. score	79.08%	85.36%
CPU time	390029 sec	247005 sec
Failure cases	4.38%	6.26%
Peak Memory	9.2 G.	26.2 G.

4   Discussions

The main reason that our algorithm is faster than Minimap2 is that we use a much small number (128) of k-mers for each read, where each k-mer corresponds to a list. If the read is of length 10k, then Minimap2 with minimizer and window size 10 needs to handle 1k k-mers and the corresponding lists. Thus Minimap2 needs more time. On the other hand, increasing the window size for minimizer approach will reduce the accuracy. Our $LCSDC_\delta$ model allows us to sample a small number of k-mers and obtain accurate results by considering the distance between two consecutive k-mers.

The main reasons that our algorithm can generate more accurate alignments are: (1) Starting with a pair of matched k-mer for ($k = 16$), we consider to match k-mers of smaller size (e.g., 7 or 9) for the two small size segments (at most 256 bps each) of the read and the genome. The length of the list for each k-mer is very short since we do not use the list for the whole genome anymore; (2) In most case, k-mers (for $k = 16, 9, 7$) are "correctly" matched; and (3) The matched k-mers decompose the whole region into many small size regions and for each small size region (≤ 15 bps in most cases) we use the exact quadratic time algorithm to give local optimal solutions. The quadratic algorithm works efficiently when the size of region is small. Moreover, our algorithm use more strict conditions to ignore cases where the two segments are not similar at all. Thus, we have slightly higher failure cases than Minimap2. A few failure cases could consume lots of computational time for our algorithm since the two sequences are not similar, the sizes of the decomposed regions are large, and the quadratic algorithm becomes very slow.

Acknowledgements. This work is supported by a GRF grant for Hong Kong Special Administrative Region, China (CityU 11256116) and a grant from the National Science Foundation of China (NSFC 61972329).

References

1. Li, H., Durbin, R.: Fast and accurate short read alignment with Burrows-Wheeler transform. Bioinformatics **25**, 1754–60 (2009)
2. Chaisson, M.J., Tesler, G.: Mapping single molecule sequencing reads using basic local alignment with successive refinement (BLASR): application and theory. BMC Bioinform. **13**(1), 238 (2012)
3. Sovic, I., Sikic, M., Wilm, A., Fenlon, S.N., Chen, S., Nagarajan, N.: Fast and sensitive mapping of Nanopore sequencing reads with GraphMap. Nat. commun. **7**, 11307 (2016)
4. Li, H.: Minimap and miniasm: fast mapping and de novo assembly for noisy long sequences. Bioinformatics **32**(14), 2103–2110 (2016)
5. Li, H.: Minimap2: fast pairwise alignment for long nucleotide sequences. Bioinformatics **34**(18), 3094–3100 (2017). arXiv:1708.01492
6. Jain, C., Dilthey, A., Koren, S., Aluru, S., Phillippy, A.M.: A fast approximate algorithm for mapping long reads to large reference databases. In: Recomb 2017, pp. 66–81 (2017)
7. Altschul, S.F., Gish, W., Miller, W., Myers, E.W., Lipman, D.J.: Basic local alignment search tool. J. Mol. Biol. **215**, 403–10 (1990)
8. Kent, W.J.: BLAT-the BLAST-like alignment tool. Genome Res. **12**, 656–64 (2002)
9. Li, R., Li, Y., Kristiansen, K., Wang, J.: SOAP: short oligonucleotide alignment program. Bioinformatics **24**, 713–4 (2008)
10. Jiang, H., Wong, W.H.: SeqMap: mapping massive amount of oligonucleotides to the genome. Bioinformatics **24**, 2395–6 (2008)
11. Hach, F., et al.: mrsFAST: a cache-oblivious algorithm for short-read mapping. Nat. Methods **7**, 576–7 (2010)
12. Campagna, D., et al.: PASS: a program to align short sequences. Bioinformatics **25**, 967–8 (2009)
13. Ma, B., Tromp, J., Li, M.: PatternHunter: faster and more sensitive homology search. Bioinformatics **18**, 440–5 (2002)
14. McKernan, K.J., et al.: Sequence and structural variation in a human genome uncovered by short-read, massively parallel ligation sequencing using two-base encoding. Genome Res. **19**, 1527–41 (2009)
15. Homer, N., Merriman, B., Nelson, S.F.: BFAST: an alignment tool for large scale genome resequencing. PLoS ONE **4**, e7767 (2009)
16. Li, H., Ruan, J., Durbin, R.: Mapping short DNA sequencing reads and calling variants using mapping quality scores. Genome Res. **18**, 1851–8 (2008)
17. Rumble, S.M., et al.: SHRiMP: accurate mapping of short color-space reads. PLoS Comput. Biol. **5**, e1000386 (2009)
18. Weese, D., Emde, A.-K., Rausch, T., Döring, A., Reinert, K.: RazerS-fast read mapping with sensitivity control. Genome Res. **19**, 1646–54 (2009)
19. Burrows, M., Wheeler, D.J.: A block-sorting lossless data compression algorithm. Digital Equipment Corporation 124 (1994)
20. Ning, Z., Cox, A.J., Mullikin, J.C.: SSAHA: a fast search method for large DNA databases. Genome Res. **11**, 1725–9 (2010)

21. Galinsky, V.L.: YOABS: yet other aligner of biological sequences-an efficient linearly scaling nucleotide aligner. Bioinformatics **28**, 1070–7 (2012)
22. Langmead, B., Trapnell, C., Pop, M., Salzberg, S.L.: Ultrafast and memory-efficient alignment of short DNA sequences to the human genome. Genome Biol. **10**, R25 (2009)
23. Liu, C.-M., et al.: SOAP3: ultra-fast GPU-based parallel alignment tool for short reads. Bioinformatics **28**, 878–9 (2012)
24. Klus, P., et al.: BarraCUDA - a fast short read sequence aligner using graphics processing units. BMC Res. Notes **5**, 27 (2012)
25. Liu, Y., Schmidt, B., Maskell, D.L.: CUSHAW: a CUDA compatible short read aligner to large genomes based on the Burrows-Wheeler transform. Bioinformatics **28**, 1830–1837 (2012)
26. Jain, M., et al.: Nanopore sequencing and assembly of a human genome with ultra-long reads. bioRxiv, 128835 (2017)
27. Ono, Y., et al.: PBSIM: pacBio reads simulator-toward accurate genome assembly. Bioinformatics **29**, 119–121 (2013)

Functional Evolutionary Modeling Exposes Overlooked Protein-Coding Genes Involved in Cancer

Nadav Brandes[1]($\boxtimes$), Nathan Linial[1], and Michal Linial[2]($\boxtimes$)

[1] School of Computer Science and Engineering, The Hebrew University of Jerusalem, Jerusalem, Israel
nadav.brandes@mail.huji.ac.il
[2] Department of Biological Chemistry, The Alexander Silberman Institute of Life Sciences, The Hebrew University of Jerusalem, Jerusalem, Israel
michall@cc.huji.ac.il

Abstract. Numerous computational methods have been developed to screening the genome for candidate driver genes based on genomic data of somatic mutations in tumors. Compiling a catalog of cancer genes has profound implications for the understanding and treatment of the disease. Existing methods make many implicit and explicit assumptions about the distribution of random mutations. We present FABRIC, a new framework for quantifying the evolutionary selection of genes by assessing the functional effects of mutations on protein-coding genes using a pre-trained machine-learning model. The framework compares the estimated effects of observed genetic variations against all possible single-nucleotide mutations in the coding human genome. Compared to existing methods, FABRIC makes minimal assumptions about the distribution of random mutations. To demonstrate its wide applicability, we applied FABRIC on both naturally occurring human variants and somatic mutations in cancer. In the context of cancer, ~3 M somatic mutations were extracted from over 10,000 cancerous human samples. Of the entire human proteome, 593 protein-coding genes show statistically significant bias towards harmful mutations. These genes, discovered without any prior knowledge, show an overwhelming overlap with contemporary cancer gene catalogs. Notably, the majority of these genes (426) are unlisted in these catalogs, but a substantial fraction of them is supported by literature. In the context of normal human evolution, we analyzed ~5 M common and rare variants from ~60 K individuals, discovering 6,288 significant genes. Over 98% of them are dominated by negative selection, supporting the notion of a strong purifying selection during the evolution of the healthy human population. We present the FABRIC framework as an open-source project with a simple command-line interface.

Keywords: Driver genes · Machine learning · TCGA · Positive selection · Cancer evolution · Single nucleotide variants · ExAC

1 Introduction

Most arising somatic mutations in cancer are considered passenger mutations, whereas only a small fraction of them have a direct role in oncogenesis, and are thus referred

© Springer Nature Switzerland AG 2020
Z. Cai et al. (Eds.): ISBRA 2020, LNBI 12304, pp. 119–126, 2020.
https://doi.org/10.1007/978-3-030-57821-3_11

to as cancer driver mutations [1, 2]. The Cancer Genome Atlas (TCGA) is a valuable resource of genomic data from cancer patients covering >10,000 samples in over 30 cancer types [3]. An ongoing effort in cancer research is compiling a comprehensive catalog of cancer genes which have a role in tumorigenesis.

Numerous computational frameworks have been designed for the purpose of identifying suspect cancer genes [4–6]. Most of these frameworks, regarded as "frequentist", are based on the premise that cancer genes are recurrent across samples and can be recognized by high numbers of somatic mutations. In contrast, passenger mutations are expected to appear at random. Assessing whether a gene shows an excessive number of mutations must be considered in view of an accurate null background model. Since cancer is characterized by order-of-magnitudes variability in mutation rates among cancer types and genomic loci [7], the frequentist approach requires complex modeling of gene mutation rates as a function of the composition of samples, cancer types and specific loci in the genomes that display extreme deviation in their mutation rates [8]. The sensitivity of the frequentist approach to modeling choices leads to lingering uncertainty and controversy [4].

An alternative to the frequentist approach, which can be regarded as "functionalist", considers the content of mutations rather than their numbers. It is based on the premise that somatic mutations in cancer genes, are subjected to positive selection and, as a result, are more damaging than expected at random. Under the functionalist approach, each gene has its own inherent background model which only depends on static properties of the gene and the number of mutations. Other variables, such as the samples or cancer types that the mutations have originated from, or the specific genomic region of the gene under study, do not need to be part of the model.

A simplistic functionalist model is based on the ratio of non-synonymous to synonymous (dN/dS) mutations [9]. This model is a common metric for studying the evolutionary selection of a gene. A richer functionalist model was recently explored by OncodriveFML [10]. It estimates the pathogenicity of mutations using CADD [11], which provides numeric scores for the clinical effects of mutations. OncodriveFML then compares the CADD effect scores of the somatic mutations observed within a gene to those of random mutations using permutation tests. OncodriveFML still uses a complex background model that includes sample identities and cancer types. As a result of its complex background model, it requires computationally demanding permutation tests and is unable to analytically calculate probabilities.

With the goal of developing an analytical functionalist model, we introduce a new framework called FABRIC (Functional Alteration Bias Recovery In Coding-regions) [12]. FABRIC is a purely functionalist framework, with a simple background model that is completely agnostic to samples, cancer-types and genomic regions. This simplicity allows calculation of precise p-values per gene. As a result, FABRIC can provide a detailed ranking of all genes by significance.

The full description of FABRIC and its demonstration to cancer is available elsewhere [12]. In this report, we iterate the highlights of that work and further demonstrate the applicability of FABRIC to broad evolutionary contexts. In particular, we show its ability to detect a trend of negative selection in the context of naturally occurring human genetic variations.

2 Methods

Framework Overview

FABRIC analyzes each protein-coding gene independently, extracting all the single nucleotide variations (SNVs) observed within the coding regions of that gene (Fig. 1A). It then uses FIRM, a machine-learning model to assign functional effect scores to each SNV, which measure the predicted effects of those variants explicitly on the protein function (Fig. 1B). Intuitively, this score can be thought of as the probability of the protein to retain its original biochemical function given the mutation. Simplistically, all synonymous mutations are assigned a score of 1 and loss of function (LoF) mutations are assigned a 0 score. Missense mutations are processed through FIRM [12] to obtain a score between 0 to 1. Notably, FIRM was trained in advance on an independent dataset.

Independently to the calculation of scores for the observed mutations, a background distribution for the expected scores is also constructed, assuming that unselected passenger mutations occur at random by a uniform distribution across the gene (Fig. 1C). This background model is precise, and calculated individually for each gene. Significant deviations between the null background distribution to the observed effect scores are then detected (Fig. 1D). Z-values measure the strengths of deviations between observed to expected scores, and used to derive exact p-values. If a gene's average z-value is significantly negative, it means that mutations are more damaging to the gene function than expected by the same number of mutations randomly distributed along the gene's coding sequence. In such case, the gene is deemed to be "alteration promoting", reflecting its tendency to harbor damaging mutations. An observed score that is significantly higher than expected indicates a gene that is more constrained than expected. We refer to these genes as "alteration rejecting".

We illustrate FABRIC's background model by analyzing *TP53*, one of the most studied cancer gene (Fig. 1E-H; see details in [12]). Importantly, we derive 12 background distributions, corresponding to the 12 possible single-nucleotide substitutions (Fig. 1F). These distributions are gene specific and are independent of the input data (Fig. 1E). Hence, the background model accounts for the exact number of mutations and their SNV frequencies as observed for the studied gene. The mixed model background distribution is gene specific (Fig. 1G). Note that we only considered SNVs (and ignored in-frame indels and splicing variants) at coding regions. Following such simplification, 93% of the somatic mutations in the analyzed dataset is considered. Note that by ignoring complex variations and effects, FABRIC underestimates the damage to gene function.

Effect Score Prediction Mode

A key component of FABRIC is a pre-trained machine-learning model for predicting the effects of missense genetic variants on protein function. This machine-learning model is called FIRM (Functional Impact Rating at the Molecular-level; Fig. 2). Different from many mutation prediction tools (e.g. CADD [11], Polyphen2 [13]) that predict clinical pathogenicity scores, FIRM seeks positive selection at the biochemical, functional level. Importantly, FIRM was pre-trained on ClinVar [14], which is independent to the datasets used in this work.

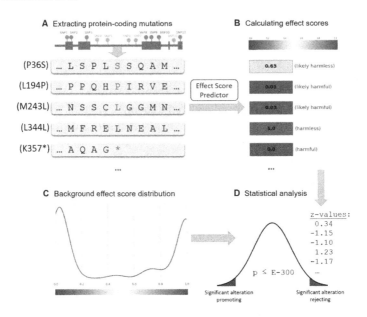

◀**Fig. 1. FABRIC framework. (A-D)** Framework overview, **(E-H)** background model (*TP53* as an example). **(A)** All somatic mutations within a particular gene are collected from a variety of samples and cancer types. SNVs within protein-coding regions are analyzed to study their effects on the protein sequence (synonymous, missense or nonsense). **(B)** Using a machine-learning model, we assign each mutation a score for its effect on the protein biochemical function, with lower scores indicating mutations that are more likely harmful. **(C)** In parallel, a precise null background score distribution is constructed (details in E-H). **(D)** By comparing the observed scores to their expected distribution, we calculate z-values for the mutations, and overall z-value and p-value for the gene. **(E)** 3,167 SNVs were observed in coding regions of *TP53* from which a 4×4 matrix of single-nucleotide substitution frequencies was derived. Note that this matrix is non-symmetric (e.g. 25.3% of the substitutions are G to A, while only 2.9% are A to G). **(F)** For each of the 12 possible nucleotide substitutions, an independent background effect score distribution was calculated, by considering all possible substitutions within the coding region of *TP53* and processing them with the same effect score prediction model used in (B). **(G)** By mixing the 12 distributions calculated in (F) with the weights of the substitution frequencies calculated in (E), we obtained the gene's final effect score distribution, used as its null background model for the analysis. **(H)** According to the null background distribution, we would expect mutations within the *TP53* gene to have a mean score of $\mu = 0.49$. However, the observed mean score of the 3,167 analyzed mutations is $\mu = 0.05$, which is 1.05 standard deviations below the mean (p-value $< E{-}300$). The observed mean (0.05) was calculated from the 3,167 SNVs observed in *TP53* which are categorized as follows: 92 synonymous mutations (effect scores of 1), 512 nonsense mutations (effect scores of 0), and 2,563 missense mutations with an average score of 0.02.

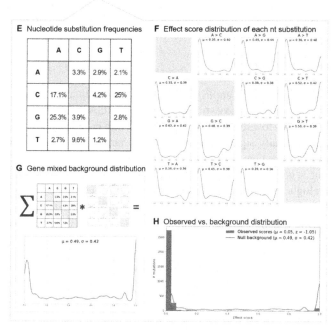

Fig. 1. (*continued*)

In order to ensure that FIRM does not capture any clinical or evolutionary information, we restricted its used features to purely biochemical properties. FIRM extracts an immense set of features (1,109 in total), aimed at capturing the rich proteomic context of each missense variant. The main classes of features include: i) the location of the variant within the protein sequence, ii) the identities of the reference and alternative amino-acids, iii) the score of the amino-acid substitution under various BLOSUM matrices, iv) an abundance of annotations extracted from UniProtKB, v) amino-acid scales (i.e.

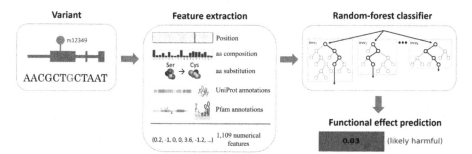

Fig. 2. Overview of FIRM. FIRM is the underlying machine-learning model used to predict the functional effects of variants, which is used by the FABRIC framework. By exploiting a rich proteomic knowledgebase, FIRM extracts features representing variants in a 1,109-dimensional space. A random-forest classifier then assigns each variant a predicted functional effect score. FIRM was pre-trained on the ClinVar dataset (which is independent of the datasets examined by FABRIC in this work).

various numeric values assigned to amino-acids [15, 16]), vi) Pfam domains and Pfam clans. More details on FIRM, including performance analysis, are described in [12].

3 Results and Discussion

Alteration Bias in Cancer We applied FABRIC on ~3 M somatic mutations from over 10,000 cancerous human samples extracted from the TCGA database [3]. Of the entire human proteome, we discovered 593 alteration promoting protein-coding genes, namely genes showing statistically significant bias towards harmful mutations [12]. To verify our results and check for new discoveries, we compared our results against prominent resources of cancer genes: COSMIC-Census catalogue [17], and the recently compiled PanSofware catalogue of 299 cancer driver genes [6]. We found a very strong and significant overlap between the discovered alteration promoting genes and those catalogues [12], although the majority of the genes (426 of 593) were not listed in them.

Alteration Bias in the Healthy Human Population
We tested the evolutionary signal that can be extracted from germline variants in healthy human population. We used the ExAC dataset [18], one of the largest and most complete contemporary catalogs of genetic variation in the healthy human population. The full dataset of ExAC contained 10,089,609 variants. We filtered out 1,054,475 low-quality variants, and among the remaining 9,035,134 variants we found 8,538,742 SNVs. Of these, 4,747,096 were found to be in coding regions, contributing to a final dataset of 4,752,768 gene effects. Applying FABRIC on this dataset, the effect scores of the variants in each gene were compared against the background distribution derived from the nucleotide substitution frequencies of the same observed variants.

We observed that variants with lower allele frequencies have lower z-values, i.e. are generally more damaging than expected (Fig. 3A). As expected, more harmful variants (with lower z-values) are less likely to become fixed in the population. We also found expected correlations between the effect score biases of genes (mean z-values) to other popular scoring techniques that measure evolutionary selection. We report Spearman's correlation of $\rho = -0.4$ (p-value $< E-300$) between the Residual Variation Intolerance Score (RVIS) [19] to the mean effect score z-values of genes. Similarly, we report Spearman's correlation of $\rho = -0.28$ (p-value $< E-300$) to the Gene Damage Index (GDI) [20]. Both metrices give higher scores to genes that are damaged more than expected, while we give lower scores to such genes, hence the expectation for negative correlation. This further confirms the evolutionary constraints reflected by the effect score biases.

Considering all ~20K protein coding genes, we discovered 6,141 significant alteration rejecting genes, and only 147 significant alteration promoting genes. In other words, almost all of the significant results (97.7%) are alteration rejecting genes, meaning that in the case of the healthy human population, most genes are under negative selection. This is the exact opposite to the trend observed in cancer (Fig. 3B). Whereas cancer is dominated by positive selection, germline variants that have undergone selective pressure throughout long-term human evolution are dominated by negative selection.

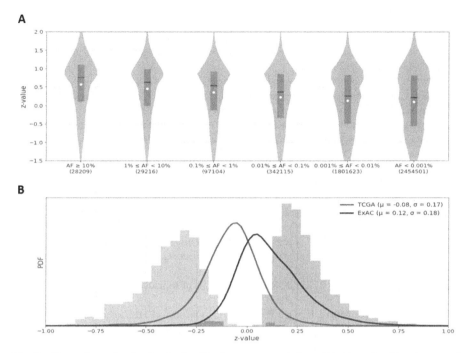

Fig. 3. Alteration rejection in the healthy human population. (**A**) Alteration bias (measured by z-value) of germline variants from ExAC across ranges of Allele Frequency (AF). The boxes represent the Q1–Q3 ranges, the middle lines the medians (Q2), and the white dots the means. Since there are ~60 k samples in the dataset, the last range (AF < 0.001%) captures only the 2,454,501 effect scores of singleton variants. (**B**) Distribution of alteration bias (measured by mean z-value) of the 17,828 and 17,946 analyzed genes in TCGA (red) and ExAC (blue), respectively. The density plots show the distribution of all analyzed genes, while the shaded histograms only the 599 and 6,288 genes with significant alteration bias in each dataset (comprised of both alteration promoting and alteration rejecting genes in both datasets). (Color figure online)

It is also interesting to note a mild overlap between the alteration promoting genes in cancer, found in the analysis of somatic mutations in TCGA, to alteration rejecting genes in the healthy human population, found in the analysis of germline variants in the ExAC dataset. Of the 17,313 genes that are shared to both analyses, 584 are significant alteration promoters in cancer, 5,995 are significant alteration rejecters in the human population, and 350 are both. According to random hyper-geometric distribution, we would expect only 202 overlapping genes ($\times 1.73$ enrichment, p-value $= 1.17E-36$). This supports the notion that cancer driver genes, which undergo positive selection during tumor evolution, are subjected to negative selection during normal human evolution.

Funding. This work was supported by the European Research Council's grant on High Dimensional Combinatorics (N.B. fellowship) [N.L. grant #339096] and a grant from Yad Hanadiv (M.L. #9660)

References

1. Vogelstein, B., Papadopoulos, N., Velculescu, V.E., et al.: Cancer genome landscapes. Science **339**(80), 1546–1558 (2013)
2. Marx, V.: Cancer genomes: discerning drivers from passengers (2014)
3. Tomczak, K., Czerwińska, P., Wiznerowicz, M.: The Cancer Genome Atlas (TCGA): an immeasurable source of knowledge. Contemp Oncol **19**, A68 (2015)
4. Tokheim, C.J., Papadopoulos, N., Kinzler, K.W., et al.: Evaluating the evaluation of cancer driver genes. Proc. Natl. Acad. Sci. **113**, 14330–14335 (2016). 201616440
5. Gonzalez-Perez, A., Deu-Pons, J., Lopez-Bigas, N.: Improving the prediction of the functional impact of cancer mutations by baseline tolerance transformation. Genome Med. **4**, 89 (2012)
6. Bailey, M.H., Tokheim, C., Porta-Pardo, E., et al.: Comprehensive characterization of cancer driver genes and mutations. Cell **173**, 371–385 (2018)
7. Lawrence, M.S., Stojanov, P., Mermel, C.H., et al.: Discovery and saturation analysis of cancer genes across 21 tumour types. Nature **505**, 495–501 (2014)
8. Zhang, J., Liu, J., Sun, J., et al.: Identifying driver mutations from sequencing data of heterogeneous tumors in the era of personalized genome sequencing. Brief. Bioinform. **15**, 244–255 (2014)
9. Greenman, C., Stephens, P., Smith, R., et al.: Patterns of somatic mutation in human cancer genomes. Nature **446**, 153–158 (2007)
10. Mularoni, L., Sabarinathan, R., Deu-Pons, J., et al.: OncodriveFML: a general framework to identify coding and non-coding regions with cancer driver mutations. Genome Biol. **17**, 128 (2016)
11. Kircher, M., Witten, D.M., Jain, P., et al.: A general framework for estimating the relative pathogenicity of human genetic variants. Nat. Genet. **46**, 310 (2014)
12. Brandes, N., Linial, N., Linial, M.: Quantifying gene selection in cancer through protein functional alteration bias. Nucleic Acids Res. **47**, 6642–6655 (2019)
13. Adzhubei, I., Jordan, D.M., Sunyaev, S.R.: Predicting functional effect of human missense mutations using PolyPhen-2. Curr. Protoc. Hum. Genet. **76**, 7–20 (2013)
14. Landrum, M.J., Lee, J.M., Benson, M., et al.: ClinVar: public archive of interpretations of clinically relevant variants. Nucleic Acids Res. **44**, D862–D868 (2015)
15. Brandes, N., Ofer, D., Linial, M.: ASAP: A machine learning framework for local protein properties. Database (2016). https://doi.org/10.1093/database/baw133
16. Ofer, D., Linial, M.: ProFET: feature engineering captures high-level protein functions. Bioinformatics **31**, 3429–3436 (2015)
17. Santarius, T., Shipley, J., Brewer, D., et al.: A census of amplified and overexpressed human cancer genes. Nat. Rev. Cancer **10**, 59–64 (2010)
18. Karczewski, K.J., Weisburd, B., Thomas, B., et al.: The ExAC browser: displaying reference data information from over 60 000 exomes. Nucleic Acids Res. **45**, D840–D845 (2017)
19. Petrovski, S., Wang, Q., Heinzen, E.L., et al.: Genic intolerance to functional variation and the interpretation of personal genomes. PLoS Genet. **9**, e1003709 (2013)
20. Itan, Y., Shang, L., Boisson, B., et al.: The human gene damage index as a gene-level approach to prioritizing exome variants. Proc. Natl. Acad. Sci. **112**, 13615–13620 (2015)

Testing the Agreement of Trees
with Internal Labels

David Fernández-Baca$^{(\boxtimes)}$ and Lei Liu

Department of Computer Science, Iowa State University, Ames, IA 50011, USA
{fernande,lliu}@iastate.edu

Abstract. The input to the agreement problem is a collection $\mathcal{P} = \{\mathcal{T}_1, \mathcal{T}_2, \ldots, \mathcal{T}_k\}$ of phylogenetic trees, called input trees, over partially overlapping sets of taxa. The question is whether there exists a tree $\mathcal{T}$, called an agreement tree, whose taxon set is the union of the taxon sets of the input trees, such that for each $i \in \{1, 2, \ldots, k\}$, the restriction of $\mathcal{T}$ to the taxon set of $\mathcal{T}_i$ is isomorphic to $\mathcal{T}_i$. We give a $\mathcal{O}(nk(\sum_{i \in [k]} d_i + \log^2(nk)))$ algorithm for a generalization of the agreement problem in which the input trees may have internal labels, where n is the total number of distinct taxa in $\mathcal{P}$, k is the number of trees in $\mathcal{P}$, and d_i is the maximum number of children of a node in $\mathcal{T}_i$.

Keywords: Phylogenetic tree · Taxonomy · Agreement · Algorithm

1 Introduction

In the *tree agreement problem* (*agreement problem*, for short), we are given a collection $\mathcal{P} = \{\mathcal{T}_1, \mathcal{T}_2, \ldots, \mathcal{T}_k\}$ of rooted phylogenetic trees with partially overlapping taxon sets. $\mathcal{P}$ is called a *profile* and the trees in $\mathcal{P}$ are the *input trees*. The question is whether there exists a tree $\mathcal{T}$ whose taxon set is the union of the taxon sets of the input trees, such that, for each $i \in \{1, 2, \ldots, k\}$, $\mathcal{T}_i$ is isomorphic to the restriction of $\mathcal{T}$ to the taxon set of $\mathcal{T}_i$. If such a tree $\mathcal{T}$ exists, then we call $\mathcal{T}$ an *agreement tree* for $\mathcal{P}$ and say that $\mathcal{P}$ *agrees*; otherwise, $\mathcal{P}$ *disagrees*. The first explicit polynomial-time algorithm for the agreement problem is in reference [16][1]. The agreement problem can be solved in $O(n^2 k)$ time, where n is the number of distinct taxa in $\mathcal{P}$ [10].

Here we study a generalization of the agreement problem, where the internal nodes of the input trees may also be labeled. These labels represent higher-order taxa; i.e., in effect, sets of taxa. Thus, for example, an input tree may contain the taxon *Glycine max* (soybean) nested within a subtree whose root is labeled Fabaceae (the legumes), itself nested within an Angiosperm subtree. Note that leaves themselves may be labeled by higher-order taxa. We present a $\mathcal{O}(nk(\sum_{i \in [k]} d_i + \log^2(nk)))$ algorithm for the agreement problem for trees

---

[1] These authors refer to what we term "agreement" as "compatibility". What we call "compatibility", they call "weak compatibility".

© Springer Nature Switzerland AG 2020
Z. Cai et al. (Eds.): ISBRA 2020, LNBI 12304, pp. 127–139, 2020.
https://doi.org/10.1007/978-3-030-57821-3_12

with internal labels, where n is the total number of distinct taxa in $\mathcal{P}$, k is the number of trees in $\mathcal{P}$, and, for each $i \in \{1, 2, \ldots, k\}$, d_i is the maximum number of children of a node in $\mathcal{T}_i$.

Background. A close relative of the agreement problem is the *compatibility problem*. The input to the compatibility problem is a profile $\mathcal{P} = \{\mathcal{T}_1, \mathcal{T}_2, \ldots, \mathcal{T}_k\}$ of rooted phylogenetic trees with partially overlapping taxon sets. The question is whether there exists a tree $\mathcal{T}$ whose taxon set is the union of the taxon sets of the input trees such that each input tree $\mathcal{T}_i$ can be obtained from the restriction of $\mathcal{T}$ to the taxon set of $\mathcal{T}_i$ through edge contractions. If such a tree $\mathcal{T}$ exists, we refer to $\mathcal{T}$ as a *compatible tree* for $\mathcal{P}$ and say that $\mathcal{P}$ is *compatible*; otherwise, $\mathcal{P}$ is *incompatible*. Compatibility is a less stringent requirement than agreement; therefore, any profile that agrees is compatible, but the converse is not true. The compatibility problem for phylogenies (i.e., trees without internal labels), is solvable in $\mathcal{O}(M_{\mathcal{P}} \log^2 M_{\mathcal{P}})$ time, where $M_{\mathcal{P}}$ is the total number of nodes and edges in the trees of $\mathcal{P}$ [9]. Note that $M_{\mathcal{P}} = \mathcal{O}(nk)$.

Compatibility and agreement reflect two distinct approaches to dealing with *multifurcations*; i.e., non-binary nodes, also known as *polytomies*. Suppose that node v is a multifurcation in some input tree of $\mathcal{P}$ and that ℓ_1, ℓ_2, and ℓ_3 are taxa in three distinct subtrees of v. In an agreement tree for $\mathcal{P}$, these three taxa must be in distinct subtrees of some node in the agreement tree. In contrast, a compatible tree for $\mathcal{P}$ may contain no such node, since a compatible tree is allowed to "refine" the multifurcation at v—that is, group two out of ℓ_1, ℓ_2, and ℓ_3 separately from the third. Thus, compatibility treats multifurcations as "soft" facts; agreement treats them as "hard" facts [15]. Both viewpoints can be valid, depending on the circumstances.

The agreement and compatibility problems are fundamental special cases of the *supertree problem*, the problem of synthesizing a collection of phylogenetic trees with partially overlapping taxon sets into a single supertree that represents the information in the input trees [2,4,18,24]. The original supertree methods were limited to input trees where only the leaves are labeled, but there has been increasing interest in incorporating internally labeled trees in supertree analysis, motivated by the desire to incorporate *taxonomies* in these analyses. Taxonomies group organisms according to a system of taxonomic rank (e.g., family, genus, and species); two examples are the NCBI taxonomy [21] and the Angiosperm taxonomy [23]. Taxonomies provide structure and completeness that can be hard to obtain otherwise [12,17,19], offering a way to circumvent one of the obstacles to building comprehensive phylogenies: the limited taxonomic overlap among different phylogenetic studies [20].

Although internally labeled trees, and taxonomies in particular, are not, strictly speaking, phylogenies, they have many of the same mathematical properties as phylogenies. Both phylogenies and internally labeled trees are *X-trees* (also called *semi-labeled trees*) [5,22]. Algorithmic results for compatibility and agreement of internally labeled trees are scarce, compared to what is available for ordinary phylogenies. To our knowledge, the first algorithm for testing compatibility of internally labeled trees is in [7] (see also [3]). The fastest known algorithm for the problem runs in $\mathcal{O}(M_\mathcal{P} \log^2 M_\mathcal{P})$ time [8]. We are unaware of any previous algorithmic results for the agreement problem for internally labeled trees.

All algorithms for compatibility and agreement that we know of are indebted to Aho et al.'s Build algorithm [1]. The time bounds for agreement algorithms are higher than those of compatibility algorithms, due to the need for agreement trees to respect the multifurcations in the input trees. To handle agreement, Build has to be modified so that certain sets of the partition of the taxa it generates are re-merged to reflect the multifurcations in the input trees, adding considerable overhead [10,16] (similar issues are faced when testing consistency of triples and fans [13]). This issue becomes more complex for internally labeled trees, in part because internal nodes with the same label, but in different trees, may jointly imply multifurcations, even if all input trees are binary.

Organization of the Paper. Section 2 provides a formal definition of the agreement problem for internally labeled trees. Section 3 studies the decomposability properties of profiles that agree. These properties allow us to reduce an agreement problem on a profile into independent agreement problems on subprofiles, leading to the agreement algorithm presented in Sect. 4. Section 5 contains some final remarks. All proofs are available in [14].

2   Preliminaries

For each positive integer r, $[r]$ denotes the set $\{1, \ldots, r\}$.

Graphs and Trees. Let G be a graph. $V(G)$ and $E(G)$ denote the node and edge sets of G. Let U be a subset of $V(G)$. Then the *subgraph of G induced by U* is the graph whose vertex set is U and whose edge set consists of all of the edges in $E(G)$ that have both endpoints in U.

A *tree* is an acyclic connected graph. All trees here are assumed to be rooted. For a tree T, $r(T)$ denotes the root of T. Suppose $u, v \in V(T)$. Then, u is an *ancestor* of v in T, denoted $u \leq_T v$, if u lies on the path from v to $r(T)$ in T. If $u \leq_T v$, then v is a *descendant* of u. Node u is a *proper ancestor* of v, denoted $u <_T v$, if $u \leq_T v$ and $u \neq v$. If $\{u, v\} \in E(T)$ and $u \leq_T v$, then u is the *parent* of v and v is a *child* of u. For each $x \in V(T)$, we use $\mathrm{parent}_T(x)$, and $\mathrm{Ch}_T(x)$, $T(x)$ to denote the parent of x, the children of x, and the subtree of T rooted at x, respectively. We extend the child notation to subsets of $V(T)$ in the natural way: for $U \subseteq V(T)$, $\mathrm{Ch}_T(U) = \bigcup_{u \in U} \mathrm{Ch}_T(u)$. Thus, if $U = \emptyset$, then $\mathrm{Ch}_T(U) = \emptyset$.

Let T be a tree and suppose $U \subseteq V(T)$. The *lowest common ancestor of U in T*, denoted $\mathrm{LCA}_T(U)$, is the unique smallest upper bound of U under $\leq_T$.

X-Trees. Throughout the paper, X denotes a set of *labels* (that is, taxa, which may be, e.g., species or families of species). An *X-tree* is a pair $\mathcal{T} = (T, \phi)$ where T is a tree and ϕ is a mapping from X to $V(T)$ such that, for every node $v \in V(T)$ of degree at most two, $v \in \phi(X)$. X is the *label set* of $\mathcal{T}$ and ϕ is the

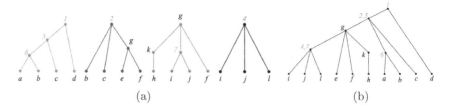

Fig. 1. (a) A profile $\mathcal{P} = \{T_1, T_2, T_3, T_4\}$. (b) An agreement tree for $\mathcal{P}$.

labeling function of T. For every node $v \in V(T)$, $\phi^{-1}(v)$ denotes the (possibly empty) subset of X whose elements map into v; these elements as the *labels of* v. If $\phi^{-1}(v) \neq \emptyset$, then v is *labeled*; otherwise, v is *unlabeled*.

By definition, every leaf in an X-tree is labeled, and any node, including the root, that has a single child must be labeled. Nodes with two or more children may be labeled or unlabeled. An X-tree $T = (T, \phi)$ is *singularly labeled* if every node in T has at most one label; T is *fully labeled* if every node in T is labeled.

X-trees, also known as *semi-labeled trees*, generalize ordinary phylogenetic trees (also known as *phylogenetic X-trees* [22]). An ordinary phylogenetic tree is a semi-labeled tree $T = (T, \phi)$ where $r(T)$ has degree at least two and ϕ is a bijection from X into leaf set of T (thus, internal nodes are not labeled).

Let $T = (T, \phi)$ be an X-tree. For each $u \in V(T)$, $X(u)$ denotes the set of all labels in the subtree of T rooted at u; that is, $X(u) = \bigcup_{v:u \leq_T v} \phi^{-1}(v)$. $X(u)$ is called a *cluster* of T. $\mathrm{Cl}(T)$ denotes the set of all clusters of T. We extend the cluster notation to sets of nodes as follows. Let U be a subset of $V(T)$. Then, $X(U) = \bigcup_{v \in U} X(v)$. If $U = \emptyset$, then $X(U) = \emptyset$.

Suppose $Y \subseteq X$ for an X-tree $T = (T, \phi)$. The *restriction* of T to Y, denoted $T|Y$, is the semi-labeled tree whose cluster set is $\mathrm{Cl}(T|Y) = \{W \cap Y : W \in \mathrm{Cl}(T) \text{ and } W \cap Y \neq \emptyset\}$. Intuitively, $T|Y$ is obtained from the minimal rooted subtree of T that connects the nodes in $\phi(Y)$ by suppressing all vertices v such that $v \notin \phi(Y)$ and v has only one child.

Let $T = (T, \phi)$ be an X-tree and $T' = (T', \phi')$ be an X'-tree such that $X' \subseteq X$. T *agrees with* T' if $\mathrm{Cl}(T') = \mathrm{Cl}(T|X')$. It is well known that the clusters of a tree determine the tree, up to isomorphism [22, Theorem 3.5.2]. Thus, T agrees with T' if T' and $T|X'$ are isomorphic.

Profiles and Agreement. Throughout the rest of this paper, $\mathcal{P}$ denotes a set $\{T_1, T_2, \ldots, T_k\}$ such that, for each $i \in [k]$, $T_i = (T_i, \phi_i)$ is a phylogenetic X_i-tree for some set X_i (Fig. 1a). We refer to $\mathcal{P}$ as a *profile*, and to the trees in $\mathcal{P}$ as *input trees*. We write $X_\mathcal{P}$ to denote $\bigcup_{i \in [k]} X_i$.

A profile $\mathcal{P}$ *agrees* if there is an $X_\mathcal{P}$-tree T that agrees with each of the trees in $\mathcal{P}$. If T exists, we refer to T as an *agreement tree for* $\mathcal{P}$. See Fig. 1b.

Given a subset Y of $X_\mathcal{P}$, the *restriction* of $\mathcal{P}$ to Y, denoted $\mathcal{P}|Y$, is the profile defined as $\mathcal{P}|Y = \{T_1|Y \cap X_1, T_2|Y \cap X_2, \ldots, T_k|Y \cap X_k\}$. The proof of the following lemma is straightforward.

Lemma 1. *Suppose a profile $\mathcal{P}$ has an agreement tree $\mathcal{T}$. Then, for any $Y \subseteq X_{\mathcal{P}}$, $\mathcal{T}|Y$ is an agreement tree for $\mathcal{P}|Y$.*

Suppose $\mathcal{P}$ contains trees that are not fully labeled. We can convert $\mathcal{P}$ into an equivalent profile $\mathcal{P}'$ of fully-labeled trees as follows. For each $i \in [k]$, let l_i be the number of unlabeled nodes in T_i. Create a set X' of $n' = \sum_{i \in [k]} l_i$ labels such that $X' \cap X_{\mathcal{P}} = \emptyset$. For each $i \in [k]$ and each $v \in V(T_i)$ such that $\phi_i^{-1}(v) = \emptyset$, make $\phi_i^{-1}(v) = \{\ell\}$, where ℓ is a distinct element from X'. We refer to $\mathcal{P}'$ as the *profile obtained by adding distinct new labels to $\mathcal{P}$* (see Fig. 1a).

The proof of the following result is analogous to that of [7, Lemma 3.4].

Lemma 2. *Let $\mathcal{P}'$ be the profile obtained by adding distinct new labels to $\mathcal{P}$. Then, $\mathcal{P}$ agrees if and only if $\mathcal{P}'$ agrees. Further, if $\mathcal{T}$ is an agreement tree for $\mathcal{P}'$, then $\mathcal{T}$ is also an agreement tree for $\mathcal{P}$.*

From this point forward, we make the following assumption.

Assumption 1. *For each $i \in [k]$, T_i is fully and singularly labeled.*

By Lemma 2, no generality is lost in assuming that all trees in $\mathcal{P}$ are fully labeled. The assumption that the trees are singularly labeled is inessential; it is only for clarity. Note that, even with the latter assumption, a tree that agrees with $\mathcal{P}$ is not necessarily singularly labeled. Figure 1b illustrates this fact.

Lemma 3. *If profile $\mathcal{P}$ agrees, then $\mathcal{P}$ has an agreement tree $\mathcal{T} = (T, \phi)$ such that $\phi^{-1}(v) \neq \emptyset$ for each node $v \in V(T)$.*

By Assumption 1, for each $i \in [k]$, there is a bijection between the labels in X_i and the nodes of $V(T_i)$. For this reason, we will often refer to nodes by their labels. In particular, given a label $\ell \in X_i$, we write $X_i(\ell)$ to denote $X_i(\phi_i(\ell))$ (the cluster of T_i at the node labeled ℓ), $\mathrm{Ch}_{T_i}(\ell)$ to denote $\phi_i(\mathrm{Ch}_{T_i}(\phi_i(\ell)))$ (the labels of children of ℓ in T_i), and, for $A \subseteq X_i$, $\mathrm{Ch}_{T_i}(A)$ to denote $\phi_i(\mathrm{Ch}_{T_i}(\phi_i(A)))$.

The following characterization of agreement generalizes a result in [10].

Lemma 4. *Let $\mathcal{P}$ be a profile and $\mathcal{T} = (T, \phi)$ be an $X_{\mathcal{P}}$-tree. Then, $\mathcal{T}$ is an agreement tree for $\mathcal{P}$ if and only if, for each $i \in [k]$, there exists a function $\phi_i : X_i \to V(T)$ such that for every label $a \in X_i$,*

(E1) $\phi_i(a) = \mathrm{LCA}_T(X_i(a))$,
(E2) for each label $b \in \mathrm{Ch}_{T_i}(a)$, $\phi_i(a) <_T \phi_i(b)$, and
(E3) for every two distinct labels $b, c \in \mathrm{Ch}_{T_i}(a)$, there exist distinct nodes $u, v \in \mathrm{Ch}_T(\phi_i(a))$ such that $\phi_i(b) \in X_{\mathcal{P}}(u)$ and $\phi_i(c) \in X_{\mathcal{P}}(v)$.

We refer to a function ϕ_i satisfying conditions (E1)–(E3) of Lemma 4 as a *topological embedding* of T_i into $\mathcal{T}$. Observe that, by transitivity, condition (E2) implies that, for any $a, b \in X_i$, if $a <_{T_i} b$, then $\phi_i(a) <_T \phi_i(b)$.

3   Positions in a Profile

A *position* in a profile $\mathcal{P}$ is a tuple $\pi = (\pi_1, \pi_2, \ldots, \pi_k)$ where, for each $i \in [k]$, either $\pi_i = \emptyset$ or $\pi_i = \{\ell\}$, for some $\ell \in X_i$. Note that the definition of a position allows for the possibility that there exist $i, j \in [k]$, $i \neq j$, such that $\ell \in \pi_i$, but $\ell \notin \pi_j$, even if $\ell \in X_i$ and $\ell \in X_j$. At any given point during its execution, our agreement algorithm focuses on testing the agreement of the subprofile of $\mathcal{P}$ determined by the subtrees associated with a specific position.

For a position π in $\mathcal{P}$, let $X_{\mathcal{P}}(\pi)$ denote the set of labels $\bigcup_{i \in [k]} X_i(\pi_i)$. A label $\ell \in X_{\mathcal{P}}(\pi)$ is *exposed in* π if $\pi_i = \{\ell\}$ for every $i \in [k]$ such that $\ell \in X_i(\pi)$. We say that position π *has an agreement tree* if $\mathcal{P}|X_{\mathcal{P}}(\pi)$ has an agreement tree.

A position π in $\mathcal{P}$ is *valid* if $X_i(\pi_i) = X_{\mathcal{P}}(\pi) \cap X_i$, for each $i \in [k]$. The *initial position* for $\mathcal{P}$ is the position π^{init}, where, for each $i \in [k]$, π_i^{init} is a singleton set consisting of the label of $r(T_i)$ (i.e., $\pi_i^{\text{init}} = \phi_i^{-1}(r(T_i))$. Clearly, π^{init} is a valid position.

Lemma 5. *A profile $\mathcal{P}$ has an agreement tree if and only if there is an agreement tree for every valid position π in $\mathcal{P}$.*

Decomposing a Position. In what follows, π denotes a valid position in $\mathcal{P}$. For each $i \in [k]$ such that $\pi_i \neq \emptyset$, let $\ell_i \in X_i$ denote the single label in π_i. Let $\text{Ch}_{\mathcal{P}}(\pi)$ denote the set of all children of some label in π; i.e., $\text{Ch}_{\mathcal{P}}(\pi) = \bigcup_{i \in [k]} \text{Ch}_{T_i}(\pi_i)$.

Let π be a valid position in $\mathcal{P}$. A *good decomposition* of π is a pair (S, Π), where S is a subset of the exposed labels in $\bigcup_{i \in \pi_i} \pi_i$ and $\Pi = \{\pi^{(1)}, \pi^{(2)}, \ldots, \pi^{(d)}\}$ is a collection of valid positions such that

(D1) $S \cup \bigcup_{j \in [d]} X_{\mathcal{P}}(\pi^{(j)}) = X_{\mathcal{P}}(\pi)$ and $S \cap \bigcup_{j \in [d]} X_{\mathcal{P}}(\pi^{(j)}) = \emptyset$, and
(D2) $X_{\mathcal{P}}(\pi^{(p)}) \cap X_{\mathcal{P}}(\pi^{(q)}) = \emptyset$, for all $p, q \in [d]$ such that $p \neq q$.

Note that we allow S or Π to be empty. We refer to the labels in S as *semi-universal labels* and to the positions in Π as *successor positions* of π. The next result is central to our agreement algorithm.

Lemma 6. *Let π be a valid position in a profile $\mathcal{P}$. Then, π has an agreement tree if and only if there exists a good decomposition (S, Π) of π such that $S \neq \emptyset$ and, for each position $\pi' \in \Pi$, π' has an agreement tree. If such a good decomposition exists, then π has an agreement tree $\mathcal{T} = (T, \phi)$ where $\phi^{-1}(r(T)) = S$.*

Good Partitions. To find a good decomposition of a position π, it is convenient to work with partitions of $\text{Ch}_{\mathcal{P}}(\pi)$. (Recall that a *partition* of a set Y is a collection Γ of nonempty subsets of Y such that every element $x \in Y$ is in exactly one set in Γ.) A good decomposition (S, Π), where $\Pi = \{\pi^{(j)}\}_{j \in [d]}$ defines a partition Γ of the set $\text{Ch}_{\mathcal{P}}(\pi)$ where, for any $a, b \in \text{Ch}_{\mathcal{P}}(\pi)$, a and b are in the same set of Γ if and only if there exists $j \in [d]$ such that $a, b \in X_{\mathcal{P}}(\pi^{(j)})$. We refer to Γ as the *partition of* $\text{Ch}_{\mathcal{P}}(\pi)$ *associated with* (S, Π). Next, we show that, conversely, certain partitions of $\text{Ch}_{\mathcal{P}}(\pi)$ define good decompositions of π.

Set $A \subseteq \text{Ch}_{\mathcal{P}}(\pi)$ is *nice* with respect to a subset S of the exposed labels in π if, for each $i \in [k]$ and each label $\ell \in \bigcup_{i \in [k]} \pi_i$ such that $\text{Ch}_{\mathcal{P}}(\ell) \cap A \neq \emptyset$,

(N1) if $\ell \in S$ and each $i \in [k]$ such that $\ell \in \pi_i$, then $|\mathrm{Ch}_{T_i}(\ell) \cap A| = 1$, and

(N2) if $\ell \notin S$, then $\mathrm{Ch}_{\mathcal{P}}(\ell) \subseteq X_{\mathcal{P}}(A)$.

Suppose A is a nice set. The *position associated with* A is the position π^A, where, for each $i \in [k]$, π_i^A is defined as follows. If $\pi_i = \emptyset$, then $\pi_i^A = \emptyset$. Otherwise, let ℓ be the single element in π_i. Then,

$$\pi_i^A = \begin{cases} \emptyset & \text{if } \mathrm{Ch}_{T_i}(\ell) \cap A = \emptyset, \\ \mathrm{Ch}_{T_i}(\ell) \cap A & \text{if } \ell \in S, \text{ and} \\ \pi_i & \text{if } \ell \notin S. \end{cases} \tag{1}$$

A partition Γ of $\mathrm{Ch}_{\mathcal{P}}(\pi)$ is *good with respect to* S if each set $A \in \Gamma$ is nice with respect to S and, for every two distinct sets $A, B \in \Gamma$, $X_{\mathcal{P}}(\pi^A) \cap X_{\mathcal{P}}(\pi^B) = \emptyset$.

Lemma 7. *There is a bijection between good decompositions of π and good partitions of $\mathrm{Ch}_{\mathcal{P}}(\pi)$. That is, the following statements hold.*

(i) *Suppose (S, Π) is a good decomposition of π. Let (S, Γ) be the partition of $\mathrm{Ch}_{\mathcal{P}}(\pi)$ associated with (S, Π). Then, (S, Γ) is a good partition of $\mathrm{Ch}_{\mathcal{P}}(\pi)$.*

(ii) *Suppose (S, Γ) is a good partition of $\mathrm{Ch}_{\mathcal{P}}(\pi)$. Let $\Pi = \{\pi^A : A \in \Gamma\}$. Then, (S, Π), a good decomposition of π.*

We refer to the good partition (S, Γ) of $\mathrm{Ch}_{\mathcal{P}}(\pi)$ obtained from a good decomposition (S, Π) of π, as described in Lemma 7 (i), as the *good partition of* $\mathrm{Ch}_{\mathcal{P}}(\pi)$ *associated with* (S, Π). Likewise, we refer to the good decomposition (S, Π) of π obtained from a good partition (S, Γ) of $\mathrm{Ch}_{\mathcal{P}}(\pi)$, as described in Lemma 7 (ii), as the *good decomposition of* $\mathrm{Ch}_{\mathcal{P}}(\pi)$ *associated with* (S, Γ).

Let $(S, \Gamma), (S', \Gamma')$ be good partitions of $\mathrm{Ch}_{\mathcal{P}}(\pi)$. We say that (S, Γ) is *finer than* (S', Γ'), denoted $(S, \Gamma) \sqsubseteq (S', \Gamma')$, if and only if, $S \supseteq S'$ and, for every $A \in \Gamma$, there exists an $A' \in \Gamma'$ such that $A \subseteq A'$. We write $(S, \Gamma) \sqsubset (S', \Gamma')$ to denote that $(S, \Gamma) \sqsubseteq (S', \Gamma')$ and $(S, \Gamma) \neq (S', \Gamma')$. We say that a partition (S, Γ) of $\mathrm{Ch}_{\mathcal{P}}(\pi)$ is *minimal* if there does not exist another partition (S', Γ') of $\mathrm{Ch}_{\mathcal{P}}(\pi)$ such that $(S', \Gamma') \sqsubset (S, \Gamma)$.

Lemma 8. *Let π be a valid position in a profile $\mathcal{P}$. Then, the minimal good partition of $\mathrm{Ch}_{\mathcal{P}}(\pi)$ is unique.*

We refer to the (unique) good decomposition (S, Π) associated with the minimal good partition of $\mathrm{Ch}_{\mathcal{P}}(\pi)$ as the *maximal good decomposition of* π.

Corollary 1. *Let π be a valid position in a profile $\mathcal{P}$ and (S, Π) be the maximal good decomposition of π. If π has an agreement tree, then $S \neq \emptyset$.*

4   Constructing an Agreement Tree

BuildAST (Algorithm 1) takes as input a profile $\mathcal{P}$ on a set of labels X and either returns an agreement tree for $\mathcal{P}$ or reports that no such tree exists. BuildAST

```
1 BuildAST(𝒫)
    Data: A profile 𝒫 = {𝒯₁, 𝒯₂, ..., 𝒯ₖ} on a set of taxa X.
    Result: Returns an agreement tree 𝒯 for 𝒫, if one exists; otherwise,
            returns disagreement.
2   𝒬.ENQUEUE(⟨π^init, null⟩)
3   while 𝒬 ≠ ∅ do
4       ⟨π, pred⟩ = 𝒬.DEQUEUE()
5       ⟨S, Π⟩ = GetDecomposition(π)
6       if S = ∅ then
7        │  return disagreement
8       Create a node r(π)
9       r(π).parent = pred
10      foreach ℓ ∈ S do
11       │  ϕ(ℓ) = r(π)
12      foreach π' ∈ Π do
13       │  𝒬.ENQUEUE(⟨π', r(π)⟩)
14  return 𝒯 = (T, ϕ), where T is the tree with root r(π^init)
```

Algorithm 1: Testing agreement

assumes the availability of an algorithm `GetDecomposition` that, given a valid position π in $\mathcal{P}$, returns a maximal good decomposition (S, Π) of π.

`BuildAST` proceeds from the top down, starting from the initial position π^{init} of $\mathcal{P}$, attempting to construct an agreement tree for $\mathcal{P}$ in a breadth-first manner. Like other algorithms based on breadth-first search, `BuildAST` uses a queue, which stores pairs $\langle \pi, \text{pred} \rangle$, where π is a position in $\mathcal{P}$ and pred is a reference to the parent of the tree node (potentially) to be created for π. At the outset, the queue contains only the pair $\langle \pi^{\text{init}}, \text{null} \rangle$, corresponding to the root of the agreement tree, which has no parent.

At each iteration of its outer **while** loop (lines 3–13), `BuildAST` extracts a pair $\langle \pi, \text{pred} \rangle$ from its queue and invokes `GetDecomposition` to obtain a maximal good decomposition (S, Π) of π. If $S = \emptyset$, then, by Corollary 1, no agreement tree for π exists. `BuildAST` reports this fact (line 7) and terminates.

If $S \neq \emptyset$, `BuildAST` creates a tree node $r(\pi)$ for π; $r(\pi)$ is the tentative root for the agreement tree for π. By Lemma 6, if π has an agreement subtree, then it has an agreement tree where $\phi(\ell) = r(\pi)$. Lines 10–11 set up the mapping ϕ accordingly. Also by Lemma 6, if π has an agreement tree, then so does each position $\pi' \in \Pi$; furthermore, the roots of the trees for each position in Π will be the children of $r(\pi)$. Thus, `BuildAST` adds $\langle \pi', r(\pi) \rangle$, for each $\pi' \in \Pi$ to the queue, to ensure that π' is processed at a later iteration and that the root of the agreement tree constructed for π' (if such a tree exists) is made to have $r(\pi)$ as its parent (lines 12–13). Therefore, if `BuildAST` terminates without reporting disagreement, then the result returned in line 14 is an agreement tree for $\mathcal{P}$. `BuildAST` indeed terminates, because there are only two possibilities at any given iteration: either the algorithm terminates reporting disagreement or (since $S \neq \emptyset$) the maximal good decomposition (S, Π) of π has the property that $\bigcup_{\pi' \in \Pi} X_{\mathcal{P}}(\pi')$ is a *proper* subset of $X_{\mathcal{P}}(\pi)$. The number of iterations of

```
1 GetDecomposition(π)
      Data: A valid position π.
      Result: Returns the maximal good decomposition (S, Π) of π.
2     S = {ℓ : ℓ is exposed in π}, K = {i : πᵢ = {ℓ} for some ℓ ∈ S}
3     Γ = {A : A =
          W ∩ Ch_𝒫(π), for some connected component W of H_𝒫(π) \ S}
4     while S contains a bad label do
5         Choose any bad label ℓ ∈ S
6         K' = {i : πᵢ = {ℓ}}
7         Γ' = {A ∈ Γ : Ch_{Tᵢ}(ℓ) ∩ A ≠ ∅ for some i ∈ K'}
8         B = ⋃_{A∈Γ'} A
9         Γ = Γ \ Γ' ∪ {B}
10        S = S \ {ℓ}, K = K \ K'
11    Π ← ∅
12    foreach A ∈ Γ do
13        foreach i ∈ [k] do πᵢ^A = ∅
14        foreach i ∈ [k] do
15            Let ℓ be the single label in πᵢ
16            if Ch_{Tᵢ}(ℓ) ∩ A ≠ ∅ then
17                if ℓ ∈ S then  πᵢ^A = Ch_{Tᵢ}(ℓ) ∩ A
18                else πᵢ^A = πᵢ
19        Π = Π ∪ π^A
20    return (S, Π)
```

Algorithm 2: Computing the maximal good decomposition.

BuildAST cannot exceed the total number of nodes in an agreement tree for $\mathcal{P}$, which is $O(n)$. Thus, we have the following result.

Theorem 1. *Given a profile* $\mathcal{P} = \{\mathcal{T}_1, \mathcal{T}_2, \ldots, \mathcal{T}_k\}$, BuildAST *returns an agreement tree* $\mathcal{T}$ *for* $\mathcal{P}$, *if such a tree exists; otherwise,* BuildAST *returns* disagreement. *The total number of iterations of* BuildAST*'s outer loop is* $O(n)$.

Finding the Maximal Good Decomposition. GetDecomposition (Algorithm 2) computes a maximal good decomposition of a position π, relying on an auxiliary graph known as *the display graph* of the input profile and denoted by $H_\mathcal{P}$ [6, 8,9]. The graph $H_\mathcal{P}$ is obtained from the disjoint union of the underlying trees $T_1, \ldots, T_k$ of the $\mathcal{P}$ by identifying nodes that have the same label. Multiple edges between the same pair of nodes are replaced by a single edge. See Fig. 2.

$H_\mathcal{P}$ has $O(nk)$ nodes and edges, and can be constructed in $O(nk)$ time. By Assumption 1, there is a bijection between the labels in X and the nodes of $H_\mathcal{P}$. Thus, from this point forward, we refer to the nodes of $H_\mathcal{P}$ by their labels. For a valid position π, $H_\mathcal{P}(\pi)$ denotes the subgraph of $H_\mathcal{P}$ induced by $X(\pi)$. Thus, $H_\mathcal{P}(\pi^{\text{init}}) = H_\mathcal{P}$.

Lines 2–10 of GetDecomposition construct the minimal good partition of $\text{Ch}_\mathcal{P}(\pi)$. Line 2 initializes S to contain all exposed labels in π, and sets K to consist of the indices of the trees in $\mathcal{P}$ that contain the labels in S. Line 3

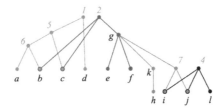

Fig. 2. The display graph $H_\mathcal{P}$ for the profile of Fig. 1a.

initializes Γ using $H_\mathcal{P}(\pi)$. We say that a label $\ell \in S$ is *bad* if there exist $i \in K$ and $A \in \Gamma$ such that $\pi_i = \{\ell\}$ and $|\mathrm{Ch}_{T_i}(\ell) \cap A| \geq 2$. Lines 4–10 construct the minimal nice partition (S, Γ) of $\mathrm{Ch}_\mathcal{P}(\pi)$ by deleting bad labels from S and merging sets in Γ accordingly. Let (S^*, Γ^*) denote the minimal good partition of $\mathrm{Ch}_\mathcal{P}(\pi)$.

Lemma 9. *Let π be a valid position in a profile $\mathcal{P}$ and let (S^*, Γ^*) be the minimal good partition of $\mathrm{Ch}_\mathcal{P}(\pi)$. Let (S_0, Γ_0) denote the initial value of (S, Γ) in* GetDecomposition *before entering the **while** loop, (S_j, Γ_j) denote the value of (S, Γ) after j executions of the body of the loop, and r denote the total number of iterations. Then, $r \leq k$ and $(S_0, \Gamma_0) \sqsubset (S_1, \Gamma_1) \sqsubset (S_2, \Gamma_2) \sqsubset \cdots \sqsubset (S_r, \Gamma_r) = (S^*, \Gamma^*)$.*

By Lemma 9, the pair (S, Γ) constructed in lines 4–10 of GetDecomposition is a minimal good partition of $\mathrm{Ch}_\mathcal{P}(\pi)$. The **for each** loop of lines 11–19 simply uses Eq. (1) to construct the maximal good decomposition (S, Π) of π from (S, Γ). We thus have the following.

Lemma 10. GetDecomposition *returns the maximal good decomposition of π.*

Implementation. Throughout its execution, BuildAST maintains the display graph $H_\mathcal{P}$. Also, for each label $\ell \in X$, it maintains a field $\ell.\mathtt{appear}$ containing every index i such that $\pi_i = \{\ell\}$ for some π in Q. Label ℓ is exposed when $|\ell.\mathtt{appear}| = k_\ell$, where k_ℓ denotes the number of input trees containing label ℓ. For each π in BuildAST's queue, the set $\mathrm{Ch}_\mathcal{P}(\pi)$ is stored as a sparse array $((i, \mathrm{Ch}_{T_i}(\pi_i)) : i \in [k]$ and $\mathrm{Ch}_{T_i}(\pi_i)) \neq \emptyset)$. This enables GetDecomposition to access the parts of $\mathrm{Ch}_\mathcal{P}(\pi)$ associated with each input tree separately. We use this representation of $\mathrm{Ch}_\mathcal{P}(\pi)$ to build similar representations of the sets in the partition Γ of $\mathrm{Ch}_\mathcal{P}(\pi)$ produced from $H_\mathcal{P}(\pi) \setminus S$ in line 3 of GetDecomposition. For each label $a \in \mathrm{Ch}_\mathcal{P}(\pi)$, we maintain a mapping that returns, in $O(1)$ time, the set $A \in \Gamma$ containing a. During the execution of GetDecomposition's **while** loop, sets in Γ may be merged, and representations of these merged sets must be produced and the mapping from $\mathrm{Ch}_\mathcal{P}(\pi)$ to Γ must be modified.

Lemma 11. *The total time needed to maintain the display graph throughout the entire execution of* BuildAST *is $\mathcal{O}(nk \log^2(nk))$.*

In the following results, d_i denotes the maximum number of children of a node in tree T_i, for each $i \in [d]$.

Lemma 12. *Excluding the time needed to maintain the display graph, Lines 2 and 3 of* GetDecomposition *take* $\mathcal{O}(\sum_{i \in [k]} d_i)$ *time.*

Lemma 13. GetDecomposition*'s* **while** *loop takes* $\mathcal{O}(k \sum_{i \in [k]} d_i)$ *time.*

Theorem 2. BuildAST *can be implemented to run in* $\mathcal{O}(nk(\sum_{i \in [k]} d_i + \log^2(nk)))$ *time, where n is the number of distinct taxa in $\mathcal{P}$, k is the number of trees in $\mathcal{P}$, and d_i is the maximum number of children of tree T_i, for $i \in [k]$.*

5   Concluding Remarks and Future Work

BuildAST may be much faster in practice than Theorem 2 suggests, since that bound assumes the unlikely scenario where every edge deletion performed in constructing $H_\mathcal{P}(\pi) \setminus S$ in GetDecomposition generates a new component and that most of these components are remerged in the GetDecomposition's **while** loop. In any case, Theorem 2 implies that BuildAST performs well if the sum of the maximum out-degrees is small relative to the number of taxa.

The running time of BuildAST can be further improved to $\mathcal{O}(nk(\sum_{i \in [k]} d_i + \log^2(nk)/\log \log(nk)))$ using the graph connectivity data structure of reference [25]. It is not clear, however, that the latter data structure is practical. In fact, our previous experimental work [11], in the context of tree compatibility, suggests that data structures much simpler than HDT (and, therefore, than [25]) perform well in practice. As part of our future work, we intend to implement BuildAST and test it on simulated and real datasets.

BuildAST can be modified to run in $\mathcal{O}(nk \log^2(nk))$ time for profiles $\mathcal{P}$ where the input trees are all binary and solely leaf-labeled. For such profiles, $|A \cap \mathrm{Ch}_{T_i}(\pi_i)| \leq 2$, for $A \in \Gamma$ and $i \in [k]$ in a position π of $\mathcal{P}$. Labels $a, a' \in \mathrm{Ch}_{T_i}(\pi_i)$ are either in the same set A or in different sets A, A' where $A, A' \in \Gamma$. In the first case, $\ell \in \pi_i$ must be bad. Bad labels can then be detected earlier in Line 3 and directly removed from S. Thus, we can skip GetDecomposition's **while** loop. Hence, maintaining graph connectivity dominates the performance of BuildAST.

BuildAST enables users to deal with hard polytomies. In applications, we may encounter both hard and soft polytomies. It would be interesting to modify BuildAST to handle a mixture of both types polytomies, as appropriate.

References

1. Aho, A., Sagiv, Y., Szymanski, T., Ullman, J.: Inferring a tree from lowest common ancestors with an application to the optimization of relational expressions. SIAM J. Comput. **10**(3), 405–421 (1981)
2. Baum, B.R.: Combining trees as a way of combining data sets for phylogenetic inference, and the desirability of combining gene trees. Taxon **41**, 3–10 (1992)

3. Berry, V., Semple, C.: Fast computation of supertrees for compatible phylogenies with nested taxa. Syst. Biol. **55**(2), 270–288 (2006)
4. Bininda-Emonds, O.R.P. (ed.): Phylogenetic Supertrees: Combining Information to Reveal the Tree of Life. Series on Computational Biology, vol. 4. Springer, Berlin (2004). https://doi.org/10.1007/978-1-4020-2330-9
5. Bordewich, M., Evans, G., Semple, C.: Extending the limits of supertree methods. Ann. Comb. **10**, 31–51 (2006)
6. Bryant, D., Lagergren, J.: Compatibility of unrooted phylogenetic trees is FPT. Theoret. Comput. Sci. **351**, 296–302 (2006)
7. Daniel, P., Semple, C.: Supertree algorithms for nested taxa. In: Bininda-Emonds, O.R.P. (ed.) Phylogenetic supertrees: combining information to reveal the tree of life, pp. 151–171. Kluwer, Dordrecht (2004)
8. Deng, Y., Fernández-Baca, D.: An efficient algorithm for testing the compatibility of phylogenies with nested taxa. Algorithms Mol. Biol. **12**, 7 (2017)
9. Deng, Y., Fernández-Baca, D.: Fast compatibility testing for rooted phylogenetic trees. Algorithmica **80**(8), 2453–2477 (2018). https://doi.org/10.1007/s00453-017-0330-4. http://rdcu.be/thB1
10. Fernández-Baca, D., Guillemot, S., Shutters, B., Vakati, S.: Fixed-parameter algorithms for finding agreement supertrees. SIAM J. Comput. **44**(2), 384–410 (2015)
11. Fernández-Baca, D., Liu, L.: Tree compatibility, incomplete directed perfect phylogeny, and dynamic graph connectivity: an experimental study. Algorithms **12**(3), 53 (2019)
12. Hinchliff, C.E., et al.: Synthesis of phylogeny and taxonomy into a comprehensive tree of life. Proc. Nat. Acad. Sci. **112**(41), 12764–12769 (2015). https://doi.org/10.1073/pnas.1423041112
13. Jansson, J., Lingas, A., Rajaby, R., Sung, W.-K.: Determining the consistency of resolved triplets and fan triplets. In: Sahinalp, S.C. (ed.) RECOMB 2017. LNCS, vol. 10229, pp. 82–98. Springer, Cham (2017). https://doi.org/10.1007/978-3-319-56970-3_6
14. Liu, L., Fernández-Baca, D.: Testing the agreement of trees with internal labels, February 2020. arXiv preprint https://arxiv.org/abs/2002.09725
15. Maddison, W.P.: Reconstructing character evolution on polytomous cladograms. Cladistics **5**, 365–377 (1989)
16. Ng, M., Wormald, N.: Reconstruction of rooted trees from subtrees. Discrete Appl. Math. **69**(1–2), 19–31 (1996)
17. Page, R.M.: Taxonomy, supertrees, and the tree of life. In: Bininda-Emonds, O.R.P. (ed.) Phylogenetic supertrees: Combining Information to Reveal the Tree of Life, pp. 247–265. Kluwer, Dordrecht (2004)
18. Ragan, M.A.: Phylogenetic inference based on matrix representation of trees. Mol. Phylogenet. Evol. **1**, 53–58 (1992)
19. Redelings, B.D., Holder, M.T.: A supertree pipeline for summarizing phylogenetic and taxonomic information for millions of species. Peer J. **5**, e3058 (2017). https://doi.org/10.7717/peerj.3058
20. Sanderson, M.J.: Phylogenetic signal in the eukaryotic tree of life. Science **321**(5885), 121–123 (2008)
21. Sayers, E.W., et al.: Database resources of the national center for biotechnology information. Nucleic Acids Res. **37**(Database issue), D5–D15 (2009)
22. Semple, C., Steel, M.: Phylogenetics. Oxford Lecture Series in Mathematics. Oxford University Press, Oxford (2003)

23. The Angiosperm Phylogeny Group: An update of the angiosperm phylogeny group classification for the orders and families of flowering plants: APG IV. Bot. J. Linn. Soc. **181**, 1–20 (2016)
24. Warnow, T.: Supertree construction: opportunities and challenges. Technical report arXiv:1805.03530 ArXiV, May 2018
25. Wulff-Nilsen, C.: Faster deterministic fully-dynamic graph connectivity. In: Proceedings of the Twenty-fourth Annual ACM-SIAM Symposium on Discrete Algorithms SODA 2013, pp. 1757–1769. Society for Industrial and Applied Mathematics, Philadelphia (2013). http://dl.acm.org/citation.cfm?id=2627817.2627943

SVLR: Genome Structure Variant Detection Using Long Read Sequencing Data

Wenyan Gu[1], Aizhong Zhou[1], Lusheng Wang[2], Shiwei Sun[3], Xuefeng Cui[1(✉)], and Daming Zhu[1(✉)]

[1] School of Computer Science and Technology,
Shandong University, Qingdao, China
`xfcui@email.sdu.edu.cn, dmzhu@sdu.edu.cn`
[2] Department of Computer Science, City University of Hong Kong,
Kowloon, Hong Kong, China
[3] Key Laboratory of Intelligent Information Processing,
Institute of Computing Technology,
Chinese Academy of Sciences, Beijing, China

Abstract. Genome structural variants have great impacts on human phenotype and diversity, and have been linked to numerous diseases. Long read sequencing technologies arise to make it possible to find structural variants of as long as ten thousand nucleotides. Thus, long read based structural variant detection has been drawing attention of many recent research projects, and many tools have been developed for long reads to detect structural variants recently.

In this article, we present a new method, called SVLR, to detect Structural Variants based on Long Read sequencing data. Comparing to existing methods, SVLR can detect three new kinds of structural variants: block replacements, block interchanges and translocations. Although these new structural variants are structurally more complicated, SVLR achieves accuracies that are comparable to those of the classic structural variants. Moreover, for the classic structural variants that can be detected by state-of-the-art methods (e.g., SVIM and Sniffles), our experiments demonstrate recall improvements of up-to 38% without harming the precisions (i.e., above 78%). We also point out three directions to further improve structural variant detection in the future.
Source codes: https://github.com/GWYSDU/SVLR.

Keywords: Genome structural variant · Genome structural variant detection · Third generation sequencing · Long-read sequencing · Single-molecule sequencing

1 Introduction

Studies show that genome variants play a major role on phenotypes [18] and contribute to large-scale chromosome evolution [26] and human disease (such

© Springer Nature Switzerland AG 2020
Z. Cai et al. (Eds.): ISBRA 2020, LNBI 12304, pp. 140–153, 2020.
https://doi.org/10.1007/978-3-030-57821-3_13

as cancer [16,22], Mendelian disorders [28], autism [10], and Alzheimer [25]). Genome variants can be categorized into single nucleotide variation (SNV) [14, 17,24], small insertion/deletion (indel) [6,27], and structural variation [20,23]. The characterization of structural variants is of major importance to genetic disorders [4] and can help to elucidate their underlying genetic and molecular processes [8]. In other organisms such as plants, structural variants can drive phenotypic variation and adaptation to different environments variation [12, 32]. From the above aspects, it is imperative to develop effective methods to identify structural variants that traditional structural variants include insertions, deletions, inversions, tandem duplications, interspersed duplications, cut&paste insertions, and translocations [21].

The development of the genome sequencing technology has highly promoted the researches on genome variant detection. Some robust algorithms [1,3,19] were developed to distinguish structural variants based on the next generation sequencing (NGS), which usually uses four different methods: pair-end mapping, split read, read depth, and de novo assembly. But due to the size and association with repeats, it is hard to efficiently detect structural variants by NGS (i.e. short reads).

Single-molecule sequencing which can produce long reads shows many advantages to characterize the full spectrum of human genetic variation. Two commercial single-molecule sequencing solutions exist to date: single-molecule real-time sequencing by Pacific Biosciences and Nanoporous sequencing by Oxford Nanoporous Technologies (ONT). Single-molecule nanopore sequencing can sequence genome with hundreds of kilobases, which has led to genome sequencing of several organisms [9]. So, long read sequencing lays the foundation for a deeper discovery of structural variants [5,31].

There are already a wide variety of tools to call structural variants on the basis of the long reads, such as Sniffles [29] and SVIM [11]. Sniffles uses signatures from split-read alignments, highmismatch regions and coverage analysis to identify structural variants [29]. SVIM consists of three components for the collection, clustering and combination of structural variant signatures from read alignments [11]. A common first step in these tools is to align the reads with the reference genome. Both Sniffles and SVIM use NGMLR [29] aligner to obtain the split read mapping. Besides NGMLR, we also use another aligner LAST [7], because LAST have been designed for long reads with high rates of insertion and deletion error, e.g. nanopore or PacBio, and it enables more accurate genome alignments, with reliability measures for local alignments and for individual aligned bases. So we can use this information to detect structural variants efficiently.

In this study, we introduce SVLR, a novel pipeline for sensitive detection and accurate classification of nine classes of structural variants (as shown in Fig. 1). Our results demonstrate that SVLR reaches substantially higher recalls and precisions than existing tools for structural variant detection from long reads. Beyond that, our method has two more prominent contributions. First, SVLR is the first tool to be able to recognize block interchange [30] and block replacement in the intra-chromosome, translocation in the inter-chromosome, and get the corresponding source and destination regions [11]. They play important roles for biologists to analyze the generation of structural variants. Another prominent contribution is that SVLR considers the situations where structural variants are interactive, and it is the reason why SVLR can detect block interchange, block replacement and translocation. We call two structural variants are interactive with each other when one's source region is covered by the destination region of the other. At the end, three directions for future structural variant detection improvements are discussed.

2   Preliminaries

The ultimate goal of the structural variant (SV) detection problem is to find all SVs between a target genome and its reference genome. For the reference genome, we have its complete sequences $G = \{g_1, g_2, \cdots, g_p\}$, where g_i represents the sequence of chromosome i. For the target genome, we have its sequencing long reads $R = \{r_1, r_2, \cdots, r_q\}$, where r_i represents a long read conducted from sequencing experiments. Before finding SVs, the reference genome G and the sequencing reads R are preprocessed and aligned by calling a read alignment tool. The alignment results are summarized as $A = \{a_1, a_2, \cdots, a_n\}$, where $a_i = (r_i, g_i, b_i^r, e_i^r, b_i^g, e_i^g)$ represents an alignment between nucleotides from b_i^r to e_i^r of read r_i and nucleotides from b_i^g to e_i^g of chromosome g_i. Given these alignments, different types of SVs can be identified. A DEL, an INS, a TDUP or an INV SV (highlighted by yellow in Fig. 1) involves a single region of the reference genome, and hence is called a monomer SV. An IDUP, a BREP, a CPI, a BINT or a TRANS SV (not highlighted by yellow in Fig. 1) involves two regions of the reference genome, and hence is called a dimer SV. Computationally, a monomer SV is denoted as $sv_i = (t_i, g_i, b_i^{sv}, e_i^{sv})$, and a dimer SV is denoted as $sv_i = (t_i, g_i, sr_i, dr_i)$. Here, t_i represents the SV type, g_i represents the associated chromosome, b_i^{sv} and e_i^{sv} represents the beginning and the ending nucleotides of the associated region on chromosome g_i, $sr_i = (b_i^{sr}, e_i^{sr})$ and $dr_i = (b_i^{dr}, e_i^{dr})$ represents the source and the destination regions on chromosome g_i. Therefore, the solution of the SV detection problem is denoted as $SV = \{sv_1, sv_2, \cdots, sv_m\}$ including all SVs between the reference genome and the target genome.

3   Method

In this manuscript, we introduce a new method, called SVLR, to detect genome Structure Variants using Long Read alignments as followings. First, SV signatures are located by analyzing the read alignments. Such signatures contain critical information to identify SVs. However, they also contain redundant and noise information. Thus, a clustering algorithm is used to eliminate such information. Based on the remaining high-confident and non-redundant signatures, the signature clusters are labeled by SV types. Finally, the labels are combined and optimized to address conflicting labels in the same reference genome regions.

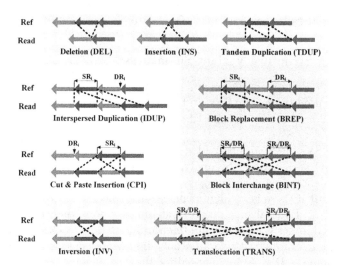

Fig. 1. Different types of structural variants between a reference genome and a long read of the target genome: the monomer structural variants are highlighted with yellow backgrounds; other than the monomer structural variants, all structural variants are dimer structural variants; **the dimer structural variants highlighted with cyan backgrounds are new structural variants that cannot be detected by previous methods.** (Color figure online)

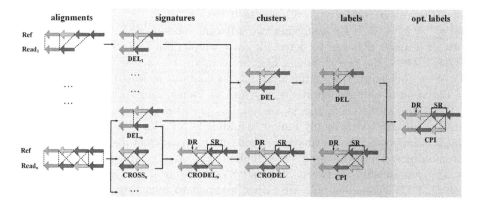

Fig. 2. A case study of the SVLR workflow: (0, white) sequencing long reads are preprocessed and aligned to the reference genome; (1, yellow) structural variant signatures are identified by analyzing long read alignments; (2, green) high-confident and non-redundant signatures are identified by clustering signatures; (3, cyan) a structural variant type is labeled for each cluster based on its signatures; (4, red) cluster labels are combined and optimized to address conflicting labels. (Color figure online)

A case study showing how the pipeline is used to identify an SV is shown in Fig. 2, and the detailed explanations of each step is provided in this section.

3.1 Finding Structure Variant Signatures

In the first step, long read alignments are analysed to find SV signatures (SGs) that are later used to identify SVs. In this section, we first define different kinds of SGs. Then, we describe an algorithm to locate SGs. Finally, we explain how SGs are related to SVs.

Similar to SVs, SGs can be divided into monomer SGs and dimer SGs. Here, a monomer SG is discovered by analyzing adjacent alignments. Recall that an alignment is denoted as $a_i = (r_i, g_i, b_i^r, e_i^r, b_i^g, e_i^g)$. Then, two alignments a_i and a_j are **adjacent** if there is no alignment a_k such that $r_i = r_j = r_k$, $g_i = g_j = g_k$ and $b_i^r < b_k^r < b_j^r$. By analyzing adjacent alignments, four kinds of monomer SVs can be detected directly, and they are temporally marked as INV, DEL, INS and TDUP SGs. For example, if two adjacent alignments contain a big gap in the reference genome, a DEL SG is marked for the gap region.

There are two special kinds of monomer SGs that are used to build up dimer SGs, and they cannot be mapped to any monomer SVs. Specifically, for two adjacent alignments, if the reference fragment of one alignment covers that of the other alignment, a COVER signature is marked. Similarly, if the reference fragment of the first alignment is in front of that of the second alignment, while the read fragment of the second alignment is in front of that of the first alignment, a CROSS SG is marked.

A dimer SG is discovered by analyzing two adjacent monomer SGs. Let $sg_i = (t_i, r_i, g_i, b_i^{sg}, e_i^{sg})$ denote a monomer SG. Then, two monomer SGs sg_i and sg_j are **adjacent** if there is no SG sg_k such that $r_i = r_j = r_k$, $g_i = g_j = g_k$ and $b_i^{sg} < b_k^{sg} < b_j^{sg}$. By analyzing two adjacent monomer SGs, COVDEL and CRODEL dimer SGs can be detected. For example, if a DEL SG is adjacent to a COVER SG, the two monomer SGs are combined as a new COVDEL dimer SG. It can be shown that a dimer SV always contains at least one dimer SG.

Based on the above definitions, SGs can be found using Algorithm 1. We set $L_{\texttt{max}} = 1$ Mbp as the maximum SV size, and $L_{\texttt{min}} = 40$ bp as the minimum SV size. Finding TRANS SVs is highly similar to finding BINT SVs with the modification that the source and the destination regions are from different instead of the same chromosomes. Thus, the details are eliminated to simply explanations.

The relationships between SGs and SVs are shown in Table 1. It can be seen that the monomer SGs are either mapped to monomer SVs or used to build dimmer SGs. Although there is no one-to-one mapping between dimer SGs and dimer SVs, additional conditions can be used in the third step to precisely distinguish different types of dimer SVs from dimer SGs. Moreover, each SV contains at least one SG. Therefore, finding SGs provides the foundation of finding SVs.

Algorithm 1. Find structural variant signatures from long read alignments.

Input: Alignments $A = \{a_1, a_2, \cdots a_p\}$, where $a_i = (r_i, g_i, b_i^r, e_i^r, b_i^g, e_i^g)$.
Output: Signatures $SG = \{sg_1, sg_2, \cdots sg_p\}$, where $sg_i = (t_i, r_i, g_i, b_i^{sg}, e_i^{sg})$ for monomer signatures,
         and $sg_i = (t_i, r_i, g_i, sr_i^{sg}, dr_i^{sg})$ for dimer signatures.

1:   $SG_1 := \emptyset$;
2: **for** adjacent triplet (a_i, a_{i+1}, a_{i+2}) with $b_i^r < b_{i+1}^r < b_{i+2}^r$ **do**
3:     **if** $((b_i^g - e_i^g) * (b_{i+1}^g - e_{i+1}^g) < 0)$ **and** $((b_{i+1}^g - e_{i+1}^g) * (b_{i+2}^g - e_{i+2}^g) < 0)$ **then**
4:        $SG_1 := SG_1 + (\texttt{INV}, r_i, g_i, e_i^g, b_{i+2}^g)$;
5:     **end if**
6: **end for**
7: **for** adjacent pair (a_i, a_{i+1}) with $b_i^r < b_{i+1}^r$ **do**
8:     $d_{i,i+1}^r := b_{i+1}^r - e_i^r$; $d_{i,i+1}^g := b_{i+1}^g - e_i^g$;
9:     **if** $L_{\min} < d_{i,i+1}^g - d_{i,i+1}^r < L_{\max}$ **then**
10:      $SG_1 := SG_1 + (\texttt{DEL}, r_i, g_i, e_i^g, b_{i+1}^g)$;
11:    **else if** $L_{\min} < d_{i,i+1}^r - d_{i,i+1}^g < L_{\max}$ **then**
12:      $SG_1 := SG_1 + (\texttt{INS}, r_i, g_i, e_i^g, b_{i+1}^g)$;
13:    **else if** $-L_{\max} < d_{i,i+1}^g - d_{i,i+1}^r < -L_{\min}$ **then**
14:      **if** $b_i^g < b_{i+1}^g < e_i^g < e_{i+1}^g$ **then**
15:        $SG_1 := SG_1 + (\texttt{TDUP}, r_i, g_i, b_{i+1}^g, e_i^g)$;
16:      **else if** $b_i^g < b_{i+1}^g < e_{i+1}^g < e_i^g$ or $b_{i+1}^g < b_i^g < e_i^g < e_{i+1}^g$ **then**
17:        $SG_1 := SG_1 + (\texttt{COVER}, r_i, g_i, b_{i+1}^g, e_i^g)$;
18:      **else if** $b_{i+1}^g < e_{i+1}^g < b_i^g < e_i^g$ **then**
19:        $SG_1 := SG_1 + (\texttt{CROSS}, r_i, g_i, b_{i+1}^g, e_i^g)$;
20:      **end if**
21:    **end if**
22: **end for**
23: $SG_2 := \emptyset$;
24: **for** adjacent pair (sg_i, sg_{i+1}) with $b_i^g < b_{i+1}^g$, where $sg_i, sg_{i+1} \in SG_1$ **do**
25:    $sr := (b_i^g, b_{i+1}^g)$; $dr := (e_i^g, e_{i+1}^g)$;
26:    **if** $(t_i = \texttt{COVER}$ and $t_{i+1} = \texttt{DEL})$ or $(t_i = \texttt{DEL}$ and $t_{i+1} = \texttt{COVER})$ **then**
27:      $SG_2 := SG_2 + (\texttt{COVDEL}, r_i, g_i, sr, dr)$;
28:    **else if** $(t_i = \texttt{CROSS}$ and $t_{i+1} = \texttt{DEL})$ or $(t_i = \texttt{DEL}$ and $t_{i+1} = \texttt{CROSS})$ **then**
29:      $SG_2 := SG_2 + (\texttt{CRODEL}, r_i, g_i, sr, dr)$;
30:    **end if**
31: **end for**
32: $SG := SG_1 \cup SG_2$;

3.2   Clustering Structure Variant Signatures

The previous step found all SGs between the reference genome and the long reads. However, there are many SGs that are highly similar to each other. Moreover, there are many false positive SGs caused by read errors. In order to address these issues, the SGs are first clustered, and the SG clusters are then scored. Since true positive SGs tend to form highly scored clusters, lowly scored clusters are

Table 1. Mappings from signatures to structural variants: monomer structural variants can be determined from monomer signatures; additional conditions can be used to distinguish dimer structural variants from dimer signatures; and all structural variants hit at least one signature.

	Monomer types					
Signatures	DEL	INS	TDUP	INV	COVER	CROSS
Structure Variants	DEL	INS	TDUP	INV	$\emptyset$	$\emptyset$

	Dimer types	
Signatures	COVDEL	CRODEL
Structure variants	{IDUP, BREP}	{IDUP, BREP, CPI, BINT}

treated as noises. Since all SGs within the same cluster are highly similar (i.e., redundant), one representative is generated for each highly scored cluster. As a result, the second step significantly reduces the number of SGs to work with, and yields high-confident and non-redundant SGs for the SVLR pipeline.

This approach was adopted from SVIM [11], and it is briefly summarized here. First, a graph is constructed, where each vertex represents an SG, and each edge represents a pair of SGs with a span-position distance below a threshold. Then, maximal cliques of the graph are found and treated as clusters. The cluster scores are evaluated based on the number of SGs, the total alignment scores and lengths, the standard deviation of the genomic span and position scores. Finally, one representative is generated for each cluster passing a cutoff score.

3.3 Labeling Signature Clusters

The purpose of this step is to classify the cluster representative SGs as SVs. Recall that the monomer SVs can be simply determined from the mapping SGs (as shown in Table 1). Hence, we focus on the dimer SG classification in this section. The idea is that we first roughly classify all dimer SGs as IDUP or BREP SVs, and then fix the two kinds of errors: CPI labeled as IDUP, and BINT labeled as BREP.

First, all dimer SGs are roughly classified as IDUP or BREP SVs. The key difference between the two SV classes is the size of the destination region. Specifically, if the destination region is small (with a size smaller than $L_{\min}$), the dimer SG (i.e., either a COVDEL or a CRODEL SG) is labeled as an IDUP SV. Otherwise, it is labeled as a BREP SV.

Since the roughly labeled SGs contain two kinds of errors, they have to be carefully checked and corrected one by one. One possible error is that a CRODEL SG is labeled as an IDUP SV, which should be a CPI SV. Such errors can be found by checking if there is a DEL SG in the source region of the CRODEL SG, and then fixed by changing the labels to CPI.

The other possible error is that two CRODEL SGs are labeled as two BREP SVs, which should be a single BINT SV. In this case, the source region of the first SG is the same as the destination region of the second SG, and the destination region of the first SG is the same as the source region of the second SG. Again, such cases can be found and fixed by combining the two SGs as a single one with a label of BINT.

3.4   Optimizing Cluster Labels

After the previous three steps, SVLR had generated a set of SV labels assigned to the SG clusters. Here, each label is correct or optimal with respect to the SGs of the corresponding cluster. However, it can still be observed that different labels are assigned to the same region of the reference genome. Thus, we need to carefully optimize the labels by picking the subset of non-conflicting labels yielding the highest total cluster score. Here, we try to explain the causes of conflicting labels because everything becomes straight forward once we understand the causes.

There are two causes of conflicting labels. First, in the error fixing process of the third step, two labels are combined to a new label without removing the original ones. The removing process cannot be done in the third step because it is not safe until all the labels are finalized. Now, it can be done safely and easily. For example, the DEL SV used to identify a CPI SV should be removed, and the two BREP SVs used to identify a BINT SV should also be removed.

The other cause of conflicting labels is the ambiguous SGs introduced by the high reading error rate of the long read sequencing technology. For example, given a reference sequence of ABCDE and a true read of ABCDbcdE (each character represents several nucleotides, and bcd represents the duplicated fragment of BCD), true read alignments of ABCD-ABCD and BCDE-bcdE should be observed, and a TDUP SG should be detected. However, due to the high reading error rate, it is possible that ambiguous read alignments of D-D and B-b are observed, and a CROSS SG is detected. Consequently, incorrect SV labels are assigned based on the ambiguous SGs. In case that the sequencing coverage is sufficiently high and both the true and the ambiguous alignments are observed, the problem can be fixed by removing the ambiguous alignments and the corresponding SGs and SVs.

4   Results

Our SVLR pipeline for SV detection has been implemented in Java, and the source codes can be downloaded from https://github.com/GWYSDU/SVLR. In order to evaluate the performance of SVLR, two datasets are used: a homozygous dataset and a heterozygous dataset. Each dataset contains a reference genome, a modified genome with implanted SVs, and simulated long reads of the genomes. One major difference between the simulated data and the real data is that the true SVs are known, and this significantly simplifies the evaluation process. Then, SVLR is used to find all SVs from the dataset. The results are compared to state-of-the-art methods, Sniffles [29] and SVIM [11]. In summary, SVLR is not only the best performing method, but also capable of finding new types of SVs (BREP, BINT and TRANS) that cannot be found by previous methods.

4.1   Simulated Data

In order to evaluate the performance of SV detection, a homozygous dataset and a heterozygous dataset is prepared. The homozygous dataset is prepared as followings. First, the hg19 genome sequences of chromosomes 21 and 22 are downloaded from the UCSD Genome Browser [15]. Both chromosomes have sizes of approximately 50Mbp, and they are used as the reference genome. Then, the reference genome is modified by implanting different SVs, including 300 DELs, 200 INS's, 200 TDUPs, 200 INVs, 100 IDUPs, 100 CPIs, 100 BINTs, 100 TRANS's and 77 BREPs. Each SV ranges between 800 bp and 10 Kbp, and they are implanted by RSVSim [2]. Afterwards, long reads of the modified genome are generated using SimLoRD [13] with default parameters. Moreover, several read sets are generated with different sequencing coverages, including 6X, 11X, 15X, 21X, 31X and 41X. Finally, the read alignments are produced using a read alignment tool, either NGMLR [29] or LAST[7] in our experiments.

In addition to the homozygous dataset, a heterozygous dataset is prepared. The only difference between preparing the two datasets is that half of the long reads are generated from the reference genome and the other half of the long reads are generated from the modified genome for the heterozygous dataset. Thus, this dataset simulates the situation that SVs do not always happen, and a frequency of 50% simulates the case with the maximum entropy.

4.2   Improving Performance by Read Aligners

The first strategy to improve the performance of SV detection is to use more accurate read aligners. Here, the importance of aligners is demonstrated by comparing the precision-recall curves of the same SV detection method (SVLR) with different aligners: NGMLR [29] and LAST [7]. As shown in Fig. 3, supporting different aligners is an important feature because different aligners might be more suitable for different sequencing technologies. For this reason, SVLR is implemented to support different read alignment formats.

The results of SV detection using 6x coverage sequencing data, SVLR and different aligners are shown in Fig. 3. It is observed that the precisions remains high and are not sensitive to the choice of aligners. Here, it is important to mention that recalls for the more challenging INS, IDUP and CPI SVs can be significantly improved by between 12% and 38% if the right aligner is used. This also suggests that improving read alignment algorithms is critical to the following biological studies based on sequence analysis's.

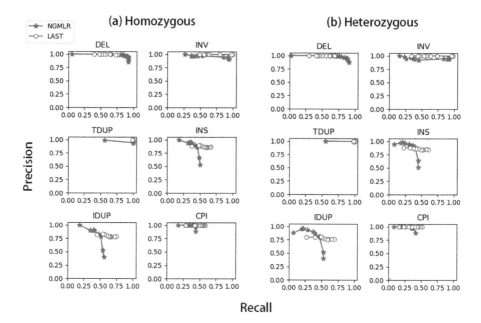

Fig. 3. Precision-recall curves for structural variant detection with different aligners: (a) the homozygous dataset and (b) the heterozygous dataset with a 6x coverage is used here; results for only SVLR are shown because SVIM and Sniffles only supports the NGMLR aligner.

4.3   Improving Recalls by SVLR

Other than the quality of the read alignments, the SV detection method is also critical to the performance. In this experiment, SVLR, SVIM [11] and Sniffles [29] SV detection methods are compared. Here, recalls of the top predictions are used instead of precision-recall curves to evaluate the performance because in practice, biologists tend to focus on studying the top scored predictions instead of trying different cutoffs to produce different number of predictions to work with. Figure 4 shows that SVLR achieved either the highest recalls in most cases or near-highest recalls in the worst case.

This experiment tries to simulate how biologists tend to use SV detection tools. In Fig. 4, the recalls of the top scored SVs found by different methods are calculated and shown. The precisions are also calculated but not shown here because the values are consistently high and they do not affect our conclusions. From Fig. 4, it can be seen that SVLR achieves near-perfect recalls when finding DEL, TDUP and INV SVs, no matter if the SVs occurred occasionally or consistently. Generally, SVLR can reliably find SVs with a recall comparable to the best performing method. Indeed, SVLR is the best performing method in most cases.

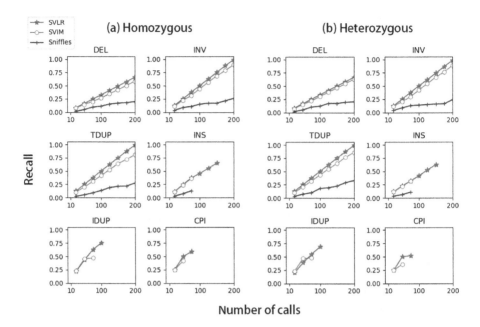

Fig. 4. Recalls for different numbers of calls (i.e., predictions): (a) the homozygous dataset and (b) the heterozygous dataset with a 6x coverage and the NGMLR aligner is used in this experiment.

4.4   Discovering New Structural Variants by SVLR

Another way to improve the SV detection is to discover new kinds of SVs that cannot be discovered by existing methods. Comparing to state-of-the-art methods, SVLR is the first method capable of discovering BREP, BINT and TRANS SVs. Moreover, the performance is similar to previously discovered SVs as shown in Fig. 5.

Recall that BINT and TRANS SVs are the most complicated SVs involving four monomer signatures and two dimer signatures (as shown in Fig. 1). However, their structural complexities are well addressed by SVLR. Specifically, Figs. 5(a–b) show that SVLR can detect BINT and TRANS SVs with accuracies comparable to those of relatively simpler IDUP and CPI SVs. Moreover, INS SVs have relativly simpler structural complexities, but all tested methods seem to have difficulties to achieve high recalls (as shown in Fig. 4). These observations imply that SV structure complexity is not strongly correlated to the accuracies. Moreover, Figs. 5(c–d) also show that sequencing coverage is not a bottleneck for SV detection.

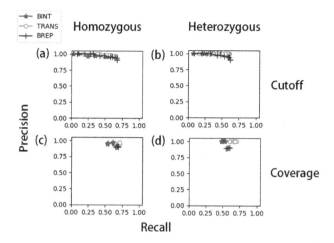

Fig. 5. Precision-recall curves for new structural variants that cannot be found by existing methods: (a) the homozygous dataset and (b) the heterozygous dataset with a 6x coverage and different score cutoffs are used; (c) the homozygous dataset and (d) the heterozygous dataset with different coverages (i.e., 6X, 11X, 15X, 21X, 31X and 41X) and the same cutoff is used.

5   Conclusion

Structural variants have drawn more and more attention with the development of genome sequencing technologies. Due to the wide variety of structural variants, it is essential to propose more sensitive tools for structural variant detection. In this manuscript, a new method, called SVLR, is introduced and implemented for Structural Variant detection using Long Read sequencing data. Our results show that SVLR is more accurate than Sniffles [29] and SVIM [11], and can discover new types of structural variants that cannot be discovered by previous tools.

References

1. Antaki, D., Brandler, W.M., Sebat, J.: SV2: accurate structural variation genotyping and de novo mutation detection from whole genomes. Bioinformatics **34**, 1774–1777 (2018)
2. Bartenhagen, C., Dugas, M.: Rsvsim: an r/bioconductor package for the simulation of structural variations. Bioinformatics **29**, 1679–1681 (2013)
3. Cameron, D.L., Stefano, L.D., Papenfuss, A.T.: Comprehensive evaluation and characterisation of short read general-purpose structural variant calling software. Nat. Commun. **10**, 3240 (2019)
4. Carvalho, C.M.B., Lupski, J.R.: Mechanisms underlying structural variant formation in genomic disorders. Nat. Rev. Genet. **17**, 224–238 (2016)
5. English, A., Salerno, W., Hampton, O., Gonzaga-Jauregui, C., Ambreth, S., Ritter, D.I., et al.: Assessing structural variation in a personal genome towards a human reference diploid genome. BMC Genom. **16**, 286–300 (2015)

6. Ferlaino, M., et al.: An integrative approach to predicting the functional effects of small indels in non-coding regions of the human genome. BMC Bioinform. **18**, 442–442 (2017)

7. Frith, M., Hamada, M., Horton, P.: Parameters for accurate genome alignment. BMC Bioinform. **11**, 80–93 (2010)

8. Gonzalez-Garay, M.: The road from next-generation sequencing to personalized medicine. Pers. Med. **11**, 523–544 (2014)

9. Goodwin, S., Gurtowski, J., Ethe-Sayers, S., Deshpande, P., Schatz, M.C., Mccombie, W.: Oxford nanopore sequencing, hybrid error correction, and de novo assembly of a eukaryotic genome. Genome Res. **25**, 1750–1756 (2015)

10. Hedges, D., Hamilton-Nelson, K., Sacharow, S., Nations, L., et al.: Evidence of novel fine-scale structural variation at autism spectrum disorder candidate loci. Mol. Autism **3**, 2–12 (2012)

11. Heller, D., Vingron, M.: SVIM: structural variant identification using mapped long reads. Bioinformatics **35**, 2907–2915 (2018)

12. Saxena, R.K., Edwards, D., Varshney, R.: Structural variations in plant genomes. Brief. Funct. Genomics **13**, 296–307 (2014)

13. Stöcker, K.B., Köster, J., Rahmann, S.: SimLORD: simulation of long read data. Bioinformatics **32**, 2704–2706 (2016)

14. Kleftogiannis, D., Punta, M., Jayaram, A., Sandhu, S., Wong, S.Q., Tandefelt, D.G., et al.: Identification of single nucleotide variants using position-specific error estimation in deep sequencing data. BMC Med. Genomics **12**, 1–12 (2019)

15. Lander, E.S., et al.: Initial sequencing and analysis of the human genome. Nature **409**, 860–921 (2001)

16. Macintyre, G., Ylstra, B.D., Brenton, J.: Sequencing structural variants in cancer for precision therapeutics. Trends Genetics **32**, 530–542 (2016)

17. Mackeh, R., Marr, A.K., Dargham, S.R., Syed, N., Fakhro, K.A., Kino, T.: Single-nucleotide variations of the human nuclear hormone receptor genes in 60,000 individuals. J. Endocrine Soc. **2**, 77–90 (2017)

18. Middelkamp, S., et al.: Prioritization of genes driving congenital phenotypes of patients with de novo genomic structural variants. Genome Med. **11**, 1–15 (2019)

19. Mohiyuddin, M., Mu, J.C., Li, J., Bani Asadi, N., Gerstein, M.B., Abyzov, A., et al.: MetaSV: an accurate and integrative structural-variant caller for next generation sequencing. Bioinformatics **31**, 2741–2744 (2015)

20. Nakagawa, H., Fujita, M.: Whole genome sequencing analysis for cancer genomics and precision medicine. Cancer Sci. **109**, 513–522 (2018)

21. Nikzainal, S., et al.: Landscape of somatic mutations in 560 breast cancer whole-genome sequences. Nature **534**, 47–54 (2016)

22. Norris, A., Workman, R., Fan, Y.R., Eshleman, J., Timp, W.: Nanopore sequencing detects structural variants in cancer. Cancer Biol. Ther. **17**, 246–253 (2016)

23. Piazza, A., Heyer, W.D.: Homologous recombination and the formation of complex genomic rearrangements. Trends Cell Biol. **29**, 135–149 (2019)

24. Poirion, O.B., Zhu, X., Ching, T., Garmire, L.X.: Using single nucleotide variations in single-cell RNA-seq to identify subpopulations and genotype-phenotype linkage. Nat. Commun. **9**, 1–13 (2018)

25. Qiang, W., Yau, W., Lu, J., Collinge, J., Tycko, R.: Structural variation in amyloid-fibrils from alzheimer's disease clinical subtypes. Nature **541**, 217–221 (2017)

26. Lupski, J.R.: Structural variation mutagenesis of the human genome: impact on disease and evolution: Mutagenesis of the human genome. Environ. Mol. Mutagenesis **56**, 419–436 (2015)

27. Ratan, A., Olson, T.L., Loughran, T.P., Miller, W.: Identification of indels in next-generation sequencing data. BMC Bioinform. **16**, 42 (2015)
28. Sanchis-Juan, A., Stephens, J., French, C.E., Gleadall, N., Mégy, K., Penkett, C., et al.: Complex structural variants in mendelian disorders: identification and breakpoint resolution using short- and long-read genome sequencing. Genome Med. **10**, 1–10 (2018)
29. Sedlazeck, F., et al.: Accurate detection of complex structural variations using single-molecule sequencing. Nat. Methods **15**, 461–468 (2018)
30. Wang, D., Wang, L.: GRSR: a tool for deriving genome rearrangement scenarios from multiple unichromosomal genome sequences. BMC Bioinform. **19**, 291–299 (2018)
31. Wenger, A., Peluso, P., Rowell, W., Chang, P.C., Hall, R., Concepcion, G., et al.: Accurate circular consensus long-read sequencing improves variant detection and assembly of a human genome. Nat. Biotechnol. **37**, 1155–1162 (2019)
32. Zhou, Y., Minio, A., Massonnet, M., Solares, E., Lv, Y., Beridze, T., et al.: The population genetics of structural variants in grapevine domestication. Nat. Plants **5**, 965–979 (2019)

De novo Prediction of Drug-Target Interaction via Laplacian Regularized Schatten-p Norm Minimization

Gaoyan Wu[1], Mengyun Yang[1,2], Yaohang Li[3], and Jianxin Wang[1(✉)]

[1] Hunan Provincial Key Lab on Bioinformatics, School of Computer Science and Engineering, Central South University, Changsha 410083, China
jxwang@mail.csu.edu.cn

[2] Provincial Key Laboratory of Informational Service for Rural Area of Southwestern Hunan, Shaoyang University, Shaoyang 422000, China

[3] Department of Computer Science, Old Dominion University, Norfolk, VA 23529, USA

Abstract. The identification of drug-target interactions plays a crucial role in drug discovery and design. However, capturing interactions between drugs and targets via traditional biochemical experiments is an extremely laborious, expensive and time-consuming procedure. Therefore, the use of computational methods for predicting potential interactions to guide the experimental verification has attracted a lot of attention. In this paper, we propose a new algorithm, named Laplacian Regularized Schatten-p Norm Minimization (LRSpNM), to predict potential target proteins for novel drugs and potential drugs for new targets. First, we take advantage of the drug and target similarity information to dynamically prefill the partial unknown interactions. Then based on the assumption that the interaction matrix is low-rank, we use Schatten-p norm minimization model to improve prediction performance in the new drug/target cases by combining the loss function with a Laplacian regularization term. Finally, we numerically solve the LRSpNM model by an efficient alternating direction method of multipliers (ADMM) algorithm. Performance evaluations on benchmark datasets show that LRSpNM achieves better and more robust performance than five state-of-the-art drug-target interaction prediction algorithms. In addition, we conduct case study in practical applications, which also illustrates the effectiveness of our proposed method.

1 Introduction

The prediction of drug-target interactions (DTIs) is an important part of pharmaceutical scientific research in drug discovery, which has various applications. According to the statistics, there are more than 90 million chemical molecules in DrugBank [1] and more than 100,000 human target proteins in UniProt [2]. However, only about 4,000 known drug-target interactions have been verified by biological experiments. Therefore, identifying more interactions is an extremely

© Springer Nature Switzerland AG 2020
Z. Cai et al. (Eds.): ISBRA 2020, LNBI 12304, pp. 154–165, 2020.
https://doi.org/10.1007/978-3-030-57821-3_14

valuable task which can bring huge breakthrough in biopharmaceutical and biomedical research.

Many computational approaches have been developed to infer novel DTIs under the advantage of lower cost and wider coverage. Yamanishi *et al.* [3] first proposed a bipartite local model (BLM) to predict target proteins of a given drug, then to predict drugs targeting a given protein. BLM uses the chemical structure similarity of drugs and the sequence similarity of targets to improve the prediction accuracy. Analogously, Laplacian Regularized Least Squares (LapRLS) [4] is another algorithm based on the BLM. LapRLS uses regularized least squares to minimize an objective function that includes an error term as well as a graph regularization term. To perform prediction, Laarhoven *et al.* [5] utilized a weighted nearest neighbor (WNN) procedure for inferring a profile of a drug by using interaction profiles of the compounds. The experimental results have shown that neighbors information is indeed beneficial to the prediction results.

It is worth noting that matrix factorization and completion methods have exhibited excellent performance for computational DTI prediction in recent years. Kernelized Bayesian Matrix Factorization with Twin Kernels (KBMF2K) [6] applies a Bayesian probabilistic matrix factorization to perform prediction. KBMF2K uses variational approximation to perform nonlinear dimensionality reduction, which can improve the computational efficiency. Collaborative Matrix Factorization (CMF) [7] employs collaborative filtering for prediction. This approach transforms the input DTI matrix into the inner product between the two feature vectors, which share the same feature dimension. Liu *et al.* [8] proposed the logistic matrix decomposition based on neighborhood regularization (NRLMF). NRLMF focuses on the probability of drug-target interaction using logistic matrix decomposition, in which the characteristics of drug and target are represented by drug-specific and target-specific potential carriers respectively. The neighborhood constraint matrix completion method (NCMC) [9] applies the similar information of drugs/targets to define the concept of neighborhood. NCMC combines nuclear norm minimization model with neighborhood constraints to captures the strong correlation between drug and target.

Although these computational methods have been achieved excellent performance for predicting DTIs, it is a challenging task to identify interactions for new drugs or new targets, which is known as *de novo* prediction. To solve the cold start problem where drugs or targets have no given interactions in *de novo* tests, the information from drugs and targets can be taken advantage to achieve further improvement. Therefore, in order to enhance the prediction accuracy in *de novo* tests, more effective computational methods can be developed to predict potential DTIs.

In this paper, we propose a Laplacian Regularized Schatten-p Norm Minimization (LRSpNM) for *de novo* prediction of DTIs. Based on the assumption that similar drugs are normally interacted with similar targets and vice versa, the DTIs matrix can be assumed to be of low rank. Accordingly, matrix completion algorithms, which efficiently construct low-rank matrix approximations consistent with known interactions, can provide tremendous help in discovering the novel DTIs. In our method, we use Schatten-p norm to approximate the matrix rank and combine the Laplacian regularized term to assist prediction. In addition, considering that many of the non-interactions in the DTIs matrix are unknown cases, we use a preprocessing step to enhance prediction. Computational results on the benchmark dataset demonstrate the effectiveness of our proposed method. Besides, LRSpNM also performs better than other five state-of-the-art methods.

2   Materials

Evaluation experiments are performed using a benchmark dataset [10], which is generally used in drug-target interaction prediction. Specifically, the dataset consists of four different sub-datasets targeting protein of Enzyme, Ion Channel, G protein-coupled receptor (GPCR), and Nuclear Receptor. Each dataset includes three matrices: an interaction matrix A between drugs and targets, a similarity matrix of drugs S_d, and a similarity matrix of targets S_t. The four datasets are publicly available at http://web.kuicr.kyoto-u.ac.jp/supp/yoshi/drugtarget/. The matrix A is the adjaceny matrix encoding the drug-target interactions, where A_{ij} is 1 if drug d_i and target t_j are known to interact and 0 otherwise. The drug similarity S_d is computed from the chemical structures of drugs by using SIMCOMP [11]. The target similarity S_t is computed according to target sequences by using a normalized Smith-Waterman score [12]. The statistical information of drug-target interaction matrix in each dataset is summarized in Table 1.

Table 1. Statistics of drug-target interactions datasets

Datasets	No. of drugs	No. of targets	No. of interactions	Sparsity
Enzyme	445	664	2926	0.010
Ion Channel	210	204	1476	0.034
GPCR	223	95	635	0.030
Nuclear Receptor	54	26	90	0.064

3   Methods

3.1   Preprocessing Step

The known DTI matrix $A \in R^{m \times n}$ has m drug rows and n target columns. Many of the non-interactions in A are unknown cases that can potentially be positive interaction. Hence, we use a preprocessing step which utilizes the similarity information between drugs and targets to estimate the interaction likelihoods for unknown cases in Y.

Firstly, for drug d_i, we select the K most similar drugs as its neighbors based on drug similarities and use $N(d_i)$ to denote the set of them. We use an adjacency matrix $\widetilde{S_d}$ to represent the drug neighborhood information, which is defined as follows:

$$\widetilde{S_d}(d_i, d_\mu) = \begin{cases} \frac{S_d(d_i, d_\mu)}{\sum_{p:d_p \in N(d_i)} S_d(d_i, d_p)} & \text{if } d_\mu \in N(d_i) \\ 0, & \text{otherwise,} \end{cases} \tag{1}$$

where $S_d(d_i, d_\mu)$ is the original similarity score between d_i and d_μ. Similarly, we use $N(t_j)$ to represent the set of t_j's neighbors, and calculate the adjacency matrix $\widetilde{S_t}$ in the same way. Based on the K nearest known neighbors information from drugs and targets, we can obtain the drug-target interaction likelihoods for partial unknown pairs, which is marked as A^N and calculated in the following equation:

$$A^N = \frac{\widetilde{S_d}A + A\widetilde{S_t}^T}{2}. \tag{2}$$

Finally, we combine the prefilling interaction probabilities with known interactions as the input matrix to be completed.

3.2 Laplacian Regularized Schatten-p Norm Minimization

Assuming a low-rank structure, the general matrix completion problem to fill out the missing entries is formulated as:

$$\begin{aligned} &\min_X rank(X) \\ &s.t. P_\Omega(X) = P_\Omega(A), \end{aligned} \tag{3}$$

where $A \in R^{m \times n}$ is the given incomplete matrix, $X \in R^{m \times n}$ is the variable matrix, $rank(X)$ denotes the rank function of X, Ω is a set containing index pairs of all known entries in A and P_Ω is the projection operator onto Ω, which is defined as:

$$(P_\Omega(X))_{ij} = \begin{cases} X_{ij}, & (i,j) \in \Omega \\ 0, & (i,j) \notin \Omega. \end{cases} \tag{4}$$

Unfortunately, the rank minimization problem (3) is known to be NP-hard. One of the solutions is that turns the rank function to a more tractable solution by minimizing the nuclear norm, which has been proven to be the convex relaxation of matrix rank [13]. Although the nuclear norm minimization model is a convex problem with a global solution, the relaxation may deviate from the problem of the original solution. Therefore, Nie *et al.* [14,15] proposed nonconvex optimization models where the Schatten-p norm of a matrix is used to replace the rank function of Eq. (3), which is defined as:

$$\|X\|_{Sp}^p = \sum_{i=1}^{\min\{n,m\}} \sigma_i^p = Tr\left((X^T X)^{\frac{p}{2}}\right), \tag{5}$$

where σ_i is the singular value of X and when $p = 1$, the Schatten 1-norm is the well-known nuclear norm. That is to say, the nuclear norm is the special case of Schatten-p norm. As a result, the baseline Schatten-p norm minimization is formulated as:

$$\min_X \|X\|_{Sp}^p + \frac{\alpha}{2} \|P_\Omega(X) - P_\Omega(A)\|_F^2 , \qquad (6)$$

where α is the harmonic parameter that balances the Schatten-p norm and the error term, We optimize the effectiveness of matrix completion by fine tuning the value of p.

Based on the assumption that similar drugs share the similar molecular pathways to interact with similar targets, the interaction matrix A is inherently low-rank. Thus, the DTI prediction problem can then be modeled as a matrix completion problem by completing the unknown elements with pharmacological space information in the interaction matrix. In this work, we first introduce Schatten-p norm to approximate the matrix rank for DTI prediction. In addition, we present a new objective function through incorporation of the Laplacian regularized terms of drugs and targets into the matrix completion framework for increasing generalization capability. Specially, a Laplacian Regularized Schatten-p Norm Minimization (LRSpNM) model is proposed for DTI prediction. The optimization problem of LRSpNM can be formulated as follows:

$$\min_X \|X\|_{Sp}^p + \frac{\alpha}{2} \|P_\Omega(X) - P_\Omega(A)\|_F^2 + \lambda_d Tr\left(X^T L_d X\right) + \lambda_t Tr\left(X L_t X^T\right), \quad (7)$$

where $L_d \in R^{m \times m}$ is the drug Laplacian matrix with $L_d = D_d - S_d$, D_d is the diagonal matrix with $D_d(i,i) = \sum_i S_d(i,j)$, $L_t \in R^{n \times n}$ is the target Laplacian matrix with $L_t = D_t - S_t$, D_t is the diagonal matrix with $D_t(i,i) = \sum_i S_t(i,j)$ and λ_d, λ_t are parameters balancing the reconstruction terms of LRSpNM model.

To solve the optimization problem in (7), we use the alternating direction methods of multipliers (ADMM) [16] framework and introduce two auxiliary variables W and Z to make the objective function separable:

$$\min_X \|W\|_{Sp}^p + \frac{\alpha}{2} \|P_\Omega(Z) - P_\Omega(A)\|_F^2 + \lambda_d Tr\left(X^T L_d X\right) + \lambda_t Tr\left(X L_t X^T\right)$$
$$s.t. X = W, X = Z. \qquad (8)$$

The corresponding augmented Lagrange function of (8) is:

$$\begin{aligned}
\mathcal{L}(W, Z, X, U, V) = &\|W\|_{Sp}^p + \frac{\alpha}{2} \|P_\Omega(Z) - P_\Omega(A)\|_F^2 \\
&+ \lambda_d Tr\left(X^T L_d X\right) + \lambda_t Tr\left(X L_t X^T\right) \\
&+ Tr\left(U^T (X - W)\right) + \frac{\mu_1}{2} \|X - W\|_F^2 \\
&+ Tr\left(V^T (X - Z)\right) + \frac{\mu_2}{2} \|X - Z\|_F^2,
\end{aligned} \qquad (9)$$

where U and V are the Lagrange multipliers, $\mu_1 > 0$ and $\mu_2 > 0$ control the penalties for violating the linear constraints. Then the variables can be approximated alternatively.

Compute W_{k+1}: The variable W can be calculated by the following equation with other variables fixed:

$$
\begin{aligned}
W_{k+1} &= \arg\min_{W} \mathcal{L}\left(W, Z_k, X_k, U_k, V_k\right) \\
&= \arg\min_{W} \|W\|_{S_p}^p + Tr\left(U_k^\top (X_k - W)\right) + \frac{\mu_1}{2}\|X_k - W\|_F^2 \\
&= \arg\min_{W} \|W\|_{S_p}^p + \frac{1}{\mu_1}\left\|W - \left(X_k + \frac{1}{\mu_1}U_k\right)\right\|_F^2,
\end{aligned}
\tag{10}
$$

where W_{k+1} can be obtained by the algorithm provided in [15], which guaranteed convergence when $0 < p < 2$.

Compute Z_{k+1}: When other variables are fixed, Z can be obtained by minimizing following function:

$$
\begin{aligned}
Z_{k+1} &= \arg\min_{W} \mathcal{L}\left(W_{k+1}, Z, X_k, U_k, V_k\right) \\
&= \arg\min_{W} \frac{\alpha}{2}\|\mathcal{P}_\Omega(Z) - \mathcal{P}_\Omega(A)\|_F^2 \\
&\quad + Tr\left(V_k^T (X_k - Z)\right) + \frac{\mu_2}{2}\|X_k - Z\|_F^2,
\end{aligned}
\tag{11}
$$

which is a convex optimization problem and can be solved by setting the derivative of Eq. (11) to zero. Referred to the solution of [17], then we directly obtain:

$$
Z_{k+1} = \frac{1}{\mu_2}V_k + \frac{\alpha}{\mu_2}\mathcal{P}_\Omega(A) + X_k - \frac{\alpha}{\alpha + \mu_2}\mathcal{P}_\Omega\left(\frac{1}{\mu_2}V_k + \frac{\alpha}{\mu_2}\mathcal{P}_\Omega(A) + X_k\right).
\tag{12}
$$

Compute X_{k+1}: When other variables are fixed, X can be solved by minimizing the following objective function:

$$
\begin{aligned}
X_{k+1} &= \arg\min_{X} \mathcal{L}\left(W_{k+1}, Z_{k+1}, X, U_k, V_k\right) \\
&= \arg\min_{X} \lambda_d\, Tr\left(X^\top L_d X\right) + \lambda_t\, Tr\left(X L_t X^\top\right) \\
&\quad + Tr\left(U_k^\top (X - W_{k+1})\right) + \frac{\mu_1}{2}\|X - W_{k+1}\|_F^2 \\
&\quad + Tr\left(V_k^\top (X - Z_{k+1})\right) + \frac{\mu_2}{2}\|X - Z_{k+1}\|_F^2,
\end{aligned}
\tag{13}
$$

By setting the derivative of Eq. (13) with respect to X to zero, we have:

$$
(2\lambda_d L_d + \mu_1 I)X + X(2\lambda_t L_t + \mu_2 I) = \mu_1 W_{k+1} + \mu_2 Z_{k+1} - U_k - V_k,
\tag{14}
$$

Formula (14) is a Sylvester equation [18] which provides the solution $X = Sylvester(A, B, C)$ of the matrix equation $AX + XB = C$. Thus X_{k+1} can be solved by the following equation:

$$
X_{k+1} = Sylvester\left(2\lambda_d L_d + \mu_1 I, 2\lambda_t L_t + \mu_2 I, \mu_1 W_{k+1} + \mu_2 Z_{k+1} - U_k - V_k\right).
\tag{15}
$$

Compute U_{k+1} and V_{k+1}: We update the multipliers by:

$$U_{k+1} = U_k + \mu_1 \left(X_{k+1} - W_{k+1} \right),$$
$$V_{k+1} = V_k + \mu_2 \left(X_{k+1} - Z_{k+1} \right). \tag{16}$$

The variables W, Z, and X are iteratively updated until convergence. Finally, we obtain the predicted DTIs based on the completed entities in matrix X.

4   Results and Discussions

4.1   Experimental Settings

In this experiment, we conduct 10-fold cross-validation (CV) to evluate the *de novo* performance of LRSpNM. In order to test the different aspect of the prediction methods, we consider two following types of *de novo* tests from new drugs and new targets aspects, respectively. The first is called CV_drug where all drugs are randomly divided into 10 subsets. Another is CV_target where all targets are randomly divided into 10 subsets. That is to say, for a given DTI prediction method, CV_drug tests its ability to predict interactions for new drugs and CV_target tests its ability to predict interactions for new targets. Each subset is treated as the testing set in turn, while the remaining nine subsets are used as the training set. Both two types of *de novo* tests are repeated five times and the average accuracy values are showed as the final results. We use Area Under the Precision-Recall curve (AUPR) [19] as the evaluation metric.

We perform the 10-fold cross-validation on the training set for setting LRSpNMs four parameters, α, p, λ_d, λ_t. Using grid search, the best parameter combination is selected from the range of values: $\alpha \in \{10^{-2}, 10^{-1}, 10^0, 10^1, 10^2\}$, $p \in \{0.25, 0.50, 0.75, 1, 1.25, 1.50, 1.75, 2\}$, λ_d and $\lambda_t \in \{10^{-4}, 10^{-3}, 10^{-2}, 10^{-1}\}$. As for the pre-filling step, the parameter $K \in \{1, 2, 3, 4, 5, 6, 7, 8, 9, 10\}$ is also set by grid search.

4.2   Performance Results

In order to measure the prediction performance, five existing state-of-the-art DTI prediction methods are used to compare with our LRSpNM model, including LapRLS [4], WNN [5], KBMF2K [6], CMF [7], and NRLMF [8]. For these competing methods, all parameters are set to their default values according to the authors' recommendation.

Table 2 shows the result of AUPR under the setting CV_drug. As shown in Table 2, LRSpNM outperforms all five competing methods on four datasets for new drug predictions. The results obtained under setting CV_target is presented in Table 3. For new target prediction, LRSpNM outperforms the competing methods except for the Enzyme dataset, where LRSpNM performs slightly worse than NRLMF algorithm. LRSpNM reports AUPR values that are 2.094%, 0.992% and 5.596% higher than the methods with second best performance in other three datasets, respectively.

These results adequately demonstrate that LRSpNM has a higher accuracy on top ranked drug-target pairs for novel prediction, which is more meaningful in drug discovery process.

Table 2. AUPR results for DTI prediction under CV_drug

AUPR	Enzyme	Ion Channel	GPCR	Nuclear Receptor
LapRLS	0.111(0.002)	0.172(0.005)	0.219(0.004)	0.370(0.020)
WNN	0.393(0.013)	0.334(0.010)	0.367(0.007)	0.540(0.020)
KBMF2K	0.254(0.010)	0.317(0.009)	0.390(0.014)	0.483(0.030)
CMF	0.386(0.008)	0.353(0.014)	0.406(0.010)	0.523(0.030)
NRLMF	0.335(0.031)	0.355(0.039)	0.353(0.028)	0.539(0.059)
LRSpNM	**0.399(0.009)**	**0.357(0.014)**	**0.408(0.012)**	**0.546(0.021)**

Best and second best AUPR results are **bold** and <u>underlined</u>, respectively. Standard deviations are given in (parentheses).

Table 3. AUPR results for DTIs prediction under CV_target

AUPR	Enzyme	Ion Channel	GPCR	Nuclear Receptor
LapRLS	0.638(0.005)	0.702(0.004)	0.310(0.011)	0.369(0.023)
WNN	0.778(0.018)	0.763(0.007)	0.574(0.021)	0.492(0.033)
KBMF2K	0.672(0.024)	0.727(0.013)	0.528(0.018)	0.406(0.021)
CMF	0.781(0.013)	0.779(0.011)	0.599(0.032)	0.475(0.016)
NRLMF	**0.810(0.017)**	0.795(0.026)	0.539(0.039)	0.523(0.082)
LRSpNM	0.803(0.017)	**0.812(0.011)**	**0.605(0.022)**	**0.554(0.047)**

Best and second best AUPR results are **bold** and <u>underlined</u>, respectively. Standard deviations are given in (parentheses).

4.3 Parameters Analysis

In this section,we analyze the parameter K from the pre-filling step in four datasets. We can see that sensitivity analyses are provided for K in Fig. 1. The result displays that the most K nearest neighbors information of drugs and targets will assist the DTI prediction.

In addition, we analyze the parameter p to explore the prediction accuracy of Schatten-p norm. From the results of Fig. 2, the AUPR values increase as the increase of the values of p and then become stable after certain value of p is reached under CV_drug setting in all datasets. Under CV_target setting, the Nuclear Receptor dataset appears a special situation, where the AUPR values fluctuate with the increase of p. From the results we can find that the Schatten-p norm-based objective can approximate the rank minimization problem much better than the nuclear norm minimization (when $p = 1$) to achieve better matrix completion results.

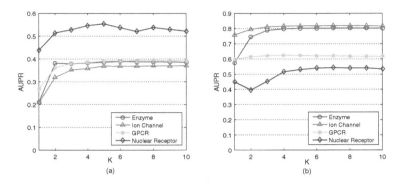

Fig. 1. Sensitivity analysis for K. (a) CV_drug; (b) CV_target.

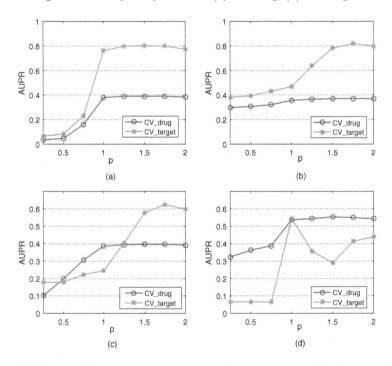

Fig. 2. AUPR with different values of p on benchmark datasets. (a) Enzyme; (b) Ion Channel; (c) GPCR; (d) Nuclear Receptor.

4.4   Case Study

In order to illustrate the prediction ability of LRSpNM in practical applications, we treat the whole known DTI matrix in the GRCR dataset as training set and regard the missing drug-target pairs as the candidate. After the prediction scores of all candidate pairs are computed by LRSpNM, we rank the unknown pairs and verify the top 10 novel DTIs by the latest version of DrugBank [1] and KEGG [20] databases. The results are shown in Table 4. The last column shows the databases evidence for each novel pair, and we can find that 70% of predictions are confirmed in latest databases. This encouraging result indicates that LRSpNM is effective for identifying new DTIs, which means it can provide reliable guidance for drug discovery. In addition, other predicted DTIs that have not been validated in public databases, which deserves to be tested by biochemical experiments in the future.

Table 4. Top 10 novel DTIs predicted by LRSpNM on GPCR dataset.

Rank	Drug Name (Drug ID)	Target Name (Target ID)	Evidence
1	Octreotide acetate (D02250)	SSTR1 (hsa6751)	KEGG
2	Metoprolol (D02358)	ADRB2 (hsa154)	DrugBank
3	Albuterol (D02147)	PTGER1 (hsa5731)	DrugBank
4	Ambuphylline (D02884)	ADORA2B (hsa136)	Unknown
5	Dinoprostone (D00079)	PTGER1 (hsa5731)	DrugBank
6	Octreotide (D00442)	SSTR5 (hsa6755)	KEGG
7	Rocuronium bromide (D00765)	CHRM1 (hsa1128)	Unknown
8	Epinephrine (D00095)	ADRA2A (hsa150)	DrugBank
9	Trospium chloride (D01103)	CHRM2 (hsa1129)	KEGG
10	Rocuronium bromide (D00765)	CHRM1 (hsa1128)	Unknown

5 Conclusions

In this paper, we proposed a novel matrix completion method named LRSpNM for *de novo* prediction of drug-target interactions. The novelty of LRSpNM comes from firstly integrating Schatten-p norm minimization with Laplacian regularization to predict the interaction probability of an unknown drug-target pair. Specifically, when p becomes a tunable parameter in completing the DTI matrix, it can be adjusted to achieve better performance than nuclear norm where $p = 1$. Moreover, we use a preprocessing step which fully utilizes the information of drug and target to fill partial unknown drug target pairs. Experiments have been conducted using two different types of cross validations for *de novo* prediction to compare our method with five state-of-the-art methods. In most of the cases, LRSpNM achieves the highest accuracy and presents the reliability of LRSpNM in *de novo* prediction. Our case study also confirms the effectiveness of our method in practical applications.

Acknowledgement. This work is supported by the National Natural Science Foundation of China (Grant No. 61972423), the Graduate Research Innovation Project of Hunan (Grant No. CX20190125), Hunan Provincial Science and technology Program (No. 2018wk4001), and 111Project (No. B18059).

References

1. Wishart, D.S., et al.: Drugbank: a knowledgebase for drugs, drug actions and drug targets. Nucleic Acids Res. **36**(Database issue), D901–D906 (2008)
2. Apweiler, R., et al.: UniProt: the universal protein knowledgebase. Nucleic Acids Res. **32**(Database issue), D115–D119 (2004)
3. Bleakley, K., Yamanishi, Y.: Supervised prediction of drug-target interactions using bipartite local models. Bioinformatics **25**(18), 2397–2403 (2009)
4. Xia, Z., Wu, L.Y., Zhou, X., Wong, S.T.: Semi-supervised drug-protein interaction prediction from heterogeneous biological spaces. BMC Syst. Biol. **4**(2), S6 (2010)
5. Van Laarhoven, T., Marchiori, E.: Predicting drug-target interactions for new drug compounds using a weighted nearest neighbor profile. PLoS One **8**(6), e66952 (2013)
6. Gönen, M.: Predicting drug-target interactions from chemical and genomic kernels using bayesian matrix factorization. Bioinformatics **28**(18), 2304–2310 (2012)
7. Zheng, X., Ding, H., Mamitsuka, H., Zhu, S.: Collaborative matrix factorization with multiple similarities for predicting drug-target interactions. In: Proceedings of the 19th ACM SIGKDD International Conference on Knowledge Discovery And Data Mining, pp. 1025–1033 (2013)
8. Liu, Y., Wu, M., Miao, C., Zhao, P., Li, X.L.: Neighborhood regularized logistic matrix factorization for drug-target interaction prediction. PLoS Comput. Biol. **12**(2), e1004760 (2016)
9. Fan, X., Hong, Y., Liu, X., Zhang, Y., Xie, M.: Neighborhood constraint matrix completion for drug-target interaction prediction. In: Phung, D., Tseng, V.S., Webb, G.I., Ho, B., Ganji, M., Rashidi, L. (eds.) PAKDD 2018. LNCS (LNAI), vol. 10937, pp. 348–360. Springer, Cham (2018). https://doi.org/10.1007/978-3-319-93034-3_28
10. Yamanishi, Y., Araki, M., Gutteridge, A., Honda, W., Kanehisa, M.: Prediction of drug-target interaction networks from the integration of chemical and genomic spaces. Bioinformatics **24**(13), i232–i240 (2008)
11. Hattori, M., Okuno, Y., Goto, S., Kanehisa, M.: Development of a chemical structure comparison method for integrated analysis of chemical and genomic information in the metabolic pathways. J. Am. Chem. Soc. **125**(39), 11853–11865 (2003)
12. Smith, T.F., Waterman, M.S., et al.: Identification of common molecular subsequences. J. Mol. Biol. **147**(1), 195–197 (1981)
13. Fazel, M.: Matrix rank minimization with applications. Ph.D. thesis, Stanford University (2002)
14. Nie, F., Huang, H., Ding, C.: Low-rank matrix recovery via efficient Schatten p-norm minimization. In: Proceedings of the 26th AAAI Conference on Artificial Intelligence (2012)
15. Nie, F., Wang, H., Huang, H., Ding, C.: Joint schatten p-norm and ℓ_p-norm robust matrix comletion for missing value recovery. Knowl. Inf. Syst. **42**(3), 525–544 (2015)
16. Chen, C., He, B., Yuan, X.: Matrix completion via an alternating direction method. IMA J. Numer. Anal. **32**(1), 227–245 (2012)
17. Yang, M., Luo, H., Li, Y., Wang, J.: Drug repositioning based on bounded nuclear norm regularization. Bioinformatics **35**(14), i455–i463 (2019)
18. Bartels, R.H., Stewart, G.W.: Solution of the matrix equation AX+XB=C [F4]. Commun. ACM **15**(9), 820–826 (1972)

19. Davis, J., Goadrich, M.: The relationship between precision-recall and roc curves. In: Proceedings of the 23rd International Conference on Machine Learning, pp. 233–240 (2006)
20. Kanehisa, M., Goto, S., Sato, Y., Furumichi, M., Tanabe, M.: KEGG for integration and interpretation of large-scale molecular data sets. Nucleic Acids Res. **40**(D1), D109–D114 (2012)

Diagnosis of ASD from rs-fMRI Images Based on Brain Dynamic Networks

Hongyu Guo[1], Wutao Yin[1], Sakib Mostafa[1], and Fang-Xiang Wu[1,2,3]($\boxtimes$)

[1] Division of Biomedical Engineering, University of Saskatchewan,
Campus Dr. 57, S7N5A9 Saskatoon, Canada
[2] Department of Mechanical Engineering, University of Saskatchewan,
Campus Dr. 57, S7N5A9 Saskatoon, Canada
[3] Department of Computer Science, University of Saskatchewan,
Campus Dr. 57, S7N5C9 Saskatoon, Canada
faw341@mail.usask.ca

Abstract. The resting-state functional magnetic resonance imaging (rs-fMRI) as a non-invasive technique with the high spatial and temporal resolution can help characterize the pathogenesis of autism spectrum disorder (ASD). Some results have been achieved with machine learning techniques to diagnose ASD with rs-fMRI data. However, most of machine learning methods have neglected the temporal dependency of the time-series fMRI data. In this study, we propose a method for diagnosing ASD based on brain dynamic networks (BDNs) which are constructed with time series rs-fMRI brain image data to describe the dynamic relationship among multiple brain regions. The least squares method with the forward model selection method was used to establish BDNs, and the Bayesian information criterion (BIC) was adopted as the model selection criteria to avoid overfitting. The resulted DBNs are weighted directed networks. Then a feature extraction method was proposed to extract representative and discriminated features from BDNs. Lastly, machine learning classifiers were trained with the whole ABIDE I cohort to diagnose ASD. The accuracy of 88.8% was achieved, which is higher than any previously reported methods.

Keywords: Autism spectrum disorder · Resting-state fMRI · Brain dynamic network · Machine learning · Time series brain image

1 Introduction

Autism spectrum disorder (ASD) is a prevalent and heterogeneous childhood neuro-developmental disease with an estimated prevalence of 1% of the global population and one in 68 children in the United States [10, 18]. The resting-state functional magnetic resonance imaging (rs-fMRI) is a non-invasive technique showing fluctuations of functional activities of a whole brain through measuring

Supported by Natural Science and Engineering Research Council of Canada (NSERC).

blood oxygen level-dependent (BOLD) signals. The lack of a task makes rs-fMRI image particularly attractive for patients who may have difficulty with task instructions. Hence, the application of rs-fMRI images in the ASD research has been growing for the past two decades [7,14,21].

A currently widely used method for fMRI study is the functional connectivity (FC) network derived from rs-fMRI image data [12] which has been extensively applied for diagnosing brain diseases [17]. FC networks are constructed by calculating Pearson's Correlation Coefficient (PCC) between pairs of ROIs resulting the FC networks only reflect the correlation of ROIs rather than their dynamic relationships. Note that the directions of the relationships between ROIs are ignored when using FC networks although directed influences between ROIs can help characterize the functional role of a brain region [2,6]. Furthermore, although rs-fMRI brain images consist of time series images, most of machine learning classifiers do not take the temporal dependency, which is a valued property of the time-series fMRI data, into consideration. In principle, the state of one ROI should be dynamically associated with multiple ROIs [2,6]. Therefore, building brain dynamic networks (BDN) would help us understand how ROIs dynamically influence each other, and thus can provide more representative and discriminate information for ASD diagnosis.

In this study, we propose a method for diagnosing ASD based on BDN which are constructed with time series rs-fMRI brain imaging data. The BDN can model directed influences among multiple ROIs across the whole brain based on the assumption that the current state of specific ROI depends on the linear combination of the previous states of multiple ROIs. Since, BDN is constructed with the temporal dependency of rs-fMRI brain image data, it can capture more complex interactions cross multiple brain ROIs than FC networks. Subsequently, graph theory and complex network analysis were applied on DBNs, which are the weighted and directed dynamic networks, to extract more representative and discriminated features. Lastly, machine learning classifiers were trained with whole ABIDE I cohort to identify ASD. The accuracy of 88.8% was achieved, which is higher than any previously reported methods (e.g.[1,10,11,13,20]).

2   Methods

2.1   Dataset and Pre-processing

The data used in this study are obtained from the Autism Brain Imaging Data Exchange I (ABIDE I), which was released in August 2012 and integrated 17 international sites. In total, the dataset includes 1111 subjects, 539 from individuals with ASD and 573 from typical developments (TDs) [3].

Brain image pre-processing includes slice-timing correction, motion correction, nuisance signal regression and temporal filtering. Firstly, functional volumes are registered to structural images, then, structural images are registered to the symmetric standard MNI152 brain atlas. Consequently, spatial smoothing and a band-pass filter of 0.01–0.1 Hz are applied to reduce the influence of heart beat and breath. The parcellation scheme used in this study is the whole-brain parcellation based on meta-analysis of fMRI, yielding 264 ROIs in MNI 152 standard space [16]. More data pre-processing details can be found in [13,20].

2.2 Brain Dynamic Networks

The construction of a Brain dynamic network is to find the relationships among multiple ROIs with the hypothesis that the state value of one ROI at one time point can be represented as a linear combination of the state values of other ROIs àt their previous time point. We apply the forward model selection approach based on the least squares method to establish one specific ROIs' dynamic model by gradually including one ROI each time until some criterion is met. The strategy for selecting the next candidate ROI is based on the correlation strengths between the target ROI and its regulating ROIs. Besides, we apply the BIC criterion to conduct the model selection and use the ridge regularization to avoid overfitting. The schematic diagram of the proposed BDN modeling algorithm with time-series rs-fMRI brain image data is depicted in Fig. 1.

Considering a time series of a target ROI, let

$$x_i = \{x_{1,i}, \cdots, x_{t,i}, \cdots, x_{T,i}\} \tag{1}$$

be the time series of the $i-$th ROI with T time points. Assuming that the observed value $x_{t,i}$ of the $i-$th ROI at time point t is a linear combination of the observed values of its regulating ROIs at their previous time point,i.e. time point $t - 1$. Formally, the model can be written as:

$$x_{t,i} = x_{t-1,1}\beta_{1,i} + x_{t-1,2}\beta_{2,i} + \cdots + x_{t-1,r}\beta_{r,i} + \cdots + x_{t-1,R}\beta_{R,i} \tag{2}$$

where $i(i = 1, \cdots, R)$ denotes the index of a ROI, and $t(t = 1, \cdots, T)$ denotes the time point, T is the length of the time series data while R represents the number of ROIs of the whole brain, $\beta_{r,i}$ corresponds to the dynamic coefficients representing the regulation strength between the objective ROI_i and its regulating ROI_r. Collecting Eq. 2 for $t = 2, \cdots, T$ yields to the following general form:

$$Y = X\beta \tag{3}$$

where

$$Y = \begin{bmatrix} x_{2,i} \\ x_{3,i} \\ \vdots \\ x_{t,i} \\ \vdots \\ x_{T,i} \end{bmatrix}, \quad X = \begin{bmatrix} x_{1,1} & \cdots & x_{1,r} & \cdots & x_{1,R} \\ x_{2,1} & \cdots & x_{2,r} & \cdots & x_{2,R} \\ \vdots & \ddots & \vdots & \ddots & \vdots \\ x_{t-1,1} & \cdots & x_{t-1,r} & \cdots & x_{t-1,R} \\ \vdots & \ddots & \vdots & \ddots & \vdots \\ x_{T-1,1} & \cdots & x_{T-1,r} & \cdots & x_{T-1,R} \end{bmatrix}, \quad \beta = \begin{bmatrix} \beta_{1,i} \\ \beta_{2,i} \\ \vdots \\ \beta_{r,i} \\ \vdots \\ \beta_{R,i} \end{bmatrix} \tag{4}$$

Given the observation data, time series fMRI brain imaging data of one subject, to estimate the parameter β using the least square method, we have the

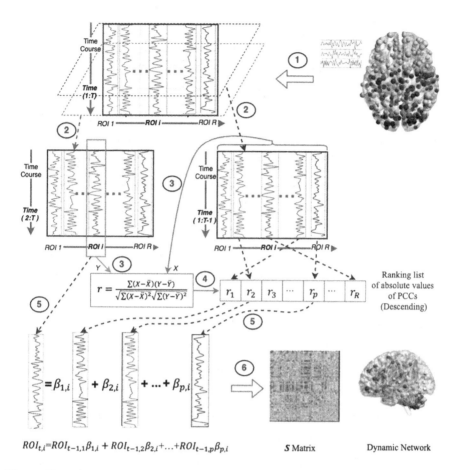

Fig. 1. The schematic diagram of BNW modeling algorithm. (1) Extract time courses of ROIs from rs-fMRI brain image data, then construct the time course matrix for a subject, which has R ROIs and T time points. (2) Construct two matrices using sliding window to $T-1$ time points from the time course matrix of the subject with one matrix represents the state of all ROIs at time points 1 to $T-1$, and the other matrix keeps the state of all ROIs at time points 2 to T. (3) Calculate Pearson Correlation Coefficients (PCCs) between the time course vector consisting of time points 2 to T of the ith ROI and the time course vector consisting of time points 1 to $T-1$ of all ROIs. (4) Rank the absolute values of PCCs in descending order. (5) Build the dynamic model for the i-th ROI using the forward model selection method initialed with containing the top ranked ROI in terms of PCC and then add one candidate ROI into the model each step until a stoping criterion is met. Thus, the state of the i-th ROI at time point t ($t = 2, \cdots, T$) may be represented by the linear combination of the state of other p ROIs (p is the number of parameters in the model) at time point $t-1$ with the regulation strengths β. (6) Collect the vectors β of all ROIs into matrix S, which is the coefficient matrix of the BDN for one subject.

objective function in the matrix format as follows:

$$S(\beta) = (Y - X\beta)^T(Y - X\beta) = \|Y - X\beta\|^2 \tag{5}$$

The least squares estimator β_{LS} is given by :

$$\beta_{LS} = [X^TX]^{-1}X^TY \tag{6}$$

In this study, the number of variables (264 ROIs) in the linear system exceeds the number of observations ($T = 120$), in this case, the ordinary least-squares problem is ill-posed. Therefore, it is hard to fit as the model has infinitely many solutions. The ridge regularization introduces further constraints to uniquely determine the solution while preventing over-fitting as follows

$$S(\beta) = (Y - X\beta)^T(Y - X\beta) = \|Y - X\beta\|^2 + \lambda\|\beta\|_2^2 \tag{7}$$

Then the estimator $\hat{\beta}^{Ridge}$ can be obtained by:

$$\hat{\beta}^{Ridge} = (X^TX + \lambda I)^{-1}X^TY \tag{8}$$

where $\lambda > 0$ is a tuning parameter, to be determined separately, I is a $R \times R$ identity matrix. Selecting a good value for λ is critical.

The forward model selection is a type of step wise regression which begins with an empty model and adds in variables one by one. In each forward step, the one 'best' variable is added which gives the single best improvement to the model. The 'best' variable is determined by some pre-determined criteria. Then, the algorithm continues adding in one variable at a time and testing at each step until the model cannot be improved [8,9]. In this study, the criteria for selecting the next 'best' is based on the rank of the values of Pearson's correlation coefficients (PCCs) between the target ROI and all other ROIs as its regulating ROIs. Firstly, we rank the absolute values of PCCs, which represent the correlation strength of the target ROI and other ROIs in descending order. The reason behind the ranking criterion is aiming to keep both the positive and negative PCCs with significant values ranked on the top so that ROIs hold higher correlation coefficients have the priority to be added into the expansion process of the model for the target ROI. Secondly, the algorithm begins with adding the top-ranked ROI into the model, then regards the remaining ROIs in the order of the rank.

When building models using the forward model approach, the likelihood can increase by adding parameters, yet it may result in overfitting. In order to avoid over-fitting, the model selection criterion is needed to be applied. The Akaike Information Criterion (AIC) and the Bayesian information criterion (BIC) are two most commonly used criteria in the model selection based on the information theory by quantifying or measuring the expected value of information [19].

The best model has the minimum value of AIC function is defined as

$$AIC(p) = -2lnL(\beta|Y) + 2p \tag{9}$$

while the BIC criterion is formally defined as Eq. 10, BIC is a criterion for model selection among a finite set of models; the model with the lowest BIC is preferred.

$$BIC(p) = -2lnL(\beta|Y) + pln(n) \tag{10}$$

where $lnL(\beta|Y)$ represents the log-likelihood of estimates of the model parameters, and the final terms in Eq. 9 and Eq. 10 stand for the penalty on the log-likelihood as a function of the number of parameters p in the model [19]. The penalty term of BIC takes the sample size into consideration and it is larger in BIC than in AIC. Hence BIC results in smaller models. We choose BIC as our model selection criterion aiming to have comparatively smaller models.

To sum up, the pseudo algorithm of dynamic modeling algorithm for fMRI brain imaging data is described in Algorithm 1.

Algorithm 1. The dynamic modeling algorithm for fMRI brain imaging data

---

Input : $Data_{T \times R}$ is the matrix of time series of one subject with column vector represents the time series vector of one ROI. T denotes the length of time course of the fMRI brain image data, and R denotes the number of ROIs in the input subject.

Output: $S_{R \times R}$ is the dynamic network of the input subject in matrix format with each row vector standing for the generated linear model of one ROI.

1: Normalise $Data$ by column to represent percent change from the average signal for that time point with having mean 0 and standard deviation 1.
2: $X \leftarrow Data[1 : T - 1, :]$ $\triangleright$ X denotes the state of $Data$ at time point t.
3: **for** $ROIid \in [1, R]$ **do**
4:     $Y = Data[2 : T, ROIid]$ $\triangleright$ Y denotes the time course of one ROI at the succeed time point $t + 1$.
5:     Calculate PCCs between the Y and each time course of X and save as PCC_{list}
6:     Rank the absolute values of PCC_{list} in descending order.
7:     **Forward stepwise selection:**
8:     Let M_1 denotes the initial model containing the time course of the top ranked ROI in PCC_{list}.
9:     $k \leftarrow 2$ $\triangleright$ k denotes the index of models
10:     **for** $k \in [2, p]$ **do** $\triangleright$ p is an empirical constant.
11:        $ROI_{selected} \leftarrow PCC_{list}(k)$ $\triangleright$ Select the k-th ROI from the ROI ranking list
12:        Expand the model in M_{k-1} with $ROI_{selected}$ using ridge least square regression to build a new model M_k.
13:     **end for**
14:     Apply BIC to select a single best model M_{best} from $(M_0, \cdots, M_p)$ by choosing the model with the minimum BIC value.
15:     $\beta_{ROIid} \leftarrow M_{best}$
16: **end for**
17: Collect the vectors $\beta_{ROIid}(ROI_{id} \in [1, R])$ into a matrix S
18: **return**= S

After obtaining the model 3 for all ROIs of one subject, collecting the vectors β of all ROIs into a matrix S, which is the coefficient matrix of the BDN for one subject.

2.3   Feature Extraction

The BDNs constructed by last step are weighted and directed graphs. Theoretically, let $G = (V, E)$ denote the BDN, where V denotes the set of R nodes (ROIs), E denotes the set of edges. The set of R nodes constitutes the columns and rows of an adjacency matrix S with the entries s_{ij} represent connections between nodes (ROIs) i and j, i.e., the coefficients of a constructed BDN. Hence, the adjacency matrix represents both topological properties and biophysics of the brain network. Figure 2 illustrates an example of the BDN as a directed and weighted graph. As shown in this figure, the DBN consists of 6 ROIs connected with weighted and directed edges. The adjacent matrix is an asymmetric matrix with each row representing the linear model of a target ROI holding the meaning that the current state of the target ROI depends on the linear combination of the previous states of its regulating ROIs. Furthermore, rows of the adjacent matix can be used as the attributes or features of the corresponding ROIs. In Fig. 2, ROI A is affected by the previous states of B and E, and the feature of ROI A is the corresponding row vector of the adjacency matrix.

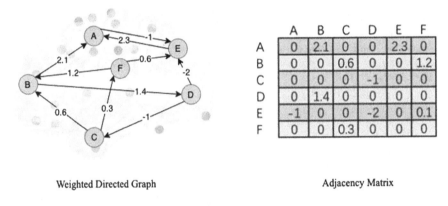

Weighted Directed Graph   Adjacency Matrix

Fig. 2. A dynamic network and its adjacency matrix.

We proposed a binary mask method based on the average network of TD group to explore the higher level features. Since we assume the variation of the BDNs in TD group is less than the those in ASD group, a comparatively consistent network structure can be found from it. Figure 3 illustrates the framework of this strategy and the major steps are described as follows:

1. Calculate the average network of TD group
   Firstly, the average dynamic network of TD group is calculated, and then we sort the absolute weights of edges in descending order. Since the absolute weights represent the regulation strength between the pair of ROIs, we set a threshold to delete those weak strength edges resulting in the average network *MeanNetwork*.

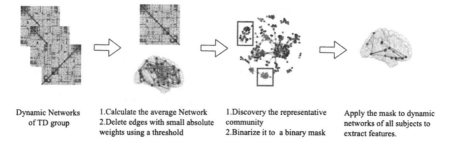

| Dynamic Networks of TD group | 1.Calculate the average Network 2.Delete edges with small absolute weights using a threshold | 1.Discovery the representative community 2.Binarize it to a binary mask | Apply the mask to dynamic networks of all subjects to extract features. |

Fig. 3. The pipeline of extracting the binary mask based on the average dynamic networks of TD group.

2. Representative modular partition discovery

   The Louvain community discovery algorithm is used to discover the representative partition of $MeanNetwork$, which refers to the partition holds the largest value of the modularity network metric.

3. Find the important ROIs using network centrality measures

   Network centrality identifies the most important vertices within a graph. We calculate several centralities in the average networks of both TD and ASD group, then select top d from each centrality to form a ROI list ROI_{list}.

   Based on the ROI list ROI_{list}, we select the sub-module of $Partition_{represent}$ holding the largest intersection set with ROI_{list}, as it can be representative if it includes more important ROIs of ROI_{list}. The selected module is named as $FeatureModule$.

4. Generate the binary mask

   Firstly, we binarize the TD average network $MeanNetwork$ by setting nonzero values as 1, then the node attributes of ROIs which belong to the selected module $FeatureModule$ are preserved through setting other node attributes as 0. Thus, the mask $BinayMask$ for feature extraction is obtained and will be used for all BDNs of both TD group and ASD group. Lastly, the weighted edges of each BDN are aggregated into a single feature vector and fed into machine learning classifiers.

2.4   Identification ASD Based on Machine Learning Algorithms

We train and evaluate machine learning classifiers SVM and Logistic Regression (LR). Evaluation metrics include the mean roc_auc score of the 10-fold cross validation, testing accuracy, F1, precision, sensitivity and specificity. The sensitivity represents the proportion of ASDs correctly predicted, while the specificity describe the proportion of TDs correctly predicted, where roc stands for the receiving operating characteristic while auc stands for the area under the roc curve.

3 Results and Discussion

3.1 Data Preparation

The lengths of time course obtained from different sites fluctuate in the range of 82 to 320 showing the heterogeneity of the ABIDE cohort. We cut the time courses of all subjects using a fixed length window $T = 120$ beginning with the first time point. For those subjects with the length of time course shorter than 120, their original lengths were kept. Then, we normalized the rs-fMRI time series matrix by rows of time series to have mean of zero and standard deviation of one in order to measure the fluctuation of BOLD signals of ROIs at each time point. The experiment data were split into two parts, 80% for training and validation, 20% for testing.

3.2 Dynamic Network Construction and Analysis

For the construction of BDNs of all subjects, the regularization parameter λ was set to 0.2. However, the structures of our constructed BDNs were inconsistent because the heterogeneity of fMRI data obtained from different sites. The minimum numbers of edges in the BDNs of TD group and ASD group are 11,934 (17.1% sparsity) and 11,845 (16.7% sparsity), respectively. In this study, sparsity refers to the ratio of the number of edges of the network to the number of all possible edges. We selected a stringent threshold (10,000 edges) to capture the main structure of the BDNs for both ASD and TD average networks.

Eigenvector centrality was used on the directed binary networks; and node strengths, betweenness, pagerank, authorities were employed to the directed weighted networks. Top 4 (1.5%) ROIs with high centralities were selected from each centrality and the unique ROIs were preserved in the ROI list ROI_{list} (1, 3, 7, 13, 17, 20, 21, 23, 30, 31, 39, 57, 59, 64, 75, 76, 118, 131, 134, 136, 138, 143, 173, 185, 196, 198, 202, 206, 224, 230, 234, 238, 240, 245, 249, 256, 262). This ROI list was used as a reference index for the feature extraction.

To discover the distinct communities within the BDNs, the Louvain modularity algorithm was applied to the average BDN of TD group. Over the course of 100 runs of the algorithm, the representative partition with 8 modules was obtained. The detailed information is shown in Table 1. As can be seen from the table, the 6th module is the largest module and the intersection with ROI_{list} is the largest as well. Therefore, we selected it as the representative module to extract features for machine learning classifiers. For comparison purpose, we also took the 5th module to do the experiments, since the module size is smaller while it contains more important ROIs.

By using the proposed feature extraction method, features of 1236 dimensions and 2721 dimensions were extracted from the 5th module and the 6th module, respectively. Consequently, we used the principal component analysis (PCA) to reduce the dimension of features and selected the principal components which preserved the 20% cumulative variance, as a result, we have the 47-dimensional

Table 1. The module sizes of the representative partition and the corresponding intersections with the important ROI list.

Index of module	1	2	3	4	5	6	7	8
Size of module	24	16	13	76	29	**77**	17	12
Intersection	4	5	5	6	6	**7**	2	2

features for the 5th module and the 57-dimensional features for the 6th module. Tables 2 and 3 illustrate the performances of ASD/TD classification based on different machine learning classifiers, such as the support vector machines (SVM) with the Radial Basis Function (RBF) kernel, SVM with the linear kernel and logistic regression (LR).

The mean roc_auc score of the 10-fold cross validation was used as the model evaluation metric. As shown from the tables, the 6th module achieved the best roc_auc value. The sensitivity metric achieved 87.5% showing that the model has strong ability to correctly detect ASD patients. Especially, the accuracy of 88.8% was achieved in this study, which is higher than any previously reported methods, as shown in Table 4.

Table 2. Performance of ASD/TD classification using the 5th module.

Classifier	Feature size	ROC_AUC	Accuracy	F1	Precision	Specificity	Sensitivity
SVM (rbf)	1236	81.7	75.0	70.1	83.9	89.0	60.3
SVM (linear)	1236	77.1	70.0	68.4	70.3	73.2	66.7
Logistic R	1236	81.7	75.6	71.9	82.0	86.6	64.1
SVM (rbf)	47	82.4	77.5	76.0	72.2	75.3	80.3
SVM (linear)	47	80.9	78.1	76.8	72.5	75.3	81.7
Logistic R	47	80.7	79.4	79.0	72.1	73.0	87.3

Table 3. Performance of ASD/TD classification using the 6th module.

Classifier	Feature size	ROC_AUC	Accuracy	F1	Precision	Specificity	Sensitivity
SVM (linear)	2721	84.8	71.5	70.9	72.2	73.3	69.7
SVM (rbf)	2721	86.4	77.8	77.4	78.4	79.2	76.5
Logistic R	2721	86.5	76.6	75.2	79.4	81.7	71.4
SVM (rbf)	57	**94.7**	**88.8**	**89.5**	**91.7**	**90.3**	**87.5**
SVM (linear)	57	92.9	85.6	84.9	84.4	85.7	85.5
Logistic R	57	93.2	85.0	86.7	84.8	80.6	88.6

Table 4. Comparison of proposed method and other methods, which used the entire ABIDE I cohort.

Methods	Accuracy (%)	Methods	Accuracy (%)
E.Wong et al. [22]	71.1	Dvornek et al. [4]	70.1
Heinsfeld et al. [10]	70.0	Eslami et al. [5]	70.1
Abraham et al. [1]	66.8	Khosla et al. [11]	73.3
Parisot et al. [15]	69.5	Mostafa et al. [13]	77.7
Proposed method	**88.8**		

4   Conclusions

In this study, we have proposed a method for diagnosing ASD based on brain dynamic networks using time series rs-fMRI brain imaging data. A constructed BDN is a weighted directed complex network which describes the dynamic relationship among multiple ROIs. Important ROIs were found through centrality measures and used as referencing index for the representative module selection. Further, the binary mask was derived from the selected module to extract features of BDNs for machine learning classification. Our proposed method have achieved the accuracy of 88.8% on the whole ABIDE I cohort, which is higher than any previously reported methods. Since our BDN construction algorithm and feature extraction method are generic methods, they could be applied to diagnose other brain diseases, which is an interesting direction for future study.

References

1. Abraham, A., et al.: Deriving reproducible biomarkers from multi-site resting-state data: an autism-based example. NeuroImage **147**, 736–745 (2017)
2. Anzellotti, S., Kliemann, D., Jacoby, N., Saxe, R.: Directed network discovery with dynamic network modelling. Neuropsychologia **99**, 1–11 (2017). https://doi.org/10.1016/j.neuropsychologia.2017.02.006. http://www.sciencedirect.com/science/article/pii/S0028393217300520
3. Di Martino, A., et al.: The autism brain imaging data exchange: towards a large-scale evaluation of the intrinsic brain architecture in autism. Mol. Psychiatry **19**(6), 659–667 (2014)
4. Dvornek, N.C., Ventola, P., Pelphrey, K.A., Duncan, J.S.: Identifying autism from resting-state fMRI using long short-term memory networks. In: Wang, Q., Shi, Y., Suk, H.-I., Suzuki, K. (eds.) MLMI 2017. LNCS, vol. 10541, pp. 362–370. Springer, Cham (2017). https://doi.org/10.1007/978-3-319-67389-9_42
5. Eslami, T., Mirjalili, V., Fong, A.C.M., Laird, A., Saeed, F.: ASD-DiagNet: a hybrid learning approach for detection of autism spectrum disorder using fMRI data. Front. Neuroinform. (2019)
6. Friston, K., Moran, R., Seth, A.K.: Analysing connectivity with Granger causality and dynamic causal modelling. Curr. Opin. Neurobiol. **23**(2), 172–178 (2013). https://doi.org/10.1016/j.conb.2012.11.010. http://www.sciencedirect.com/science/article/pii/S0959438812001845

7. Greicius, M.: Resting-state functional connectivity in neuropsychiatric disorders. Curr. Opin. Neurol. **21**(4), 424–430 (2008)
8. Gutlein, M., Frank, E., Hall, M., Karwath, A.: Large-scale attribute selection using wrappers. In: 2009 IEEE Symposium on Computational Intelligence and Data Mining, pp. 332–339, March 2009. https://doi.org/10.1109/CIDM.2009.4938668
9. Hall, M.A., Holmes, G.: Benchmarking attribute selection techniques for discrete class data mining. IEEE Trans. Knowl. Data Eng. **15**(6), 1437–1447 (2003). https://doi.org/10.1109/TKDE.2003.1245283
10. Heinsfeld, A.S., Franco, A.R., Craddock, R.C., Buchweitz, A., Meneguzzi, F.: Identification of autism spectrum disorder using deep learning and the ABIDE dataset. NeuroImage: Clin. **17**, 16–23 (2018)
11. Khosla, M., Jamison, K., Kuceyeski, A., Sabuncu, M.R.: 3D convolutional neural networks for classification of functional connectomes. In: Stoyanov, D., et al. (eds.) DLMIA/ML-CDS -2018. LNCS, vol. 11045, pp. 137–145. Springer, Cham (2018). https://doi.org/10.1007/978-3-030-00889-5_16
12. Liu, Y., et al.: Regional homogeneity, functional connectivity and imaging markers of Alzheimer's disease: a review of resting-state fMRI studies. Neuropsychologia **46**(6), 1648–1656 (2008). https://doi.org/10.1016/j.neuropsychologia.2008.01.027
13. Mostafa, S., Tang, L., Wu, F.: Diagnosis of autism spectrum disorder based on eigenvalues of brain networks. IEEE Access **7**, 128474–128486 (2019)
14. Murphy, K., Birn, R.M., Bandettini, P.A.: Resting-state fMRI confounds and cleanup. NeuroImage **80**, 349–359 (2013)
15. Parisot, S., et al.: Spectral graph convolutions for population-based disease prediction. In: Descoteaux, M., Maier-Hein, L., Franz, A., Jannin, P., Collins, D.L., Duchesne, S. (eds.) MICCAI 2017. LNCS, vol. 10435, pp. 177–185. Springer, Cham (2017). https://doi.org/10.1007/978-3-319-66179-7_21
16. Power, J.D., et al.: Functional network organization of the human brain. Neuron **72**(4), 665–678 (2011)
17. Price, T., Wee, C.-Y., Gao, W., Shen, D.: Multiple-network classification of childhood autism using functional connectivity dynamics. In: Golland, P., Hata, N., Barillot, C., Hornegger, J., Howe, R. (eds.) MICCAI 2014. LNCS, vol. 8675, pp. 177–184. Springer, Cham (2014). https://doi.org/10.1007/978-3-319-10443-0_23
18. Quesnel-Vallières, M., Weatheritt, R.J., Cordes, S.P., Blencowe, B.J.: Autism spectrum disorder: insights into convergent mechanisms from transcriptomics. Nat. Rev. Genet. **20**(1), 51–63 (2019)
19. Schwarz, G.: Estimating the dimension of a model. Ann. Stat. **6**, 461 (1978). https://doi.org/10.1214/aos/1176344136
20. Tang, L., Mostafa, S., Liao, B., Wu, F.X.: A network clustering based feature selection strategy for classifying autism spectrum disorder. BMC Med. Genomics **12**(Suppl 7), 153–153 (2019)
21. Van Essen, D.C., Ugurbil, K.: The future of the human connectome. NeuroImage **62**(2), 1299–1310 (2012)
22. Wong, E., Anderson, J.S., Zielinski, B.A., Fletcher, P.T.: Riemannian regression and classification models of brain networks applied to autism. In: Wu, G., Rekik, I., Schirmer, M.D., Chung, A.W., Munsell, B. (eds.) CNI 2018. LNCS, vol. 11083, pp. 78–87. Springer, Cham (2018). https://doi.org/10.1007/978-3-030-00755-3_9

MiRNA-Disease Associations Prediction Based on Negative Sample Selection and Multi-layer Perceptron

Na Li[1], Guihua Duan[1(✉)], Cheng Yan[1,2], Fang-Xiang Wu[3], and Jianxin Wang[1]

[1] Hunan Provincial Key Lab on Bioinformatics, School of Computer Science and Engineering, Central South University, Changsha 410083, China
duangh@csu.edu.cn
[2] School of Computer and Information, Qiannan Normal University for Nationalities, Duyun 558000, Guizhou, China
[3] Division of Biomedical Engineering and Department of Mechanical Engineering, University of Saskatchewan, Saskatoon, SK S7N5A9, Canada

Abstract. MicroRNAs (miRNAs) are a class of non-coding RNAs of approximately 22 nucleotides. Cumulative evidence from biological experiments has confirmed that miRNAs play a key role in many complex human diseases. Therefore, the accurate identification of potential associations between miRNAs and diseases is beneficial to understanding the mechanisms of diseases, developing drugs and treating complex diseases. We propose a new method to predict miRNA-disease associations based on a negative sample selection strategy and multi-layer perceptron (called NMLPMDA). For obtaining more similarity information, NMLPMDA integrates the miRNA functional similarity and the Gaussian interaction profile (GIP) kernel similarity of miRNAs as the final miRNA similarity, and integrates the disease semantic similarity and the GIP kernel similarity of diseases as the final disease similarity. In particular, we propose a negative sample selection strategy based on common gene information to select more reliable negative samples from unknown miRNA-disease associations. The 5-fold cross validation is used to evaluate the performance of NMLPMDA and other competing methods. On four datasets (HMDD2.0-Yan, HMDD2.0-Lan, HMDD2.0-You, HMDD3.0), the AUC values of NMLPMDA are 0.9278, 0.9206, 0.9301 and 0.9350, respectively. In addition, we also illustrate the prediction ability of NMLPMDA in Lymphoma. As a result, 28 of the top 30 miRNAs associated with the disease have been validated experimentally in dbDEMC and previous studies, respectively. These experimental results indicate that NMLPMDA is a reliable model for predicting associations between miRNAs and diseases.

Keywords: miRNA-disease associations · Negative sample selection · Multi-layer perceptron

© Springer Nature Switzerland AG 2020
Z. Cai et al. (Eds.): ISBRA 2020, LNBI 12304, pp. 178–188, 2020.
https://doi.org/10.1007/978-3-030-57821-3_16

1 Introduction

MicroRNAs (miRNAs) are small, endogenous, single-stranded, non-coding RNAs (about ~22 nucleotides), which are involved in many important biological processes, such as viral infections, immune responses, tumor invasion, signal transduction, cell proliferation, cell growth and cell death. In addition, human diseases may be caused by miRNA abnormalities and abnormal regulation of disease gene expression, such as cancer, hereditary diseases, etc. With the synergy of bioinformatics, more and more miRNA-disease associations have been revealed. For example, miR-145 is a candidate tumor suppressor miRNA, which may be involved in the regulation of human hepatocarcinoma cell differentiation [1]. However, the traditional experimental methods are of small scale, as well as time-consuming and costly. Therefore, developing more efficient computational models is particularly urgent to achieve large-scale and reliable prediction of associations between miRNAs and diseases.

More and more computational methods have been proposed to predict miRNA-disease associations. These computational methods can be roughly divided into two categories: similarity-based methods and machine learning-based methods. The similarity-based methods predict the probability scores that a miRNA interacts with a disease. You et al. [2] developed the prediction model PBMDA to discover miRNA-disease associations, which constructed a heterogeneous graph with paths. Lan et al. [3] developed KBMF-MDI, in which kernelized Bayesian matrix factorization method was employed to infer potential miRNA-disease associations by integrating different data sources. Yan et al. [4] proposed DNRLMF-MDA to predict miRNA-disease associations based on dynamic neighborhood regularized logistic matrix factorization.

Prediction models based on machine learning apply the machine learning classification algorithm to predict potential miRNA-disease associations. Compared to predicting miRNA-disease associations by measuring the association strength between nodes in the miRNA and disease network, machine learning-based methods can better predict new miRNAs (have no known miRNA-disease associations). Xu et al. [5] proposed a miRNA target-dysregulated network model MTDN based on support vector machine (SVM) to prioritize candidate disease-related miRNAs for prostate cancer. Chen et al. [6] developed a prediction model ABMDA, which was able to integrate weak classifiers to form a strong classifier based on corresponding weights. Chen et al. [7] proposed a model of Extreme Gradient Boosting machine for miRNA-disease association prediction.

In this study, we present a new miRNA-disease association prediction method. We integrate the miRNA functional similarity and the GIP kernel similarity of miRNAs as the final miRNA similarity, and integrate the disease semantic similarity and the GIP kernel similarity of diseases as the final disease similarity. Furthermore, a negative sample selection strategy is proposed based on miRNA-gene associations and disease-gene associations. Unlike existing methods, our negative sample selection is based on the reasoning that if a miRNA and a disease are not associated with common genes, this miRNA-disease pair is likely to be a reliable negative sample, otherwise it is an unreliable negative sample. Finally, a multi-layer perceptron (MLP) neural network is trained for miRNA-disease association prediction using feature vectors through concatenating similarities of miRNAs and diseases. In order to comprehensively evaluate the performance of our model, NMLPMDA compares with current state-of-the-art methods on four datasets (HMDD2.0-Yan, HMDD2.0-Lan, HMDD2.0-You, HMDD3.0) through the 5-fold cross validation. Our method has the AUC values of 0.9278, 0.9206, 0.9301, 0.9350 on four datasets HMDD2.0-Yan, HMDD2.0-Lan, HMDD2.0-You, HMDD3.0, respectively. Furthermore, we validate the proposed model against Lymphoma. Ultimately, 28 of top 30 miRNA candidates associated with the disease predicted by NMLPMDA were confirmed in dbDEMC and previous studies.

2   Materials and Methods

2.1   Data Description

Four datasets are used to evaluate our method, including HMDD2.0-You, HMDD2.0-Yan, HMDD2.0-Lan and HMDD3.0. All these datasets are downloaded from HMDD database [8]. HMDD (the Human microRNA Disease Database) is a database that curates experiment-supported evidence for human miRNA and disease associations. Some miRNAs were judged to be unreliable by the public database miRBase, which were removed from HMDD3.0 [9]. HMDD2.0-You, HMDD2.0-Lan and HMDD2.0-Yan are downloaded from HMDD2.0 dataset at different times. The statistics of four datasets are shown in Table 1, where N_m is the number of miRNAs, N_d is the number of diseases, and N_{md} is the number of known miRNA-disease associations.

Table 1. MiRNAs, diseases and miRNA-disease associations in each dataset.

Dataset	N_m	N_d	N_{md}
HMDD2.0-You	495	383	5430
HMDD2.0-Lan	550	383	6084
HMDD2.0-Yan	576	356	6391
HMDD3.0	1057	850	32226

On this basis, matrix $Y \in R^{N_m \times N_d}$ is defined as adjacency matrix of N_m miRNAs and N_d diseases. The value of element y_{ij} is 1 if the association between miRNA m_i and disease d_j has been confirmed in database, otherwise 0. The disease-gene associations and the miRNA-gene associations used in the negative sample selection strategy are from DisGeNET database [10] and miRTArBase database [11], respectively. In our study, we remove those genes that have no association with diseases or miRNAs.

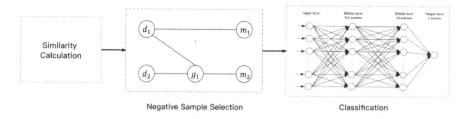

Fig. 1. The framework of NMLPMDA.

2.2 Method

As shown in Fig. 1, our method mainly consists of three steps to predict miRNA-disease associations. First, we calculate the miRNA similarity and disease similarity. Second, given a miRNA, disease and gene network, reliable negative samples are selected from unlabeled miRNA-disease pairs. Third, we generate feature vectors for miRNA-disease pairs based on these similarities and a multi-layer perceptron network is constructed to predict the associations between miRNAs and diseases based on the representation vectors.

2.2.1 Similarity Calculation

Mesh database is used to calculate the disease semantic similarity [12], which describes the relationship between different diseases by a direct acyclic graph (DAG). Given disease d_1 and disease d_2, their semantic similarity is calculated as follows:

$$D_{ss}(d_1, d_2) = \frac{\sum_{t \in T_{d_1} \cap T_{d_2}} (SV_{d_1}(t) + SV_{d_2}(t))}{Sem(d_1) + Sem(d_2)} \quad (1)$$

where T_{d_i} is the ancestor node set of disease d_i and itself. $Sem(d_1)$ represents the semantic value of disease d_1 in DAG, which is calculated as follows:

$$Sem(d_1) = \sum_{t \in T_{d_1}} SV_{d_1}(t) \quad (2)$$

where $SV_{d_1}(t)$ is defined as the semantic value between diseases t and d_1, and t is a common ancestor of diseases d_1 and d_2. $SV_{d_1}(t)$ is calculated as follows:

$$SV_{d_1}(t) = \begin{cases} 1, & t = d_1 \\ \Delta^K, & t = the\ smallest\ K\ layer\ ancestor\ node\ of\ d_1 \end{cases} \quad (3)$$

where Δ is defined as the layer contribution factor between disease t and its direct ancestor disease. The value of Δ is 0.5 in our study [12]. Similarly, $Sem(d_2)$ and $SV_{d_2}(t)$ can be calculated in the same way.

We also compute the GIP kernel similarity of diseases. The GIP kernel similarity between disease d_1 and disease d_2 is calculated as follows [13]:

$$K_{GIP,d}(d_1, d_2) = exp(-\gamma_d ||yd_1 - yd_2||^2) \quad (4)$$

where $yd_1 = \{y_{11}, y_{21}...y_{N_m1}\}$ is the association profile of disease d_1. γ_d is the regulation parameter of kernel bandwidth, which is calculated as follows [13]:

$$\gamma_d = \gamma_d'/(\frac{1}{N}\sum_{i=1}^{N_d} ||yd_i||^2) \tag{5}$$

where γ_d' is set to be 1 in this study.

To gain more similarity information, the final disease similarity of diseases d_1 and d_2 is calculated as follows [14]:

$$S_d(d_1, d_2) = \begin{cases} D_{ss}(d_1, d_2), & if\ d_1, d_2\ has\ semantic\ similarity \\ K_{GIP,d}(d_1, d_2), & otherwise \end{cases} \tag{6}$$

Wang et al. proposed a model to calculate the functional similarity between different miRNAs [13]. Given miRNA m_1 and miRNA m_2, their functional similarity is calculated as follows:

$$M_{fs}(m_1, m_2) = \frac{\sum_{1\leq i\leq n_1} S(dt_{1i}, DT_2) + \sum_{1\leq j\leq n_2} S(dt_{2j}, DT_1)}{n_1 + n_2} \tag{7}$$

DT_1 and DT_2 are diseases sets associated with miRNAs m_1 and m_2, respectively. n_1 and n_2 are their cardinality. $S(dt_{1i}, DT_2)$ is the semantic similarity of disease dt_{1i} and disease set DT_2, which is calculated as follows:

$$S(dt_{1i}, DT_2) = \max_{1\leq i\leq n_2}(D_{ss}(dt_{1i}, dt_{2j})) \tag{8}$$

Similarly, $S(dt_{2j}, DT_1)$ can be calculated in the same way. The final miRNA similarity of miRNAs m_1 and m_2 is calculated as follows:

$$S_m(m_1, m_2) = \begin{cases} M_{fs}(m_1, m_2), & if\ m_1, m_2\ has\ functional\ similarity \\ K_{GIP,m}(m_1, m_2), & otherwise \end{cases} \tag{9}$$

2.2.2 Selecting Reliable Negative Samples from Unlabeled MiRNA-Disease Pairs

In HMDD database, any miRNA-disease association entries without any relationship have not been provided. Traditional methods treat the non-association samples as negative samples which is unreasonable as those non-association samples may be undetected miRNA-disease associations. Therefore, a negative sample selection strategy is designed to find more reliable negative samples from unlabeled samples (unknown miRNA-disease pairs) in this work.

We build a network (MGDN) including miRNA-disease associations, miRNA-gene associations, disease-gene associations and select a group of miRNA-disease pairs from MGDN as reliable negative samples. In this network, each node represents a disease, a disease (miRNA)-associated gene or a miRNA and each node is connected with its associated nodes. If a disease node and a miRNA node share the same genes, there is a potential association between them. From a biological perspective, if a miRNA shares certain genes with a disease, the miRNA may have the potential to cause this disease. On the contrary, if a miRNA does not share any gene with a disease, it can assume that the miRNA can't cause the disease.

To illustrate, the second part of Fig. 1 depicts a sub network of MGDN with 2 diseases (d_1, d_2), 2 miRNAs (m_1, m_2) and a gene (g_1). The connection between disease node and miRNA node in the MGDN represents their relationships. A disease node and a miRNA node are connected $(d_1$ and $m_1)$, which indicates the miRNA is associated with the disease. A disease node and a miRNA node share the same genes $(d_1$ and m_2; d_2 and $m_2)$, which indicates they have a potential association. If there is no path between a miRNA node and a disease node $(d_2$ and $m_1)$, the miRNA-disease pair is a reliable negative sample. It is worth noting that some of the miRNA names in miRTarBase are different from the miRNA names in HMDD, so we make a mapping on two datasets for miRNA names based on miRBase.

2.2.3 Multilayer Perceptron Neural Network

Each miRNA-disease pair is represented by concatenating integrated similarities of miRNAs and diseases as a feature vector. Each disease is described by a N_d-dimensional feature vector. For example, disease d_i is described by a feature vector as follows:

$$S_d(d_i) = (a_1, a_2, a_3 ... a_{N_d}) \tag{10}$$

where $S_d(d_i, :)$ is the i-th row vector of matrix S_d, and a_j is the integrated similarity value between disease d_i and d_j. Similarly, miRNA can be represented by a N_m-dimensional feature vector as follows:

$$S_m(m_u) = (b_1, b_2, b_3 ... b_{N_m}) \tag{11}$$

Therefore, each miRNA-disease pair can be described by an $(N_d + N_m)$-dimensional vector based on integrated similarities of miRNA and disease as follows:

$$F_{m_u d_i} = (S_m(m_u), S_d(d_i)) \tag{12}$$

Multi-layer perceptron (MLP) neural networks are the most commonly used feedforward neural networks because of their fast operation, ease of implementation and smaller training set requirements. The MLP consists of three sequential layers, including input layer, hidden layer and output layer. MLP model with insufficient or excessive number of neurons in the hidden layer most likely causes the problems of bad generalization and overfitting, respectively. There is no analytical method for determining the number of neurons in the hidden layer. Therefore, it is only found by trial and error. In the study, an MLP neural network is trained with two hidden layers of 512, 64 hidden neurons to get the final prediction of the association between each miRNA-disease pair. Sigmoid function and Adam algorithm are used as the activation function and the optimization method, respectively. In addition, since the number of reliable negative samples are more than that of positive ones, we randomly select a subset from reliable negative samples with the size equal to the positive samples to train the model.

3   Results and Discussion

3.1   Experimental Settings

In order to evaluate the performance of NMLPMDA, we conduct the 5-fold cross validation, which is widely used in inferring miRNA-disease associations. In each round, the balanced data is divided into the five sets, four of which are used as the training set and the rest one as the testing sets. The AUC (area under the receiver operating curve) value is used as an evaluation metric. The AUC value of a model may lie between 0.5 and 1, and it is less than or equal to 0.5 when the prediction model has no predictive ability while an optimal model has the value of AUC near 1.0.

3.2   Comparison with Other Methods

NMLPMDA is compared with other two competing methods, namely, ABMDA [6] and EGBMMDA [7], because both models are built using balanced data. This makes the data structure of the testing sets consistent and makes the comparison of AUC more fair and interpretable. ABMDA was able to integrate weak classifiers to form a strong classifier based on corresponding weights and balanced the positive and negative samples by performing random sampling based on k-means clustering on negative samples. EGBMMDA was a model of Extreme Gradient Boosting machine for miRNA-disease association prediction by integrating the miRNA functional similarity, the disease semantic similarity, and known miRNA-disease associations.

3.2.1   The 5-fold Cross Validation

We perform the 5-fold cross validation experiments on four datasets. Figure 2(a) plots the ROC curves and shows the AUC values of ABMDA, EGBMMDA and NMLPMDA on HMDD2.0-You dataset. In terms of AUC, NMLPMDA is the best because of its AUC value is 0.9301, while the other results are 0.9007(ABMDA), 0.7433(EGBMMDA). We also conduct the experiment on HMDD2.0-Yan dataset. Figure 2(b) shows the ROC curves of ABMDA, EGB-MMDA and NMLPMDA on HMDD2.0-Yan dataset. Compared with the other results of 0.9050(ABMDA), 0.7283(EGBMMDA), NMLPMDA achieves 0.9278, which is superior to other methods. Similarly, as shown in Fig. 2(c), NMLP-MDA also outperforms other methods on HMDD2.0-Lan dataset. The AUC value of NMLPMDA is 0.9206, which is superior to the other results of 0.9031(ABMDA), 0.7316(EGBMMDA). Finally, the AUC of NMLPMDA is 0.9350 on the HMDD3.0 dataset, while those of ABMDA and EGBMMDA are 0.8693, 0.7941, respectively. Obviously, NMLPMDA performs the best among three methods in terms of AUC, as shown in Fig. 2(d). In summary, these experiments show that NMLPMDA can achieve improvement in predicting miRNA-disease associations compared to some state-of-the-art approaches.

In this study, the original miRNA-disease adjacency matrix is updated in each repetition of the 5-fold cross validation experiment, as miRNA functional similarity and GIP kernel similarities of miRNAs and diseases calculation is based on the adjacency matrix that is different in each repetition of 5-fold cross validation experiment. We follow the same way to reproduce ABMDA and EGB-MMDA and find that the AUC values in our experiment is lower than the AUC values of the original text. The reason may be that the authors neglected the fact that the adjacency matrix should be updated when computing miRNA functional similarity and GIP kernel similarities of miRNAs and diseases.

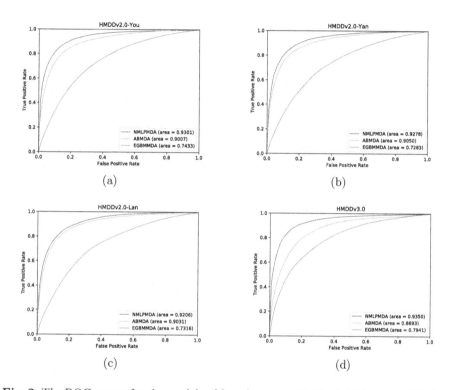

Fig. 2. The ROC curves for the models of four datasets with the 5-fold cross validation.

3.2.2 Denovo MiRNA Validation

In addition, we adopt the denovo miRNA validation on four datasets. We randomly select 50 miRNAs on each dataset and remove all their known associations in turn as the test set to evaluate the prediction performance for new miRNAs (have no known miRNA-disease associations). Table 2 shows the AUC value of denovo miRNA validation. The AUC values of NMLPMDA on four datasets are 0.8937, 0.9054, 0.7679 and 0.8974, respectively, which are superior to the other results of denovo miRNA validation.

Table 2. The AUC value of denovo miRNA validation.

AUC	NMLPNDA	ABMDA	EGBMMDA
HMDD2.0-You	0.8937	0.8842	0.8384
HMDD2.0-Yan	0.9054	0.8900	0.7683
HMDD2.0-Lan	0.7679	0.7584	0.7575
HMDD3.0	0.8974	0.8827	0.7393

3.3   Case Studies

In order to further illustrate the performance of NMLPMDA, we evaluate the prediction ability of NMLPMDA on Lymphoma. Lymphoma is a malignant tumor that originates in the lymphoid hematopoietic system [15]. More and more studies have reported that many miRNAs are closely related to the disease. Therefore, Lymphoma are chosen as the case study to evaluate the prediction ability of NMLPMDA. In this experiment, the model is trained using all known miRNA-disease pairs as the training set in the HMDD3.0 dataset to predict miRNAs related with Lymphoma. After obtaining the predicted results, we pick up the 30 miRNAs with the highest scores and verify them with dbDEMCv2.0 database.

Table 3. The result of top 30 related miRNAs of Lymphoma predicted by NMLPMDA on the HMDD3.0 dataset.

Top1-15miRNA	Evidence	Top16-30miRNA	Evidence
hsa-mir-874	dbDEMC 2.0	hsa-mir-9-3	PMID:22310293
hsa-mir-663a	Unknown	hsa-mir-20a	dbDEMC 2.0
hsa-mir-21	dbDEMC 2.0	hsa-mir-31	dbDEMC 2.0
hsa-mir-155	dbDEMC 2.0	hsa-mir-15a	dbDEMC 2.0
hsa-mir-146a	dbDEMC 2.0	hsa-mir-222	dbDEMC 2.0
hsa-mir-223	dbDEMC 2.0	hsa-mir-9-2	PMID:22310293
hsa-mir-34a	dbDEMC 2.0	hsa-mir-92a	dbDEMC 2.0
hsa-mir-145	dbDEMC 2.0	hsa-mir-489	dbDEMC 2.0
hsa-mir-126	dbDEMC 2.0	hsa-mir-146b	dbDEMC 2.0
hsa-mir-7-1	Unknown	hsa-mir-9	dbDEMC 2.0
hsa-mir-221	dbDEMC 2.0	hsa-mir-210	dbDEMC 2.0
hsa-mir-150	dbDEMC 2.0	hsa-mir-19a	dbDEMC 2.0
hsa-mir-16	dbDEMC 2.0	hsa-mir-29a	dbDEMC 2.0
hsa-mir-17	dbDEMC 2.0	hsa-mir-205	dbDEMC 2.0
hsa-mir-143	dbDEMC 2.0	hsa-mir-451	dbDEMC 2.0

Table 3 demonstrates the result of top 30 miRNAs related-Lymphoma predicted by NMLPMDA on the HMDD3.0 dataset. It shows that 26 of top 30 are confirmed in the dbDEMC database. In addition, there are 2 novel associations that are validated in literature. For example, inhibition of miR-9 de-represses HuR and DICER1 and impairs Hodgkin lymphoma tumour outgrowth in vivo [16]. With these recent literature and database evidences, 28 out of the top 30 potentially Lymphoma-related miRNAs were verified.

4   Conclusion

In this study, we propose a new method to predict miRNA-disease associations based on negative sample selection strategy and MLP, in which reliable negative samples are used to train the multi-layer perceptron neural network. On four datasets (HMDD2.0-Yan, HMDD2.0-Lan, HMDD2.0-You, HMDD3.0), the AUC values of 5-fold cross validation for NMLPMDA are 0.9278, 0.9206, 0.9301 and 0.9350, respectively. In addition, 28 of top 30 miRNAs associated with Lymphoma were validated experimentally in various case studies, respectively. Nowadays, although some important progress in predicting miRNA-disease associations have been made, the lack of non-associated miRNA-disease pairs and the limitation of data volume of known miRNA-disease associations also influence the further development of miRNA-disease association prediction methods. Therefore, to further development of prediction methods for miRNA-disease associations, larger scale of experimentally validated data should be collected for associated data and non-associated data. In addition, more biological data related to miRNAs and diseases should be explored to further predict miRNA-disease associations, such as miRNA family and cluster information, neighborhood information, etc.

Acknowledgement. This work is supported in part by the National Natural Science Foundation of China (No. 61772552, No. 61832019 and No. 61962050), 111 Project (No. B18059), Hunan Provincial Science and Technology Program (No. 2018WK4001), the Science and Technology Foundation of Guizhou Province of China under Grant NO. [2020]1Y264.

References

1. Gao, P., Wong, C.C.L., Tung, E.K.K., Lee, J.M.F., Wong, C.M., Ng, I.O.L.: Deregulation of microrna expression occurs early and accumulates in early stages of hbv-associated multistep hepatocarcinogenesis. J. Hepatol. **54**(6), 1177–1184 (2011)
2. You, Z.H., et al.: PBMDA: a novel and effective path-based computational model for mirna-disease association prediction. PLoS Comput. Biol. **13**(3), e1005455 (2017)
3. Lan, W., Wang, J., Li, M., Liu, J., Wu, F.X., Pan, Y.: Predicting microRNA-disease associations based on improved microRNA and disease similarities. IEEE/ACM Trans. Comput. Biol. Bioinf. **15**(6), 1774–1782 (2018)

4. Yan, C., Wang, J., Ni, P., Lan, W., Wu, F.X., Pan, Y.: DNRLMF-MDA: predicting microrna-disease associations based on similarities of micrornas and diseases. IEEE/ACM Trans. Comput. Biol. Bioinf. **16**(1), 233–243 (2017)

5. Giardine, B., et al.: Galaxy: a platform for interactive large-scale genome analysis. Genome Res. **15**(10), 1451–1455 (2005)

6. Zhao, Y., Chen, X., Yin, J.: Adaptive boosting-based computational model for predicting potential mirna-disease associations. Bioinformatics **35**(22), 4730–4738 (2019)

7. Chen, X., Huang, L., Xie, D., Zhao, Q.: EGBMMDA: extreme gradient boosting machine for mirna-disease association prediction. Cell Death Dis. **9**(1), 3 (2018)

8. Li, Y., et al.: HMDD v2.0: a database for experimentally supported human microRNA and disease associations. Nucleic Acids Res. **42**(D1), D1070–D1074 (2013)

9. Huang, Z., et al.: HMDD v3.0: a database for experimentally supported human microrna-disease associations. Nucleic Acids Res. **47**(D1), D1013–D1017 (2018)

10. Pinero, J., et al.: DisGeNET: a discovery platform for the dynamical exploration of human diseases and their genes. Database **2015**, bav028 (2015)

11. Hsu, S.D., et al.: Mirtarbase: a database curates experimentally validated microrna-target interactions. Nucleic Acids Res. **39**((suppl-1)), D163–D169 (2010)

12. Wang, D., Wang, J., Lu, M., Song, F., Cui, Q.: Inferring the human microrna functional similarity and functional network based on microrna-associated diseases. Bioinformatics **26**(13), 1644–1650 (2010)

13. van Laarhoven, T., Nabuurs, S.B., Marchiori, E.: Gaussian interaction profile kernels for predicting drug-target interaction. Bioinformatics **27**(21), 3036–3043 (2011)

14. Yu, H., Chen, X., Lu, L.: Large-scale prediction of microrna-disease associations by combinatorial prioritization algorithm. Sci. Rep. **7**, 43792 (2017)

15. Siegel, R.L., Miller, K.D., Jemal, A.: Cancer statistics, 2016. CA Cancer J. Clin. **66**(1), 7–30 (2016)

16. Leucci, E., et al.: Inhibition of mir-9 de-represses hur and dicer1 and impairs hodgkin lymphoma tumour outgrowth in vivo. Oncogene **31**(49), 5081 (2012)

Checking Phylogenetic Decisiveness in Theory and in Practice

Ghazaleh Parvini, Katherine Braught, and David Fernández-Baca$^{(\boxtimes)}$

Department of Computer Science, Iowa State University, Ames, IA 50011, USA
{ghazaleh,kbraught,fernande}@iastate.edu

Abstract. Suppose we have a set X consisting of n taxa and we are given information from k loci from which to construct a phylogeny for X. Each locus offers information for only a fraction of the taxa. The question is whether this data suffices to construct a reliable phylogeny. The decisiveness problem expresses this question combinatorially. Although a precise characterization of decisiveness is known, the complexity of the problem is open. Here we relate decisiveness to a hypergraph coloring problem. We use this idea to (1) obtain lower bounds on the amount of coverage needed to achieve decisiveness, (2) devise an exact algorithm for decisiveness, (3) develop problem reduction rules, and use them to obtain efficient algorithms for inputs with few loci, and (4) devise an integer linear programming formulation of the decisiveness problem, which allows us to analyze data sets that arise in practice.

Keywords: Phylogenetic tree · Taxon coverage · Algorithms

1 Introduction

Missing data poses a challenge to assembling phylogenetic trees. The question we address here is how much data can one afford to miss without compromising accuracy. We focus on data sets assembled by concatenating data from many (sometimes thousands) of loci [13,14,25]. Such data sets are used to construct phylogenetic trees by either (i) combining the data from all the loci into a single *supermatrix* that is then used as input to some standard phylogeny construction method (e.g., [10,20]) or (ii) taking phylogenetic trees computed separately for each locus and combining them into a single *supertree* that summarizes their information [4,18,24]. For various reasons, the *coverage density* of concatenated datasets — i.e., the ratio of the amount of available data to the maximum possible amount — is often much less than 1 [16]. Reference [6] examines a wide range of phylogenetic analyses using concatenated data sets, and reports coverage densities ranging from 0.06 to 0.98, with the majority being under 0.5.

Low coverage density can give rise to ambiguity [17,21,26]. In supertree analyses, ambiguity manifests itself in multiple supertrees that are equivalent with respect to the method upon which they are based. In super-matrix analyses, it is manifested in multiple topologically different, but co-optimal (in terms of

© Springer Nature Switzerland AG 2020
Z. Cai et al. (Eds.): ISBRA 2020, LNBI 12304, pp. 189–202, 2020.
https://doi.org/10.1007/978-3-030-57821-3_17

parsimony or likelihood scores) trees. Note that high coverage density does not, by itself, guarantee lack of ambiguity. More important is the coverage pattern itself. The question is whether one can identify conditions under which a given coverage pattern guarantees a unique solution. Sanderson and Steel [17,23] have proposed a formal approach to studying this question, which we explain next.

A *taxon coverage pattern* for a taxon set X is a collection of sets $\mathcal{S} = \{Y_1, Y_2, \ldots, Y_k\}$, where, for each $i \in \{1, 2, \ldots, k\}$, Y_i is a subset of X consisting of the taxa for which locus i provides information. $\mathcal{S}$ is *decisive* if it satisfies the following property: Let T and T' be two binary phylogenetic trees for X such that, for each $i \in \{1, 2, \ldots, k\}$, the restrictions of T and T' to Y_i are isomorphic (restriction and isomorphism are defined in Sect. 2). Then, it must be the case that T and T' are isomorphic. The *decisiveness problem* is: Given a taxon coverage pattern $\mathcal{S}$, determine whether or not $\mathcal{S}$ is decisive. Intuitively, if a taxon coverage pattern $\mathcal{S}$ is *not* decisive, we have ambiguity. That is, there are at least two trees that cannot be distinguished from each other by the subtrees obtained when these trees are restricted to the taxon sets in $\mathcal{S}$.

The complexity of the decisiveness problem has been surprisingly hard to settle, and, to our knowledge, remains an open question (but see Sect. 2, Remark 1). A necessary and sufficient condition — the four-way partition property — for a coverage pattern to be decisive is known [17,23] (see also Sect. 2). However, it is not clear how to test this condition efficiently. On the positive side, the *rooted* case, where at least one taxon for which every locus offers data, is known to be polynomially solvable, and software for it is available [27]. *Groves* [1,7] are a related, but not identical, notion. For a discussion on the relationship between groves and decisiveness, see [16].

Contributions. In Sect. 2, we define decisiveness precisely, and review some earlier results, including the four-way partition property. In Sect. 3, we show that checking decisiveness is equivalent to checking for the *non-existence* of a "no-rainbow" coloring in a hypergraph associated with the given coverage pattern. The no-rainbow coloring problem is conjectured to be NP-complete. If this conjecture holds true, then checking decisiveness is co-NP complete. In Sect. 4 we derive a lower bound on the amount of coverage needed to achieve decisiveness. This bound can be used to quickly rule out the decisiveness of certain coverage patterns. Next, we turn our attention to developing algorithms to check decisiveness. Using the four-way partition property naïvely leads to a $O(4^n)$ algorithm for decisiveness. In Sect. 5, we give a considerably faster exact algorithm, which checks decisiveness in $O(2.8^n)$ time by exploiting the connection with no-rainbow 4-colorings. Section 6 studies reduction rules that allow us to compress an instance of the decisiveness problem to a smaller, but equivalent, instance. One consequence is that we can show that the decisiveness problem is fixed-parameter tractable in the number of loci. The performance in practice of the algorithms of Sects. 5 and 6 remains to be tested. Nevertheless, in Sect. 7, we show that checking decisiveness in practice may not be as difficult as the conjectured complexity of the problem would suggest. In that section, we present an integer linear programming (ILP) formulation of the decisiveness problem, along with experimental results using this formulation on real data sets. In all cases, an ILP solver was able to check decisiveness within seconds; indeed, an answer was found within a fraction of a second in most cases. We also show that the ILP approach can be used to obtain subsets of taxa for which the given data is decisive. Section 8 gives some concluding remarks.

2 Preliminaries

Throughout the rest of this paper, X denotes a set of taxa, n denotes $|X|$, and, for any positive integer q, $[q]$ denotes the set $\{1, 2, \ldots, q\}$.

Phylogenetic Trees. A *phylogenetic X-tree* [19,22] is a tree T with leaf set X, where every internal vertex has degree at least three. Biologists are often interested in *rooted* trees, where the root is considered as the origin of species and edges are viewed as being directed away from the root, indicating direction of evolution. Note, however, that most phylogeny construction methods produce unrooted trees.

A *split* of taxon set X is a bipartition $A|B$ of X such that $A, B \neq \emptyset$. Let T be a phylogenetic X-tree. Each edge e in T defines a split $\sigma_T(e) = A|B$, where A and B are the subsets of X lying in each component of $T - e$. Spl(T) denotes the set $\{\sigma_e : e \in E(T)\}$. It is well-known that a phylogenetic X-tree T is completely determined by Spl(T) [19, Theorem 3.5.2]. Two X-trees T and T' are *isomorphic* if Spl(T) = Spl(T').

Let T be a phylogenetic X-tree, and suppose $Y \subseteq X$. The *restriction* of T to Y, denoted by $T|Y$, is the phylogenetic Y-tree where Spl($T|Y$) = $\{A \cap Y|B \cap Y : A|B \in \text{Spl}(T) \text{ and } A \cap Y, B \cap Y \neq \emptyset\}$. Equivalently, $T|Y$ is obtained from the minimal subtree of T that connects Y by suppressing all vertices of degree two that are not in Y.

Decisiveness. A taxon coverage pattern S for X is *phylogenetically decisive* if it satisfies the following property: If T and T' are binary phylogenetic X-trees, with $T|Y = T'|Y$ for all $Y \in S$, then $T = T'$. In other words, for any binary phylogenetic X-tree T, the collection $\{T|Y : Y \in S\}$ uniquely determines T (up to isomorphism). The *decisiveness problem* is the problem of determining whether a given coverage pattern is decisive.

Let Q_S denote the set of all quadruples from X that lie in at least one set in S. That is: $Q_S = \bigcup_{Y \in S} \binom{X}{4}$. A collection S of subsets of X satisfies the *four-way partition property* (for X) if, for all partitions of X into four disjoint, nonempty sets A_1, A_2, A_3, A_4 (with $A_1 \cup A_2 \cup A_3 \cup A_4 = X$) there exists $a_i \in A_i$ for $i \in \{1, 2, 3, 4\}$ for which $\{a_1, a_2, a_3, a_4\} \in Q_S$.

Theorem 1 [23]. *A taxon coverage pattern S for X is phylogenetically decisive if and only if S satisfies the four-way partition property for X.*

Corollary 1. *The decisiveness problem is in co-NP.*

Proof. A certificate for *non*-decisiveness is a partition of X into four disjoint, nonempty sets A_1, A_2, A_3, A_4, such that there is no quadruple $\{a_1, a_2, a_3, a_4\} \in Q_S$ where $a_i \in A_i$ for each $i \in \{1, 2, 3, 4\}$.   □

Conjecture 1. The decisiveness problem is co-NP-complete.

Remark 1. A recent preprint [28] presents a proof that the no-rainbow 3-coloring problem (defined in Sect. 3) is NP-complete. This result would imply that decisiveness is indeed co-NP complete.

Note that Theorem 1 implies that a taxon coverage pattern S for X such that $X \in S$ (that is, one set in S contains all the taxa) is trivially decisive.

Theorem 2 [23]. *Let S be a taxon coverage pattern for X.*

(i) *If S is decisive, then for every set $A \in \binom{X}{3}$, there exists a set $Y \in S$ such that $A \subseteq Y$.*
(ii) *If $\bigcap_{Y \in S} Y \neq \emptyset$, then, S is decisive if and only if for every set $A \in \binom{X}{3}$, there exists a set $Y \in S$ such that $A \subseteq Y$.*

Part (ii) of Theorem 2 implies that decisiveness is polynomially solvable in the rooted case [22].

3   Hypergraphs, No-Rainbow Colorings, and Decisiveness

Hypergraphs. A *hypergraph* H is a pair $H = (X, E)$, where X is a set of elements called *nodes* or *vertices*, and E is a set of non-empty subsets of X called *hyperedges* or *edges* [2,3]. Two nodes $u, v \in V$ are *neighbors* if $\{u, v\} \subseteq e$, for some $e \in E$. A hypergraph $H = (X, E)$ is *r-uniform*, for some integer $r > 0$, if each hyperedge of H contains exactly r nodes.

A *chain* in a hypergraph $H = (X, E)$ is an alternating sequence $v_1, e_1, v_2, \ldots,$ e_s, v_{s+1} of nodes and edges of H such that: (1) $v_1, \ldots, v_s$ are all distinct nodes of H, (2) $e_1, \ldots, e_s$ are all distinct edges of H, and (3) $\{v_j, v_{j+1}\} \in e_j$ for $j \in \{1, \ldots, s\}$. Two nodes $u, v \in X$ are *connected* in H, denoted $u \equiv v$, if there exists a chain in H that starts at u and ends at v. The relation $u \equiv v$ is an equivalence relation [2]; the equivalence classes of this relation are called the *connected components* of H. H is *connected* if it has only one connected component; otherwise H is *disconnected*.

No-Rainbow Colorings and Decisiveness. Let $H = (X, E)$ be a hypergraph and r be a positive integer. An *r-coloring* of H is a mapping $c : X \to [r]$. For node $v \in X$, $c(v)$ is the *color* of v. Throughout this paper, r-colorings are assumed to be *surjective*; that is, for each $i \in [r]$, there is at least one node $v \in X$ such that $c(v) = i$. An edge $e \in E$ is a *rainbow edge* if, for each $i \in [r]$, there is at least one $v \in e$ such that $c(v) = i$. A *no-rainbow r-coloring* of H is a surjective r-coloring of H such that H has no rainbow edge.

Given an r-uniform hypergraph $H = (X, E)$, the *no-rainbow r-coloring problem* (r-NRC) asks whether H has a no-rainbow r-coloring [5]. r-NRC is clearly in NP, but it is unknown whether the problem is NP-complete [5].

Let S be a taxon coverage pattern for X. We associate with S a hypergraph $H(S) = (X, S)$, and with Q_S, we associate a 4-uniform hypergraph $H(Q_S) = (X, Q_S)$. The next result states that r-NRC is equivalent to the complement of the decisiveness problem.

Proposition 1. *Let S be a taxon coverage pattern. The following statements are equivalent.*

(i) S is not decisive.
(ii) $H(Q_S)$ admits a no-rainbow 4-coloring.
(iii) $H(S)$ admits a no-rainbow 4-coloring.

Proof. (1) $\Leftrightarrow$ (2): By Theorem 1, it suffices to show that S fails to satisfy the 4-way partition property if and only if $H(Q_S)$ has a no-rainbow 4-coloring. S does not satisfy the 4-way partition property if and only if there exists a 4-way partition A_1, A_2, A_3, A_4 of X such that, for every $q \in Q_S$, there is an $i \in [4]$ such that $A_i \cap q = \emptyset$. This holds if and only if the coloring c, where $c(v) = i$ if and only if $v \in A_i$, is a no-rainbow 4-coloring of $H(Q_S)$.

(2) $\Leftrightarrow$ (3): It is clear that if c is a no-rainbow 4-coloring of $H(S)$, then c is a no-rainbow 4-coloring of $H(Q_S)$. We now argue that if c is a no-rainbow 4-coloring of $H(Q_S)$, then c is a no-rainbow 4-coloring of $H(S)$. Suppose, to the contrary, that there a rainbow edge $Y \in S$. Let q be any 4-tuple $\{v_1, v_2, v_3, v_4\} \subseteq Y$ such that $c(v_i) = i$, for each $i \in [4]$. Then, q is a rainbow edge in Q_S, a contradiction. $\square$

Proposition 2. *Let $H = (X, E)$ be a hypergraph and r be a positive integer.*

(i) If H has at least r connected components, then H admits a no-rainbow r-coloring.
(ii) If $r = 2$, then H admits a no-rainbow r-coloring if and only if H is disconnected.

Proof. (i) Suppose the connected components of H are $C_1, \ldots, C_q$, where $q \geq r$. For each $i \in \{1, \ldots, r-1\}$, assign color i to all nodes in C_i. For $i = \{r, \ldots, q\}$, assign color r to all nodes in C_i. Thus, no edge is rainbow-colored.

(ii) By part (i), if H is disconnected, it admits a no-rainbow 2-coloring. To prove the other direction, assume, for contradiction that H admits a no-rainbow 2-coloring but it is connected. Pick any two nodes u and v such that $c(u) = 1$ and $c(v) = 2$. Since H is connected, there is a (u, v)-chain in H. But this chain must contain an edge with nodes of two different colors; i.e., a rainbow edge. $\square$

Part (ii) of Proposition 2 implies the following.

Corollary 2. *2-NRC $\in P$.*

Lemma 1. *Let $H = (X, E)$ be an r-uniform hypergraph. Suppose that there exists a subset A of X such that $2 \leq |A| \leq r-1$ and $A \not\subseteq e$ for any $e \in E$. Then, H has a no-rainbow r-coloring.*

Proof. Let c be the coloring where each of the nodes in A is assigned a distinct color from the set $[|A|]$ and the remaining nodes are assigned colors from the set $\{|A|+1, \ldots, r\}$. Then, c is a no-rainbow r-coloring of H. □

4   A Tight Lower Bound on the Coverage

The next result provides a tight lower bound on the minimum amount of coverage that is needed to achieve decisiveness.

Theorem 3. *Let S be a taxon coverage pattern for X and let $n = |X|$. If S is decisive, then $|Q_S| \geq \binom{n-1}{3}$. This lower bound is tight. That is, for each $n \geq 4$, there exists a decisive taxon coverage pattern S for X such that $|Q_S| = \binom{n-1}{3}$.*

To prove Theorem 3, for every pair of integers n, r such that $n \geq r \geq 1$ let us define the function $A(n, r)$ as follows.

$$A(n,r) = \begin{cases} 1 & \text{if } r = 1 \text{ or } n = r \\ A(n-1, r-1) + A(n-1, r) & \text{otherwise.} \end{cases} \tag{1}$$

Lemma 2. *Let n and r be integers such that $n \geq r \geq 1$ and let $H = (X, E)$ be an n-vertex r-uniform hypergraph. If $|E| < A(n, r)$, then H admits a no-rainbow r-coloring. If $|E| \geq A(n, r)$, then H may or may not admit a no-rainbow r-coloring. Furthermore, there exist n-vertex r-uniform hypergraphs with exactly $A(n, r)$ edges that do not admit a no-rainbow r-coloring.*

Proof. For $r = 1$ or $n = r$, H has at most one hyperedge. If H has exactly one hyperedge, then any coloring that uses all r colors contains a rainbow edge. If H contains no hyperedges, then H trivially admits a no-rainbow r-coloring.

Let us assume that for any i and j with $1 \leq i < n$ and $1 \leq j \leq r$, $A(i, j)$ equals the minimum number of hyperedges an i-node, j-uniform hypergraph H that does not admit a no-rainbow r-coloring. We now prove the claim for $i = n$ and $j = r$.

Pick an arbitrary node $v \in X$. There are two mutually disjoint classes of colorings of H: (1) the colorings c such that $c(v) \neq c(u)$ for any $u \in X \setminus \{v\}$, and (2) the colorings c such that $c(v) = c(u)$ for some $u \in X \setminus \{v\}$.

For the colorings in class 1, we need hyperedges that contain node v, since in the absence of such hyperedges, any coloring is a no-rainbow coloring. Assume, without loss of generality, that $c(v) = r$. The question reduces to finding the number of hyperedges in an $(n-1)$-node $(r-1)$-uniform hypergraph (since v's color, r, is unique). The minimum number of hyperedges needed to avoid a no-rainbow $(r-1)$-coloring for an $(n-1)$-node hypergraph is $A(n-1, r-1)$.

To find the minimum number of hyperedges needed to cover colorings of class 2, we ignore v, since v is assigned a color that is used by other nodes as well. The number of hyperedges needed for this class is $A(n-1, r)$.

To obtain a lower bound, we add the lower obtained for the two disjoint classes of colorings. Thus, $A(n, r) = A(n-1, r-1) + A(n-1, r)$. □

Lemma 3. $A(n,r) = \binom{n-1}{r-1}$.

Proof. For $r = 1$, $A(n,r) = \binom{n-1}{0} = 1$ and for $n = r$, $A(n,r) = A(r,r) = \binom{r-1}{r-1} = 1$. Now, assume that $A(i,j) = \binom{i-1}{j-1}$, for $1 \leq i \leq n-1$ and $1 \leq j \leq r$. Then, $A(n,r) = A(n-1,r) + A(n-1,r-1) = \binom{n-2}{r-1} + \binom{n-2}{r-2} = \binom{n-1}{r-1}$.   □

Proof (of Theorem 3). Follows from Lemmas 2 and 3, by setting $r = 4$, and Proposition 1(ii).   □

5 An Exact Algorithm for Decisiveness

The naïve way to use Theorem 1 to test whether a coverage pattern $\mathcal{S}$ is decisive is to enumerate all partitions of X into four non-empty sets A_1, A_2, A_3, A_4 and verify that there is a set $Y \in \mathcal{S}$ that intersects each A_i. Equivalently, by Proposition 1, we can enumerate all surjective colorings of $H(S)$ and check if each of these colorings yields a rainbow edge. In either approach, the number of options to consider is given by a Stirling number of the second kind, namely $\left\{ {n \atop 4} \right\} \sim \frac{4^n}{4!}$ [9]. The next result is a substantial improvement over the naïve approach.

Theorem 4. *Let $\mathcal{S}$ be a taxon coverage pattern for a taxon set X. Then, there is an algorithm that, in $O^*(2.8^n)$ time[1] determines whether or not $\mathcal{S}$ is decisive.*

The proof of Theorem 4 relies on the following result.

Lemma 4. *There exists an algorithm that, given a 4-uniform hypergraph $H = (X, E)$, determines if H has a no-rainbow 4-coloring in time $O^*(2.8^n)$.*

Proof. We claim that algorithm `FindNRC` (Algorithm 1) solves 4-NRC in $O^*(2.8^n)$ time. `FindNRC` relies on the observation that if H has a no-rainbow 4-coloring c, then (1) there must exist a subset $A \subseteq X$ where $|A| \leq \lfloor \frac{n}{4} \rfloor$, such that all nodes in A have the same color, which is different from the colors used for $X \setminus A$, and (ii) there must exist a subset $B \subseteq X \setminus A$, where $|B| \leq \lfloor \frac{n-|A|}{3} \rfloor$, such that all nodes in B have the same color, which is different from the colors used for the nodes in $X \setminus B$. `FindNRC` tries all possible choices of A and B and, without loss of generality, assigns $c(v) = 1$, for all $v \in A$ and $c(v) = 2$, for all $v \in B$. We are now left with the problem of determining whether we can assign colors 3 and 4 to the nodes in $X \setminus (A \cup B)$ to obtain a no-rainbow 4-coloring for H.

Let c be the current coloring of H. For each $e \in E$ and each $i \in [4]$, $m_e^c(i)$ denotes the number of nodes $v \in e$ such that $c(v) = i$. Consider the situation after `FindNRC` assigns colors 1 and 2 to the nodes in A and B. There are two cases, both of which can be handled in polynomial time.

1. *There is no $e \in E$, such that, for each $i \in [2]$, $m_e^c = 1$.* Then, if we partition the nodes of $X \setminus (A \cup B)$, arbitrarily into subsets C and D and assign $c(v) = 3$ for each $v \in C$ and $c(v) = 4$ for each $v \in D$, we obtain a no-rainbow 4-coloring of H.

---

[1] The O^*-notation is a variant of O-notation that ignores polynomial factors [8].

```
1 FindNRC(H)
      Input: A 4-uniform hypergraph H = (X, E) such that |X| ≥ 4.
      Output: A no-rainbow 4-coloring of H, if one exists; otherwise, fail.
2     for i = 1 to ⌊n/4⌋ do
3         foreach v ∈ X do c(v) = uncolored
4         foreach A ⊆ X such that |A| = i do
5             foreach v ∈ A do c(v) = 1
6             for j = 1 to ⌊(n-i)/3⌋ do
7                 foreach B ⊆ X \ A such that |B| = j do
8                     foreach v ∈ B do c(v) = 2
9                     if there is no e ∈ E such that, for each i ∈ [2], m_e^c(i) = 1
                      then
10                        Arbitrarily split X \ (A ∪ B) into nonempty sets C, D
11                        foreach v ∈ C do c(v) = 3
12                        foreach v ∈ D do c(v) = 4
13                        return c
14                    else
15                        Choose any e ∈ E such that m_e^c(i) = 1 for each i ∈ [2]
16                        foreach uncolored node x ∈ e do c(x) = 3
17                        while there exists e ∈ E s.t. m_e^c(i) = 1 for each i ∈ [3]
                          do
18                            Pick any e ∈ E s.t. m_e^c(i) = 1 for each i ∈ [3]
19                            Let x be the unique uncolored node in e
20                            c(x) = 3
21                        if X contains no uncolored node then return fail
22                        else
23                            foreach uncolored vertex u ∈ X do c(u) = 4
24                            return c
25    return fail
```

Algorithm 1: No-rainbow 4-coloring of H

2. *There exists* $e \in E$, *such that, for each* $i \in [2]$, $m_e^c = 1$. Let e be any such edge. Then e must exactly contain two uncolored nodes, x and y. To avoid e becoming a rainbow edge, we must set $c(x) = c(y) \notin [2]$. Without loss of generality, make $c(x) = c(y) = 3$. Next, as long as there exists any hyperedge e such that $m_e^c(i) = 1$ for each $i \in [3]$, the (unique) uncolored node x in e must be assigned $c(x) = 3$, because setting $c(x) = 4$ would make e a rainbow hyperedge. Once no such hyperedges remain, we have two possibilities:
   (a) X *does not contain uncolored nodes.* Then, there does not exist a no-rainbow 4-coloring, given the current choice of A and B.
   (b) X *contains uncolored nodes.* Then, there is no $e \in E$ such that $m_e^c(i) = 1$ for each $i \in [3]$. Thus, if we set $c(u) = 4$ for each uncolored node u, we obtain a no-rainbow 4-coloring for H.

The total number of pairs (A, B) considered throughout the execution of FindNRC is at most $\sum_{i=1}^{\lfloor n/4 \rfloor} \binom{n}{i} \sum_{j=1}^{\lfloor (n-i)/3 \rfloor} \binom{n-i}{j}$. We have estimated this sum numerically to be $O(2.8^n)$. The time spent per pair (A, B) is polynomial in n; hence, the total running time of `FindNRC` is $O^*(2.8^n)$.   □

Proof (of Theorem 4). Given $\mathcal{S}$, we construct the hypergraph $H(Q_\mathcal{S})$, which takes time polynomial in n, and then run `FindNRC`$(H(Q_\mathcal{S}))$, which, by Lemma 4, takes $O^*(2.8^n)$ time. If the algorithm returns a no-rainbow 4-coloring c of $H(Q_\mathcal{S})$, then, by Proposition 1, $\mathcal{S}$ is not decisive; if `FindNRC`$(H(Q_\mathcal{S}))$ returns `fail`, then $\mathcal{S}$ is decisive.   □

6   Reduction Rules and Fixed Parameter Tractability

A *reduction rule* for the decisiveness problem is a rule that replaces an instance $\mathcal{S}$ of the problem by a smaller instance $\widetilde{\mathcal{S}}$ such that $\mathcal{S}$ is decisive if and only if $\widetilde{\mathcal{S}}$ is. Here we present reduction rules that can reduce an instance of the decisiveness problem into a one whose size depends only on k. This size reduction is especially significant for taxon coverage patterns where the number of loci, k, is small relative to the number of taxa. Such inputs are not uncommon in the literature — examples of such data sets are studied in Sect. 7.

We need to introduce some definitions and notation. Let $H = (X, E)$ be a hypergraph where $X = \{x_1, x_2, \ldots, x_n\}$ and $E = \{e_1, e_2, \ldots, e_k\}$. The *incidence matrix* of H is the $n \times k$ binary matrix where $M_H[i, j] = 1$ if $x_i \in e_j$ and $M_H[i, j] = 0$ otherwise. Two rows in M_H are *copies* if the rows are identical when viewed as 0-1 strings; otherwise, they are *distinct*.

Let $\widetilde{M_H}$ denote the matrix obtained from M_H by striking out duplicate rows, so that $\widetilde{M_H}$ retains only one copy of each row in M_H. Let $\widetilde{n}$ denote the number of rows of $\widetilde{M_H}$. Then, $\widetilde{n} \le 2^k$. $\widetilde{M}$ is the incidence matrix of a hypergraph $\widetilde{H} = (\widetilde{X}, \widetilde{E})$, where $\widetilde{X} \subseteq X$, and each $v \in \widetilde{X}$ corresponds to a distinct row of M_H. For each $v \in X$, $X(v) \subseteq X$ consists of all nodes $u \in X$ that correspond to copies of the row of M_H corresponding to v.

Given two binary strings s_1 and s_2 of length k, $s_1 \,\&\, s_2$ denotes the bitwise *and* of s_1 and s_2; $\mathbf{0}$ denotes the all-zeroes string of length k.

The next result is a direct consequence of Lemma 1.

Proposition 3. *If $\widetilde{M_H}$ has two rows r_1 and r_2 such that $r_1 \,\&\, r_2 = \mathbf{0}$ or three rows r_1, r_2 and r_3 such that $r_1 \,\&\, r_2 \,\&\, r_3 = \mathbf{0}$, then $\widetilde{H}$ and H admit no-rainbow 4-colorings.*

Corollary 3. *If $\widetilde{M_H}$ has more than 2^{k-1} rows, where k is the number of columns, then H admits a no-rainbow 4-coloring.*

Proof. Suppose $\widetilde{n} \ge 2^{k-1}$. Then, there are at least two rows r_1 and r_2 in $\widetilde{M_H}$ that are complements of each other (that is, r_2 is is obtained by negating each bit in r_1) and, thus, $r_1 \,\&\, r_2 = \mathbf{0}$. The claim now follows from Proposition 3.   □

Theorem 5. *Suppose $n \ge \widetilde{n} + 2$. H admits a no-rainbow 4-coloring if and only if $\widetilde{H}$ admits a no-rainbow r-coloring for some $r \in \{2, 3, 4\}$.*

Proof. (*If*) Suppose $\widetilde{H} = (\widetilde{X}, \widetilde{E})$ admits a no-rainbow r-coloring $\widetilde{c}$ for some $r \in \{2, 3, 4\}$. Let c be the coloring for H obtained by setting $c(u) = \widetilde{c}(v)$, for each $v \in \widetilde{X}$ and each $u \in X(v)$. If $\widetilde{c}$ is a no-rainbow 4-coloring of $\widetilde{H}$, then c is also one for H, and we are done. Suppose $\widetilde{c}$ is a 3-coloring. Since $n \geq \widetilde{n} + 2$, there must exist $v \in \widetilde{X}$ such that $|X(v)| \geq 2$. We choose one node $u \in X(v) \setminus \{v\}$, and set $c(u) = 4$, making c a no-rainbow 4-coloring for H. Suppose $\widetilde{c}$ is a no-rainbow 2-coloring. If there exists $v \in \widetilde{X} \setminus \{v\}$ such that $|X(v)| \geq 3$, we pick any $u, w \in \widetilde{X} \setminus \{v\}$, and set $c(u) = 3$ and $c(w) = 4$. If there is no $v \in \widetilde{X}$ such that $|X(v)| \geq 3$, there must exist $v_1, v_2 \in \widetilde{X}$ such that $|X(v_i)| \geq 2$ for $i \in \{1, 2\}$. For $i \in \{1, 2\}$, choose any $u_i \in X(v_i) \setminus \{v_i\}$ and set $c(u_i) = i + 2$.

(*Only if*) Suppose H has a no-rainbow 4-coloring c. Let $\widetilde{c}$ be the coloring of $\widetilde{H}$ where, for each $v \in \widetilde{X}$, $\widetilde{c}(v) = c(u)$, for some arbitrarily chosen node in $u \in X(v)$. If $\widetilde{c}$ is a surjective 4-coloring of v, $\widetilde{c}$ must be a no-rainbow 4-coloring of $\widetilde{H}$, and we are done. In the extended version of the paper [15], we show that any r-coloring of $\widetilde{H}$, where $r \in \{1, 2, 3, 4\}$ can be converted into a no-rainbow 4-coloring of $\widetilde{H}$ by altering some of the colors assigned by $\widetilde{c}$.   □

Theorem 6. *Decisiveness is fixed-parameter tractable in k.*

Proof. Let $\mathcal{S}$ be the input coverage pattern. First, in $O^*(2^k)$ time, we construct $\widetilde{H}(\mathcal{S})$. By Theorem 2, we need to test if $\widetilde{H}(\mathcal{S})$ admits a non-rainbow r-coloring for any $r \in \{2, 3, 4\}$. If the answer is "yes" for any such r, then $\mathcal{S}$ is not decisive; otherwise $\mathcal{S}$ is decisive. By Corollary 2, the test for $r = 2$ takes polynomial time. We perform the steps for $r = 3$ and $r = 4$ using the algorithm of Sect. 5. The total time is $O^*(2.8^{\widetilde{n}})$, which is $O^*(2.8^{2^k})$.   □

7   An Integer Linear Programming Formulation

Let $\mathcal{S} = \{Y_1, Y_2, \ldots, Y_k\}$ be a taxon coverage pattern for X. Here we formulate a 0-1 integer linear program (ILP) that is feasible if and only if $\mathcal{S}$ is non-decisive.[2] We use the equivalence between non-decisiveness of $\mathcal{S}$ and the existence of a no-rainbow 4-coloring of hypergraph $H(\mathcal{S})$ (Proposition 1).

Suppose $X = \{a_1, a_2, \ldots, a_n\}$. For each $i \in [n]$ and each color $q \in [4]$, define a binary *color variable* x_{iq}, where $x_{iq} = 1$ if taxon i is assigned color q. To ensure that each $i \in X$ is assigned only one color, we add constraints

$$\sum_{q \in [4]} x_{iq} = 1, \quad \text{for each } i \in X. \tag{2}$$

The following constraints ensure that each color $q \in [4]$ appears at least once.

$$\sum_{i \in X} x_{iq} \geq 1, \quad \text{for each } q \in [4]. \tag{3}$$

---

[2] For an introduction to the applications of integer linear programming, see [12].

To ensure that, for each $j \in [k]$, Y_j is not rainbow colored, we require that there exist at least one color that is not used in Y_j; i.e, that $\sum_{i \in Y_j} x_{iq} = 0$, for some $q \in [4]$. To express this condition, for each $j \in [k]$ and each $q \in [4]$, we define a binary variable z_{jq}, which is 1 if and only if $\sum_{i \in Y_j} x_{iq} = 0$. We express z_{jq} using the following linear constraints.

$$(1 - z_{jq}) \leq \sum_{i \in Y_j} x_{iq} \leq n \cdot (1 - z_{jq}), \quad \text{for each } j \in [k] \text{ and each } q \in [4] \quad (4)$$

The requirement that Y_j not be rainbow-colored is expressed as

$$\sum_{q \in [4]} z_{jq} \geq 1, \quad \text{for each } j \in [k]. \quad (5)$$

Proposition 4. *S is non-decisive if and only if the 0-1 ILP with variables x_{iq} and z_{jq} and constraints* (2), (3), (4), *and* (5) *is feasible.*

Experimental Results. Here we summarize our computational results using ILP on the data sets studied in [6].

We wrote a Python script that given a taxon coverage pattern, generates an ILP model. Table 1 shows the time taken to generate the ILP models for the data sets analyzed in [6] (see the latter reference for full citations of the corresponding phylogenetic studies). The models were generated on a Linux server.

We generated ILP models for all the data sets in [6] and used Gurobi [11] to solve 9 of these models. All but one of these models were solved in under 0.1 seconds. Table 2 shows the time taken to solve the models. Only one of the data sets, Insects, is decisive (and its ILP took the longest to solve). Indeed, the Insects data set is trivially decisive, as one locus spans all the taxa.

For the non-decisive data sets, we used a simple heuristic to identify a subset of the taxa for which the data is decisive. If the data set is non-decisive, we remove the taxon covered by the fewest loci, breaking ties in favor of the first taxon in the input. After removing a taxon, we update the ILP model and run it again. When the model becomes infeasible, the remaining data set is decisive.

For two data sets (Saxifragales and Mammals), the heuristic yielded trivially decisive coverage patterns. We obtained non-trivial results for three data sets. For the complete Birds data set, the largest of all, the heuristic took 1.1 h. Although the heuristic retained only 2.5% of the original taxa, every family of taxa except one from the original data set is represented in the final result. For Bats, the heuristic took 70 seconds and achieved 4.3% coverage, but had sparse coverage across the families. For Primates, the heuristic took 33 seconds and achieved 50.3% coverage, distributed over most families. For all data sets, the most time-consuming step was attempting to solve the final, infeasible, ILP.

Table 1. Running times for generating ILPs for data sets studied in [6]

Data set	Execution time (seconds)	Number of taxa	Number of loci
Allium	0.051037	57	6
Asplenium	0.047774	133	6
Bats	0.152805	815	29
Birds (complete)	4.688950	7000	32
Birds	2.723334	5146	32
Caryophyllaceae	0.068084	224	7
Chameleons	0.059073	202	6
Eucalyptus	0.058591	136	6
Euphorbia	0.061188	131	7
Ficus	0.063072	112	5
Fungi	0.223971	1317	9
Insects	7.649374	144	479
Iris	0.055743	137	6
Mammals	0.110263	169	26
Primates	0.363623	372	79
Primula	0.064607	185	6
Scincids	0.071276	213	6
Ranunculus	0.059699	170	7
Rhododendron	0.052903	117	7
Rosaceae	0.092148	529	7
Solanum	0.062660	187	7
Saxifragales	0.173522	946	51
Szygium	0.051021	106	5

Table 2. Solution times for a subset of ILPs listed in Table 1

Data set	Execution time (seconds)
Bats	0.098
Birds (complete)	0.03
Eucalyptus	0.002
Ficus	0.001
Insects	5.902
Iris	0.002
Mammals	0.091
Primates	0.013
Saxifragales	0.004

8   Discussion

Despite its apparent complexity, the decisiveness problem appears to be quite tractable in practice. Since real data sets are likely to be non-decisive, testing for decisiveness can only be considered a first step. Indeed, if we determine that a data set is not decisive, it is useful to find a subset of the data that is decisive. In Sect. 7, we have taken some preliminary steps in that direction, using a simple heuristic. This heuristic could potentially be improved upon, perhaps relying on the data reduction ideas of Sect. 6. One open problem is whether the doubly-exponential algorithm of Theorem 6 can be improved.

Acknowledgements. Mike Steel pointed out the connection between decisiveness and hypergraph coloring. We thank Mike Sanderson for useful discussions. He and Barbara Dobrin provided the data studied in Sect. 7.

References

1. Ané, C., Eulenstein, O., Piaggio-Talice, R., Sanderson, M.J.: Groves of phylogenetic trees. Ann. Comb. **13**(2), 139–167 (2009)
2. Berge, C.: Graphs and Hypergraphs. North-Holland, Amsterdam (1973)
3. Berge, C.: Hypergraphs: Combinatorics of Finite Sets. North-Holland Mathematical Library, vol. 45. Elsevier (1984)
4. Bininda-Emonds, O.R.P. (ed.): Phylogenetic Supertrees: Combining Information to Reveal the Tree of Life, Series on Computational Biology, vol. 4. Springer, Berlin (2004)
5. Bodirsky, M., Kára, J., Martin, B.: The complexity of surjective homomorphism problems–a survey. Discrete Appl. Math. **160**(12), 1680–1690 (2012)
6. Dobrin, B.H., Zwickl, D.J., Sanderson, M.J.: The prevalence of terraced treescapes in analyses of phylogenetic data sets. BMC Evol. Biol. **18**, 46 (2018)
7. Fischer, M.: Mathematical aspects of phylogenetic groves. Ann. Comb. **17**(2), 295–310 (2013)
8. Fomin, F.V., Kratsch, D.: Exact Exponential Algorithms. Springer, Heidelberg (2010)
9. Graham, R.L., Knuth, D.E., Patashnik, O.: Concrete Mathematics: A Foundation for Computer Science. Addison-Wesley, Reading (1989)
10. Guindon, S., Gascuel, O.: A simple, fast, and accurate algorithm to estimate large phylogenies by maximum likelihood. Syst. Biol. **52**(5), 696–704 (2003)
11. Gurobi Optimization, LLC: Gurobi optimizer reference manual (2019). http://www.gurobi.com
12. Gusfield, D.: Integer Linear Programming in Computational and Systems Biology: An Entry-level Text and Course. Cambridge University Press, New York (2019)
13. Hinchliff, C.E., et al.: Synthesis of phylogeny and taxonomy into a comprehensive tree of life. Proc. Natl. Acad. Sci. **112**(41), 12764–12769 (2015). https://doi.org/10.1073/pnas.1423041112
14. Jarvis, E.D., et al.: Whole-genome analyses resolve early branches in the tree of life of modern birds. Science **346**(6215), 1320–1331 (2014)
15. Parvini, G., Braught, K., Fernández-Baca, D.: Checking phylogenetic decisiveness in theory and in practice, February 2020. arXiv preprint https://arxiv.org/abs/2002.09722

16. Sanderson, M.J., McMahon, M.M., Steel, M.: Phylogenomics with incomplete taxon coverage: the limits to inference. BMC Evol. Biol. **10**, 155 (2010). https://doi.org/10.1186/1471-2148-10-155

17. Sanderson, M.J., McMahon, M.M., Steel, M.: Terraces in phylogenetic tree space. Science **333**(6041), 448–450 (2011). https://doi.org/10.1126/science.1206357, http://science.sciencemag.org/content/333/6041/448

18. Scornavacca, C.: Supertree methods for phylogenomics. Ph.D. thesis, Univ. of Montpellier II, Montpellier, France, December 2009

19. Semple, C., Steel, M.: Phylogenetics. Oxford Lecture Series in Mathematics. Oxford University Press, Oxford (2003)

20. Stamatakis, A.: RAxML version 8: a tool for phylogenetic analysis and post-analysis of large phylogenies. Bioinformatics **30**(9), 1312–1313 (2014)

21. Stamatakis, A., Alachiotis, N.: Time and memory efficient likelihood-based tree searches on phylogenomic alignments with missing data. Bioinformatics **26**(12), i132–i139 (2010)

22. Steel, M.: Phylogeny: Discrete and Random Processes in Evolution. CBMS-NSF Conference Series in Applied Mathematics, vol. 89. SIAM, Philadelphia (2016)

23. Steel, M., Sanderson, M.J.: Characterizing phylogenetically decisive taxon coverage. Appl. Math. Lett. **23**(1), 82–86 (2010)

24. Warnow, T.: Supertree construction: opportunities and challenges. Technical report, arXiv:1805.03530, ArXiV, May 2018. https://arxiv.org/abs/1805.03530

25. Wickett, N.J., et al.: Phylotranscriptomic analysis of the origin and early diversification of land plants. Proc. Natl. Acad. Sci. **111**(45), E4859–E4868 (2014)

26. Wilkinson, M.: Coping with abundant missing entries in phylogenetic inference using parsimony. Syst. Biol. **44**(4), 501–514 (1995)

27. Zhbannikov, I.Y., Brown, J.W., Foster, J.A.: decisivatoR: an R infrastructure package that addresses the problem of phylogenetic decisiveness. In: Proceedings of the International Conference on Bioinformatics, Computational Biology and Biomedical Informatics, pp. 716–717 (2013)

28. Zhuk, D.: No-rainbow problem is NP-hard, March 2020. arXiv preprint https://arxiv.org/abs/2003.11764

TNet: Phylogeny-Based Inference of Disease Transmission Networks Using Within-Host Strain Diversity

Saurav Dhar[1], Chengchen Zhang[1,3], Ion Mandoiu[1,2], and Mukul S. Bansal[1,2(✉)]

[1] Department of Computer Science and Engineering,
University of Connecticut, Storrs, USA
{ion.mandoiu,mukul.bansal}@uconn.edu
[2] Institute for Systems Genomics, University of Connecticut, Storrs, USA
[3] Computer Science and Engineering, University of California, San Diego, USA

Abstract. The inference of disease transmission networks from genetic sequence data is an important problem in epidemiology. One popular approach for building transmission networks is to reconstruct a phylogenetic tree using sequences from disease strains sampled from (a subset of) infected hosts and infer transmissions based on this tree. However, most existing phylogenetic approaches for transmission network inference cannot take within-host strain diversity into account, which affects their accuracy, and, moreover, are highly computationally intensive and unscalable.

In this work, we introduce a new phylogenetic approach, TNet, for inferring transmission networks that addresses these limitations. TNet uses multiple strain sequences from each sampled host to infer transmissions and is simpler and more accurate than existing approaches. Furthermore, TNet is highly scalable and able to distinguish between ambiguous and unambiguous transmission inferences. We evaluated TNet on a large collection of 560 simulated transmission networks of various sizes and diverse host, sequence, and transmission characteristics, as well as on 10 real transmission datasets with known transmission histories. Our results show that TNet outperforms two other recently developed methods, phyloscanner and SharpTNI, that also consider within-host strain diversity using a similar computational framework. TNet is freely available open-source from https://compbio.engr.uconn.edu/software/TNet/.

1 Introduction

The accurate inference of disease transmission networks is fundamental to understanding and containing the spread of infectious diseases [2,10,16]. A key challenge with inferring transmission networks, particularly those of rapidly evolving RNA and retroviruses [7], is that they exist in the host as "clouds" of closely related sequences. These variants are referred to as *quasispecies* [6,22], and the resulting genetic diversity of the strains circulating within a host has important implications for efficiency of transmission, disease progression, drug/vaccine

© Springer Nature Switzerland AG 2020
Z. Cai et al. (Eds.): ISBRA 2020, LNBI 12304, pp. 203–216, 2020.
https://doi.org/10.1007/978-3-030-57821-3_18

resistance, etc. The availability of quasispecies, or sequences from multiple strains per infected host, also has direct relevance for inferring transmission networks and has the potential to make such inference easier and far more accurate [18, 20, 23]. Yet, while the advent of next-generation sequencing technologies has revolutionized the study of quasispecies, most existing transmission network inference methods cannot use multiple distinct strain sequences per host.

Existing methods for inferring transmission networks can be classified into two categories: Those based on constructing and analyzing sequence similarity or relatedness graphs, and those based on constructing and analyzing phylogenetic trees for the infecting strains. Many methods based on sequence similarity or relatedness graph analysis exist and several recently developed methods in this category are also able to take into account multiple distinct strain sequences per host [9, 14, 19]. However, similarity/relatedness based methods can suffer from a lack of resolution and are often unable to infer transmission directions or complete transmission histories. Phylogeny-based methods [5, 11, 13, 16, 23] attempt to overcome these limitations by constructing and analyzing phylogenies of the infecting strains. We refer to these strain phylogenies as *transmission phylogenies*. These phylogeny-based methods infer transmission networks by computing a host assignment for each node of the transmission phylogeny, where this phylogeny is either first constructed independently or is co-estimated along with the host assignment. Leaves of the transmission phylogeny are labelled by the host from which they are sampled, and an ancestral host assignment is then inferred for each node/edge of the phylogeny. This ancestral host assignment defines the transmission network, where transmission is inferred along any edge connecting two nodes labeled with different hosts.

Several sophisticated phylogeny-based methods have been developed over the last few years. These include BEASTlier [11], SCOTTI [4], phybreak [13], TransPhylo [5], and phyloscanner [23], BadTrIP [3]. Among these, only SCOTTI [4], BadTrIP [3], and phyloscanner [23] can explicitly consider multiple strain sequences per host. BEASTlier also allows for the presence of multiple sequences per host, but requires that all sequences from the same host be clustered together on the phylogeny, a precondition that is often violated in practice. Among the methods that explicitly consider multiple strain sequences per host, SCOTTI, BadTrIP, and BEASTlier are model-based and highly computationally intensive, relying on the use of Markov Chain Monte Carlo (MCMC) algorithms for inference. These methods also require several difficult-to-estimate epidemiological parameters, such as infection times, and make several strong assumptions about pathogen evolution and the underlying transmission network. Thus, phyloscanner [23] is the only previous method that takes advantage of multiple sequences per host and that is also computationally efficient, easy to use, and scalable to large datasets.

In this work, we introduce a new phylogenetic approach, TNet, for inferring transmission networks. TNet uses multiple strain sequences from each sampled host to infer transmissions and is simpler and more accurate than existing approaches. TNet uses an extended version of the classical Sankoff algorithm [17]

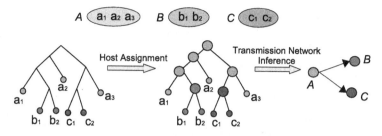

Fig. 1. Phylogeny-based transmission network inference. The figure shows a simple example with three infected individuals A, B, and C, represented here by the three different colors, where A has three viral variants while B and C have two each. The tree on the left depicts the transmission phylogeny for the seven sampled strains, with each of these strains colored by the host from which it was sampled. The tree in the middle shows a hypothetical assignment of hosts to ancestral nodes of the transmission phylogeny. This ancestral host assignment can then be used to infer the transmission network shown on the right.

from the phylogenetics literature for ancestral host assignment, where the extension makes it possible to efficiently compute support values for individual transmission edges based on a sampling of optimal host assignments where the number of back-transmissions (or reinfections by descendant disease strains) is minimized. TNet is parameter-free and highly scalable and can be easily applied within seconds to datasets with hundreds of strain sequences and hosts. In recent independent work, Sashittal et al. [18] developed a new method called SharpTNI that is based on similar ideas to TNet. SharpTNI is based on an NP-hard problem formulation that seeks to find parsimonious ancestral host assignments minimizing the number of co-transmissions [18].The authors provide an efficient heuristic for this problem that is based on uniform sampling of parsimonious ancestral host assignments (not necessarily minimizing co-transmissions) and subsequently filtering them to only keep those assignments among the samples that minimize co-transmissions [18]. Thus, both TNet and SharpTNI are based on the idea of parsimonious ancestral host assignments and on aggregating across the diversity of possible solutions obtained through some kind of sampling of optimal solutions. The primary distinction between the two methods is the strategy employed for sampling of the optimal solutions, with SharpTNI minimizing co-transmissions and TNet minimizing back-transmissions.

We evaluated TNet, SharpTNI, and phyloscanner on a large collection of 560 simulated transmission networks of various sizes and representing a wide range of host, sequence, and transmission characteristics, as well as on 10 real transmission datasets with known transmission histories. We found that both TNet and SharpTNI significantly outperformed phyloscanner under all tested conditions and all datasets, yielding more accurate transmission networks for both simulated and real datasets. Between TNet and SharpTNI, we found that both methods performed similarly on the real datasets but that TNet showed better accuracy on the simulated datasets. TNet is freely available open-source from: https://compbio.engr.uconn.edu/software/TNet/.

2   Basic Definitions and Preliminaries

Given a rooted tree T, we denote its node set, edge set, and leaf set by $V(T)$, $E(T)$, and $Le(T)$ respectively. The root node of T is denoted by $rt(T)$, the parent of a node $v \in V(T)$ by $pa_T(v)$, its set of children by $Ch_T(v)$, and the (maximal) subtree of T rooted at v by $T(v)$. The set of *internal nodes* of T, denoted $I(T)$, is defined to be $V(T) \setminus Le(T)$. A rooted tree is *binary* if all of its internal nodes have exactly two children. In this work, the term *tree* refers to a rooted binary tree.

2.1   Problem Formulation

Let T denote the transmission phylogeny constructed from the genetic sequences of the infecting strains (i.e., pathogens) sampled from the infected hosts under consideration. Note that such trees can be easily constructed using standard phylogenetic methods such as RAxML [21]. These trees can also be rooted relatively accurately using either standard phylogenetic rooting techniques or by using a related sequence from a previous outbreak of the same disease as an outgroup. Let $H = \{h_1, h_2, \ldots, h_n\}$ denote the set of n hosts under consideration. We assume that each leaf of T is labeled with the host from H from which the corresponding strain sequence was obtained. Figure 1 shows an example of such a tree and its leaf labeling, where the labeling is depicted using the different colors.

Observe that each internal node of T represents an ancestral strain sequence that existed in some infected host. Moreover, each internal node (or bifurcation) represents either intra-host diversification and evolution of that ancestral strain or a transmission event where that ancestral strain is transmitted from one host to another along one of the child edges. Thus, each node of T is associated with an infected host. Given $t \in V(T)$, we denote the host associated with node t by $h(t)$. Note that internal nodes may represent strains from hosts that do not appear in H, i.e., strains from unsampled hosts, and so there may be $t \in I(T)$ for which $h(t) \notin H$. Given an ancestral host assignment for T, i.e., given $h(t)$ for each $t \in I(T)$, the implied transmission network can be easily inferred as follows: A transmission edge is inferred from host x to host y if there is an edge $(pa(t), t) \in E(T)$, where $h(pa(t)) = x$ and $h(t) = y$. Note that each transmission edge in the reconstructed transmission network may represent either direct transmission or indirect transmission through one or more unsampled hosts. Thus, to reconstruct transmission networks it suffices to compute $h(t)$ for each $t \in I(T)$.

TNet (along with SharpTNI) is based on finding ancestral host assignments that minimize the number of inter-host transmission events on T. The utility of such parsimonious ancestral host assignment for transmission network inference when multiple strain sequences per host are available was first systematically demonstrated by Romero-Severson et al. [16] and later developed further by Wymant et al. [23] in their phyloscanner method. The basic computational problem under this formulation can be stated as follows:

Problem 1 (Optimal ancestral host assignment). *Given a transmission phylogeny T on strain sequences sampled from a set $H = \{h_1, h_2, \ldots, h_n\}$ of n infected hosts, compute $h(t)$ for each $t \in I(T)$ such that the number of edges $(t', t'') \in E$ for which $h(t') \neq h(t'')$ is minimized.*

Problem 1 is equivalent to the well-known small parsimony problem in phylogenetics and can be solved efficiently using the classical Fitch [8] and Sankoff [17] algorithms. In TNet, we solve a modified version of the problem above that considers all possible optimal ancestral host assignments and samples greedily among them to minimize the number of back-transmissions (or reinfection by a descendant disease strain). To accomplish this goal efficiently, TNet uses an extended version of Sankoff's algorithm.

3   Algorithmic Details

A primary methodological and algorithmic innovation responsible for the improved accuracy of TNet (and also of SharpTNI) is the explicit and principled consideration of variability in optimal ancestral host assignments. More precisely, TNet recognizes that there are often a very large number of distinct optimal ancestral host assignments and it samples the space of all optimal ancestral host assignments in a manner that preferentially preserves optimal ancestral host assignments (described in detail below). TNet then aggregates across these samples to compute a support value for each edge in the final transmission network. This approach is illustrated in Fig. 2. Thus, the core computational problem solved by TNet can be formulated as follows:

Definition 1 (Back-Transmission). *Given a transmission network N on n infected hosts $H = \{h_1, h_2, \ldots, h_n\}$, we say that there exists a back-transmission for host h_i if there exists a directed cycle containing h_i in N. The total number of back-transmissions implied by N equals the number of hosts with back-transmissions.*

Problem 2 (Minimum back-transmission sampling). *Given a transmission phylogeny T on strain sequences sampled from a set $H = \{h_1, h_2, \ldots, h_n\}$ of n infected hosts, let $\mathcal{O}$ denote the set containing all distinct ancestral host assignments for T. Further, let $\mathcal{O}'$ denote the subset of $\mathcal{O}$ that implies the fewest back-transmissions in the resulting transmission network. Compute an optimal ancestral host assignment from $\mathcal{O}'$ such that each element of $\mathcal{O}'$ has an equal probability of being computed.*

Observe that the actual number of optimal ancestral host assignments (both $\mathcal{O}$ and $\mathcal{O}'$) can grow exponentially in the number of hosts n. By addressing the sampling problem above instead, TNet seeks to efficiently account for

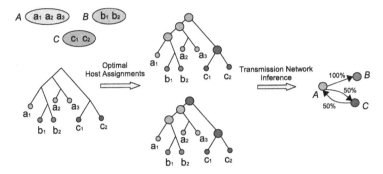

Fig. 2. Accounting for multiple optima in transmission network inference. The tree on the left depicts the transmission phylogeny for the seven strains sampled from three infected individuals A, B, and C, represented here by the three different colors. This tree admits two distinct optimal ancestral host assignments as shown in the figure. These two optimal ancestral host assignments can then be together used to infer a transmission network, as shown on the right, in which each edge has a support value. The support value of a transmission edge is define to be the percentage of optimal ancestral host assignments that imply that transmission edge.

the diversity within optimal ancestral host assignments with minimum back-transmissions, without explicitly having to enumerate them all.

Note that SharpTNI performs a similar sampling among all optimal ancestral host assignments, but employs a different optimality objective. Specifically, SharpTNI seeks to sample optimal ancestral host assignments that minimize the number of *co-transmissions*, i.e., minimize the number of inter-host edges in the transmission network.

3.1 Minimum Back-Transmission Sampling of Optimal Host Assignments

TNet approximates minimum back-transmission sampling by combining uniform sampling of ancestral host assignments with a greedy procedure to assign specific hosts to internal nodes. This is accomplished by suitably extending and modifying Sankoff's algorithm. This extended Sankoff algorithm computes, for each $t \in V(T)$ and $h_i \in H$, the number of distinct optimal host assignments for the subtree $T(t)$ under the constraint that $h(t) = h_i$, denoted by $N(t, h_i)$. After all $N(\cdot, \cdot)$ numbers have been computed, we perform our greedy sampling procedure using probabilistic backtracking. The basic idea is to perform a pre-order traversal of T and make final host assignment at the current node based on the number of optimal ancestral host assignments available for each optimal choice at that node, while preferentially preserving the parent host assignment. This is described in detail in Procedure *GreedyProbabilisticBacktracking* below.

Procedure *GreedyProbabilisticBacktracking*

1: Let $\alpha = \min_i\{C(rt(T), h_i)\}$.
2: **for** each $t \in I(T)$ in a pre-order traversal of T **do**
3: **if** $t = rt(T)$ **then**
4:   Let $X = \{h_i \in H \mid C(rt(T), h_i) = \alpha\}$.
5:   For each $h_i \in X$, assign $h(t) = h_i$ with probability $\frac{N(t,h_i)}{\sum_{h_j \in X} N(t,h_j)}$.
6: **if** $t \neq rt(T)$ **then**
7:   Let $X = \{h_i \in H \mid C(t, h_i) + p(h(pa(t)), h_i) \text{ is minimized}\}$.
8:   **if** $h(pa(t)) \in X$ **then**
9:    Assign $h(t) = h(pa(t))$.
10: **if** $h(pa(t)) \notin X$ **then**
11:   For each $h_i \in X$, assign $h(t) = h_i$ with probability $\frac{N(t,h_i)}{\sum_{h_j \in X} N(t,h_j)}$.

The procedure above preferentially assigns each internal node the same host assignment as that node's parent, if such an assignment is optimal. This strategy is based on the following straightforward observation: If the host assignment of an internal node t *could* be the same as that of its parent (while remaining optimal), i.e., $h(t) = h(pa(t))$ is optimal, then assigning a different optimal mapping $h(t) \neq h(pa(t))$ can result in a transmission edge back to $h(pa(t))$, effectively implying a reinfection of host $h(pa(t))$ by a descendant disease strain. Thus, the goal of TNet's sampling strategy is to strike a balance between sampling the diversity of optimal ancestral host assignments but avoiding sampling solutions with unnecessary back-transmissions.

3.2 Additional Methodological Details

Aggregation Across Multiple Optimal Ancestral Host Assignments. As illustrated in Fig. 2, aggregating across the sampled optimal ancestral host assignments can be used to improve transmission network inference by distinguishing between high-support and low-support transmission edges. Specifically, each directed edge in the transmission network can be assigned a support value based on the percentage of sampled optimal ancestral host assignments that imply that transmission edge. By executing TNet multiple times on the same transmission phylogeny (100 times per tree in our experimental study), these support values for edges can be estimated very accurately.

Accounting for Phylogenetic Inference Error. In addition to capturing the uncertainty of minimum back-transmission ancestral host assignments, which we show how to handle above, a second key source of inference uncertainty is phylogenetic error, i.e., errors in the inferred transmission phylogeny. Phyloscanner [23] accounts for such phylogenetic error by aggregating results across multiple transmission phylogenies (e.g., derived from different genomic regions of the samples strains, bootstrap replicates, etc.). We employ the same approach with TNet, aggregating the transmission network across multiple transmission phylogenies, in addition to the aggregation across multiple optimal ancestral host assignments per transmission phylogeny.

4   Datasets and Evaluation Methodology

Simulated Datasets. To evaluate the performance of TNet, SharpTNI, and phyloscanner, we generated a number of simulated viral transmission data sets across a variety of parameters. These datasets were generated using FAVITES [15], which can simultaneous simulate transmission networks, phylogenetic trees, and sequences. The simulated contact networks consisted of 1000 individuals, with each individual connected to other individuals through 100 outgoing edges preferentially attached to high-degree nodes using the Barabasi-Albert model [1]. On these contact networks, we simulated datasets with (i) four types of transmission networks using both Susceptible-Exposed-Infected-Recovered (SEIR) and Susceptible-Infected-Recovered (SIR) [12] models with two different infection rates for each, (ii) number of viruses sampled per host (5, 10, and 20), (iii) three different nucleotide sequence lengths (1000nt, 500nt, and 250nt), and (iv) three different rates of with-in host sequence evolution (normal, half, and double). This resulted in 560 different transmission network datasets representing 28 different parameter combinations. Further details on the construction and specific parameters used for these simulated datasets appear in [20]. These 560 simulated datasets had between 35 and 1400 sequences (i.e., leaves in the corresponding transmission phylogeny), with an average of 287.44 leaves. The maximum number of hosts per tree was 75, with an average of 26.72.

Data from Real HCV Outbreaks. We also evaluated the accuracies of TNet, SharpTNI, and phyloscanner on real datasets of HCV outbreaks made available by the CDC [19]. This collection consists 10 different datasets, each representing a separate HCV outbreak. Each of these outbreak data sets contains between 2 and 19 infected hosts and a few dozen to a few hundred strain sequences. The approximate transmission network is known for each of these datasets through CDC's monitoring and epidemiological efforts. In each of the 10 cases, this estimated transmission network consists of a single known host infecting all the other hosts in that network.

Evaluating Transmission Network Inference Accuracy. For all simulated and real datasets, we constructed transmission phylogenies using RAxML and used RAxML's own balanced rooting procedure to root them [21]. Note that TNet, SharpTNI, and phyloscanner all require rooted transmission phylogenies. To account for phylogenetic uncertainty and error, we computed 100 bootstrap replicates for each simulated and real dataset. For SharpTNI we used the efficient heuristic implementation for evaluation (not the exponential-time exact solution). All TNet results were based on aggregating across 100 sampled optimal host assignments per transmission phylogeny, and all SharpTNI results were aggregated across that subset of 100 samples that had minimum co-transmission cost, per transmission phylogeny. Results for all methods were aggregated across the different bootstrap replicates to account for phylogenetic uncertainty and yield edge-weighted transmission networks. To convert such edge-weighted transmission networks into unweighted transmission networks, we used the same 0.5 (or 50%) tree-support threshold used by phyloscanner in [23]. Thus, all directed edges with an edge-weight of at least 0.5 (or 50%) tree-support were retained in the final inferred transmission network and other edges were deleted. For a fair evaluation, none of the methods were provided with any epidemiological information such as sampling times or infection times. Finally, since both TNet and SharpTNI build upon uniform sampling procedures for optimal ancestral host assignments (minimizing the total number of inter-host transmissions), we also report results for uniform random sampling of optimal ancestral host assignments as a baseline.

To evaluate the accuracies of these final inferred transmission networks, we computed *precision* (i.e., the fraction of inferred edges in the transmission network that are also in the true network), *recall* (i.e., the fraction of true transmission network edges that are also in the inferred network), and *F1 scores* (i.e., harmonic mean of precision and recall).

5   Experimental Results

5.1   Simulated Data Results

Accuracy of Single Samples. We first considered the impact of inferring the transmission network using only a single optimal solution, i.e., without any aggregation across samples or bootstrap replicates. Figure 3 shows the results of this analysis. As the figure shows, TNet has by far the best overall accuracy, with precision, recall, and F1 scores of 0.72, 0.75, and 0.73, respectively. Phyloscanner showed the greatest precision at 0.828 but had significantly lower recall and F1 at 0.522 and 0.626, respectively. SharpTNI performed slightly better than a random optimal solution (uniform sampling), with precision, recall, and F1 scores of 0.68, 0.71, and 0.694, respectively, compared to 0.67, 0.71, and 0.687, respectively, for a randomly sampled optimal solution.

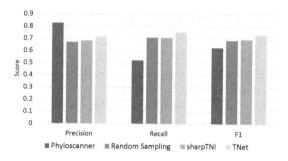

Fig. 3. Accuracy of methods using single samples. This figure plots precision, recall, and F1 scores for the different methods without any aggregation of results across multiple samples or bootstrap replicates. Results are averaged across the 560 simulated datasets.

Impact of Sampling Multiple Optimal Solutions on TNet and Sharp TNI. For improved accuracy, both TNet and SharpTNI rely on aggregation across multiple samples per transmission phylogeny. Note that, when aggregating across multiple optimal ancestral host assignments, the final transmission network is obtained by applying a cutoff for the edge support values. For example, in Fig. 2, at a cutoff threshold of 100%, only a single transmission from $(A \rightarrow B)$ would be inferred, while with a cutoff threshold of 50%, all three transmission edges shown in the figure would be inferred. We studied the impact of multiple sample aggregation by considering two natural sampling cutoff thresholds: 50% and 100%. As Fig. 4 shows, results improve as multiple optimal are considered. Specifically, for the 50% sampling cutoff threshold, we found that the overall accuracy of all methods improves as multiple samples are considered. For TNet, precision, recall, and F1 score all increase to 0.73, 0.75, and 0.74, respectively. For SharpTNI, precision and F1 score increase significantly to 0.76 and 0.72, respectively, while recall decreases slightly to 0.706. Surprisingly, we found that uniform random sampling outperformed SharpTNI, with precision, recall, and F1 score of 0.77, 0.70, and 0.73, respectively.

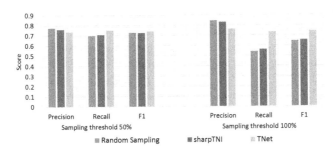

Fig. 4. Accuracy of methods using multiple samples on a single transmission phylogeny. This figure plots average precision, recall, and F1 scores for random sampling, sharpTNI, and TNet when 100 samples are used on a single transmission phylogeny. Values reported are averaged across all 560 simulated datasets, and results are shown for both 50% and 100% sampling cutoff thresholds.

We also see a clear tradeoff between precision and recall as the sampling cutoff threshold is increased. Specifically, for the 100% sampling cutoff threshold, the precision of all methods increases significantly, but overall F1 score falls to 0.65 and 0.64 for SharpTNI and random sampling, respectively. Surprisingly, recall only decreases slightly for TNet, and its overall F1 score remains 0.74 even for the 100% sampling cutoff threshold.

Accuracy on Multiple Bootstrapped Transmission Phylogenies. To further improve inference accuracy, results can be aggregated across the different bootstrap replicates to account for phylogenetic uncertainty. We therefore ran phyloscanner, TNet, and SharpTNI with 100 transmission phylogeny estimates (bootstrap replicates) per dataset. (We tested for the impact of using varying numbers of bootstrap replicates, trying 25, 50, and 100, but found that results were roughly identical in each case. We therefore report results for only the

100 bootstrap analyses.) As Fig. 5 shows, for the 50% sampling cutoff threshold, the accuracies of all methods improve over the corresponding single-tree results, with particularly notable improvements in precision. For the 100% sampling cutoff threshold, the precision of all methods improves further, but for phyloscanner and SharpTNI this comes at the expense of large reductions in recall. TNet continues to be best performing method overall for both sampling cutoff thresholds, with precision, recall, and F1 score of 0.79, 0.73, and 0.76, respectively, at the 50% sampling cutoff threshold, and 0.82, 0.71, and 0.754, respectively at the 100% sampling cutoff threshold.

Precision-Recall Characteristics of SharpTNI and TNet. The results above shed light on the differences between the sampling strategies (i.e, objective functions) used by SharpTNI and TNet, revealing that SharpTNI tends to have higher precision but much lower recall. Thus, depending on use case, either SharpTNI or TNet may be the method of choice. We also note that random sampling shows similar accuracy and precision-recall characteristics as SharpTNI, suggesting that SharpTNI may not offer much improvement over the much simpler random sampling strategy.

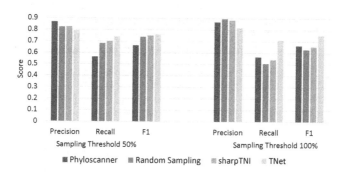

Fig. 5. Transmission network inference accuracy when multiple transmission phylogenies are used. This figure plots average precision, recall, and F1 scores for phyloscanner, random sampling, sharpTNI, and TNet when 100 bootstrap replicate transmission phylogenies are used for transmission network inference. Values reported are averaged across all 560 simulated datasets, and results are shown for both 50% and 100% sampling cutoff thresholds.

5.2 Real Data Results

We applied TNet, SharpTNI, and phyloscanner to the 10 real HCV datasets using 100 bootstrap replicates per dataset. We found that both TNet and SharpTNI performed almost identically on these datasets, and that both dramatically outperformed phyloscanner on the real datasets in terms of both precision and recall (and, consequently, F1 scores). Figure 6 shows these results averaged across the 10 real datasets. As the figure shows, both TNet and SharpTNI have identical F1 scores for the 50% and 100% sampling cutoff thresholds, with both methods showing F1 scores of 0.57 and 0.56, respectively. In contrast, phyloscanner shows much lower precision and recall, with an F1 score of only 0.22. Random sampling had slightly worse performance than TNet and SharpTNI at both the 50% and 100% sampling cutoff thresholds. At the 100% sampling cutoff threshold, we observe the same precision-recall characteristics seen in the simulated datasets, with SharpTNI showing higher precision but lower recall.

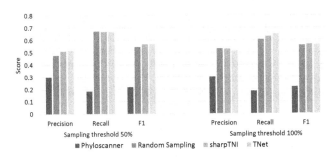

Fig. 6. Transmission network inference accuracy across the 10 real HCV datasets. This figure plots average precision, recall, and F1 scores for phyloscanner, random sampling, sharpTNI, and TNet on the 10 real HCV datasets with known transmission histories. Results are shown for both 50% and 100% sampling cutoff thresholds.

6   Discussion

In this paper, we introduced TNet, a new method for transmission network inference when multiple strain sequences are sampled from the infected hosts. TNet has two distinguishing features: First, it systematically accounts for variability among different optimal solutions to efficiently compute support values for individual transmission edges and improve transmission inference accuracy, and second, its objective function seeks to find those optimal host assignments that minimize the number of back-transmissions. TNet is based on a relatively simple parsimony-based formulation and is parameter-free and highly scalable. It can be easily applied within seconds to datasets with many hundreds of strain sequences and hosts. As our experimental results on both simulated and real datasets show, TNet is highly accurate and significantly outperforms phyloscanner. We find that TNet also outperforms SharpTNI, a distinct but very similar method developed independently and published recently.

Going forward, several aspects of TNet can be tested and improved further. The simulated datasets used in our experimental study assume that all infected hosts have been sampled. It would be useful to test how accuracy decreases as fewer and fewer infected hosts are sampled. Phyloscanner employs a simple technique to estimate if an ancestral host assignment may be to an unsampled host, and a similar technique could be used in TNet. Currently, TNet does not use branch lengths or overall strain diversity within hosts, and these could be used to further improve the accuracy of ancestral host assignment and transmission network inference. Finally, our results suggest that, despite their conceptual similarities, SharpTNI and TNet, show different precision-recall characteristics. It may be possible to meaningfully combine the objective functions of SharpTNI and TNet to create a more accurate hybrid method.

Acknowledgements. The authors wish to thank Dr. Pavel Skums (Georgia State University) and the Centers for Disease Control for sharing their HCV outbreak data. We also thank Samuel Sledzieski for creating and sharing the simulated transmission network datasets used in this work.

Funding. This work was supported in part by NSF award CCF 1618347 to IM and MSB.

References

1. Albert, R., Barabási, A.L.: Statistical mechanics of complex networks. Rev. Mod. Phys. **74**, 47–97 (2002). https://doi.org/10.1103/RevModPhys.74.47
2. Clutter, D., et al.: Trends in the molecular epidemiology and genetic mechanisms of transmitted human immunodeficiency virus type 1 drug resistance in a large US clinic population. Clin. Infect. Dis. **68**(2), 213–221 (2018). https://doi.org/10.1093/cid/ciy453
3. De Maio, N., Worby, C.J., Wilson, D.J., Stoesser, N.: Bayesian reconstruction of transmission within outbreaks using genomic variants. PLoS Comp. Biol. **14**(4), 1–23 (2018). https://doi.org/10.1371/journal.pcbi.1006117
4. De Maio, N., Wu, C.H., Wilson, D.J.: Scotti: Efficient reconstruction of transmission within outbreaks with the structured coalescent. PLoS Comp. Biol. **12**(9), 1–23 (2016). https://doi.org/10.1371/journal.pcbi.1005130
5. Didelot, X., Fraser, C., Gardy, J., Colijn, C., Malik, H.: Genomic infectious disease epidemiology in partially sampled and ongoing outbreaks. Mol. Biol. Evol. **34**(4), 997–1007 (2017). https://doi.org/10.1093/molbev/msw275
6. Domingo, E., Holland, J.: RNA virus mutations and fitness for survival. Annu. Rev. Microbiol. **51**, 151–178 (1997)
7. Drake, J.W., Holland, J.J.: Mutation rates among RNA viruses. Proc. Natl. Acad. Sci. USA **96**(24), 13910–13913 (1999)
8. Fitch, W.: Towards defining the course of evolution: minimum change for a specified tree topology. Syst. Zool. **20**, 406–416 (1971)
9. Glebova, O., Knyazev, S., Melnyk, A., Artyomenko, A., Khudyakov, Y., Zelikovsky, A., Skums, P.: Inference of genetic relatedness between viral quasispecies from sequencing data. BMC Genom. **18**(suppl. 10), 918 (2017)
10. Grulich, A., et al.: A10 Using the molecular epidemiology of HIV transmission in New South Wales to inform public health response: Assessing the representativeness of linked phylogenetic data. Virus Evol. **4**(suppl. 1), April 2018. https://doi.org/10.1093/ve/vey010.009
11. Hall, M., Woolhouse, M., Rambaut, A.: Epidemic reconstruction in a phylogenetics framework: transmission trees as partitions of the node set. PLoS Comp. Biol. **11**(12), e1004613 (2015). https://doi.org/10.1371/journal.pcbi.1004613

12. Kermack, W.O., McKendrick, A.G., Walker, G.T.: A contribution to the mathematical theory of epidemics. Proc. Roy. Soc. Lond. Ser. A **115**(772), 700–721 (1927). https://doi.org/10.1098/rspa.1927.0118

13. Klinkenberg, D., Backer, J.A., Didelot, X., Colijn, C., Wallinga, J.: Simultaneous inference of phylogenetic and transmission trees in infectious disease outbreaks. PLoS Comp. Biol. **13**, 1–32 (2017). https://doi.org/10.1371/journal.pcbi.1005495

14. Kosakovsky Pond, S.L., Weaver, S., Leigh Brown, A.J., Wertheim, J.O.: HIV-TRACE (TRAnsmission Cluster Engine): a tool for large scale molecular epidemiology of HIV-1 and other rapidly evolving pathogens. Mol. Biol. Evol. **35**(7), 1812–1819 (2018). https://doi.org/10.1093/molbev/msy016

15. Moshiri, N., Wertheim, J.O., Ragonnet-Cronin, M., Mirarab, S.: FAVITES: simultaneous simulation of transmission networks, phylogenetic trees and sequences. Bioinformatics **35**(11), 1852–1861 (2019). https://doi.org/10.1093/bioinformatics/bty921

16. Romero-Severson, E.O., Bulla, I., Leitner, T.: Phylogenetically resolving epidemiologic linkage. Proc. Natl. Acad. Sci. **113**(10), 2690–2695 (2016). https://doi.org/10.1073/pnas.1522930113

17. Sankoff, D.: Minimal mutation trees of sequences. SIAM J. Appl. Math. **28**(1), 35–42 (1975). http://www.jstor.org/stable/2100459

18. Sashittal, P., El-Kebir, M.: SharpTNI: counting and sampling parsimonious transmission networks under a weak bottleneck. bioRxiv (2019). https://doi.org/10.1101/842237

19. Skums, P., et al.: QUENTIN: reconstruction of disease transmissions from viral quasi species genomic data. Bioinformatics **34**(1), 163–170 (2018). https://doi.org/10.1093/bioinformatics/btx402

20. Sledzieski, S., Zhang, C., Mandoiu, I., Bansal, M.S.: TreeFix-TP: phylogenetic error-correction for infectious disease transmission network inference. bioRxiv (2019). https://doi.org/10.1101/813931

21. Stamatakis, A.: RAxML version 8: a tool for phylogenetic analysis and post-analysis of large phylogenies. Bioinformatics **30**(9), 1312–1313 (2014). https://doi.org/10.1093/bioinformatics/btu033

22. Steinhauer, D., Holland, J.: Rapid evolution of RNA viruses. Annu. Rev. Microbiol. **41**, 409–433 (1987)

23. Wymant, C., et al.: PHYLOSCANNER: inferring transmission from within- and between-host pathogen genetic diversity. Mol. Biol. Evol. **35**(3), 719–733 (2017). https://doi.org/10.1093/molbev/msx304

Cancer Breakpoint Hotspots Versus Individual Breakpoints Prediction by Machine Learning Models

Kseniia Cheloshkina[1], Islam Bzhikhatlov[2], and Maria Poptsova[1]([✉])

[1] Laboratory of Bioinformatics, Faculty of Computer Science, National Research University Higher School of Economics, 11 Pokrovsky boulvar, Moscow 101000, Russia
mpoptsova@hse.ru
[2] Faculty of Control Systems and Robotics, ITMO University,
49 Kronverksky Pr., St. Petersburg 197101, Russia

Abstract. Genome rearrangement is a hallmark of all cancers. Cancer breakpoint prediction appeared to be a difficult task, and various machine learning models did not achieve high prediction power. We investigated the power of machine learning models to predict breakpoint hotspots selected with different density thresholds and also compared prediction of hotspots versus individual breakpoints. We found that hotspots are considerably better predicted than individual breakpoints. While choosing a selection criterion, the test ROC AUC only is not enough to choose the best model, the lift of recall and lift of precision should be taken into consideration. Investigation of the lift of recall and lift of precision showed that it is impossible to select one criterion of hotspot selection for all cancer types but there are three to four distinct groups of cancer with similar properties. Overall the presented results point to the necessity to choose different hotspots selection criteria for different types of cancer.

Keywords: Cancer genome rearrangements · Cancer breakpoints · Cancer breakpoint hotspots · Machine learning · Random forest

1 Introduction

Cancer genome rearrangement is a hallmark of all cancers and hundreds of thousands of cancer breakpoints has been documented for different types of cancers [1–3]. Heterogeneity of cancer mutations has been noticed long ago [4] and the accumulated data on cancer genome mutations was termed as cancer genome landscapes [5]. Thousands of cancer full genome data became available to researchers by the International Cancer Genome Consortium (ICGC) [6]. Later, instead of individual cancer genomes a notion of pan cancer has emerged [7, 8] revealing common and individual properties of cancer genome mutations. Recently, the Pan-Cancer Analysis of Whole Genomes (PCAWG) Consortium [9] of the International Cancer Genome Consortium (ICGC) [6] and The Cancer Genome Atlas (TCGA) [10] reported the integrative analysis of more than 2,500

© Springer Nature Switzerland AG 2020
Z. Cai et al. (Eds.): ISBRA 2020, LNBI 12304, pp. 217–228, 2020.
https://doi.org/10.1007/978-3-030-57821-3_19

whole-cancer genomes across 38 tumour types [11]. Apart from [11–13] point muta-tions, a genome rearrangement with creation of structural elements is often an early event in cancer evolution sometimes preceding point mutation accumulation.

Machine learning methods were successful in finding regularities in studying cancer genome mutations. The most successful machine learning models was shown to be in predicting densities of somatic mutations [14, 15]. In [14] the density of somatic point mutations was predicted by densities of HDNase and histone modifications with determination coefficient of 0.7–0.8. The relative contribution of non-B DNA structures and epigenetic factors in predicting the density of cancer point mutations was studied in [15]. It was shown that taking both groups of factors into account increased prediction power of models.

Despite success in prediction of densities of somatic point mutations, cancer break-point prediction models showed low or moderate power [15, 16]. This fact could be explained both by the lack of causal determinants in the models and constrains of the machine learning algorithms.

Previously we showed that the breakpoint density distribution varies across different chromosomes in different cancer types [16]. Also, we showed that determination of hotspot breakpoints depends on a threshold, and the choice of the threshold could vary between different types of cancer and could affect the results of machine learning models. Here we aimed at conducting a systematic study of how breakpoint hotspots density thresholds influence prediction power of machine learning models. We also posed a question whether the prediction power of machine learning models will be different whether we predict individual breakpoints or breakpoint hotspots.

2   Methods

2.1   Data

Data on cancer breakpoints were downloaded from the International Cancer Genome Consortium (ICGC) [6]. The dataset comprises more than 652 000 breakpoints of 2803 samples from more than 40 different types cancers that we grouped in 10 groups of cancer according to tissue types and further refer as cancer types. We cut the genome into non-overlapping windows of 100 KB of length and excluded regions from centromeres, telomeres, blacklisted regions and Y chromosome. Then for each window we estimated breakpoint density as the ratio of the number of breakpoints in the window to the total number of breakpoints in a given chromosome. We used the density metric to designate hotspots, i.e. genomic regions with a relatively high concentration of breakpoints. In the study, we investigate three labeling types of hotspots - 99%, 99.5% and 99.9% percentiles of breakpoint density distribution. Besides, we assigned "individual breakpoints" label to windows containing at least one breakpoint. The proportion of the number of these windows from the total number of windows varied from 2.8% to 90% for different cancer types.

In the study we used the most comprehensive set of predictors, available as of today mostly from next-generation sequencing experiments. The features include genomic regions, TAD boundaries, secondary structures, transcription factor binding sites and a set of epigenetic factors (chromatin accessibility, histone modifications, DNA methylation). These data were collected from The Encode, DNA Punctuation, Non-B DB projects, UCSC Genome Browser. The data were transformed into feature vectors by calculating window coverage of each characteristic.

2.2 Machine Learning Models

After data collection we got 30 datasets that comprise genomic features' coverage and binary target labeling (3 hotspots labeling with different quantile threshold of 99%, 99.5% and 99.9% per each of 10 cancer types).

The hotspot prediction power was evaluated through the train-test splits with stratification by a chromosome with proportion of 70–30, retaining 30% of data for testing. To get a reliable estimate of quality metrics we performed train-test splits 30 times for each dataset because of high class-imbalance (very small ratio of positive examples). For this reason, we also applied the class balancing technique (oversampling) when training a machine learning model. We selected Random Forest as one of the most performing and popular classification algorithm for table data to assess hotspots and breakpoints prediction power. For the model we estimated the best hyperparameters (the number of trees, number of features to grow a tree, minimal number of examples in a terminal node, maximal number of nodes in a tree) by averaging performance metrics among all cancer types.

2.3 Evaluation Metrics

We used several metrics for model evaluation: ROC AUC (Area Under Receiver Operating Curve), precision, recall, lift of precision and lift of recall. As we were dealing with a high class imbalance for each dataset we averaged the results from 30 random train-test splits by taking mean (or median) ROC AUC on the test set and controlling for its standard deviation, which demonstrates how strongly the results depend on the distribution of examples in train and test sets. We calculated recall and precision for different probability percentiles – from 0.5% to 50%. To get an estimate of how well a model performs in comparison with a random choice we used the lift of recall and lift of precision. The lift of recall for a given probability percentile shows how many times the recall of the model (estimated on examples labeled as the positive class according to a probability percentile threshold) is higher than a random choice (it is equal to the recall of the model divided by the probability percentile). Similarly, the lift of precision demonstrates how many times precision for a given probability percentile is higher than a random choice (equal to the precision of the model divided by the proportion of positive examples in a dataset).

3   Results

3.1 Distribution of Test and Train ROC AUC for All Cancer Types by Hotspot Labeling Type

We train Random Forest model on all 30 datasets for hotspot prediction and 10 datasets for individual breakpoint prediction. The distribution of ROC AUC on test set by cancer type and labeling type is given in Fig. 1. It could be seen that for the half of cancer types including blood, brain, breast, pancreatic and skin cancer the higher the hotspot labeling threshold the higher the median of test ROC AUC. For bone, liver, uterus cancer there is no monotonically increasing median test quality but for the highest labeling type this value is higher than for the lowest while for the rest of cancers (ovary and prostate) there is no significant difference between labeling types.

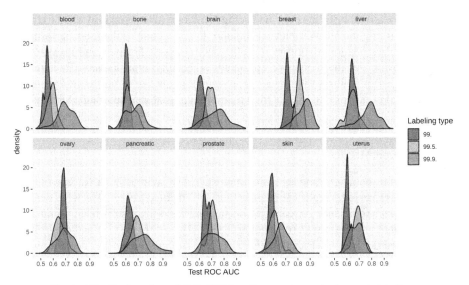

Fig. 1. Distribution of test ROC AUC for all cancer types by hotspots labeling.

Median values of the test ROC AUC by cancer type and labeling type are presented in Fig. 2 and Table 1. The highest quality in terms of considered metric belongs to the breast cancer for all labeling types while the lowest – to the blood and skin cancer for 99% labeling type. The difference greater than 0.10 between the median test ROC AUC for models of the 99% and 99.9% hotspot labeling types is observed for the blood, brain, breast, liver and pancreatic cancer. For the half of the cancer types the highest labeling type implies significantly higher quality according to the median test ROC AUC.

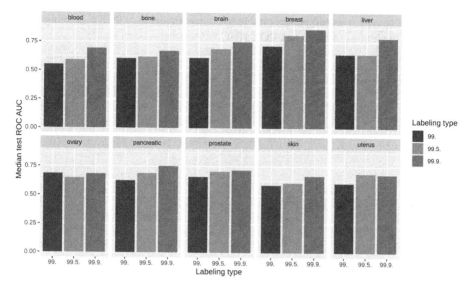

Fig. 2. Median test ROC AUC for each cancer and labeling type.

Table 1. Median test ROC AUC for each cancer and labeling type.

Cancer type	Median test ROC AUC (99%)	Median test ROC AUC (99.5%)	Median test ROC AUC (99.9%)
Blood	0,552	0,593	0,693
Bone	0,606	0,621	0,673
Brain	0,612	0,689	0,75
Breast	0,715	0,81	0,861
Liver	0,645	0,646	0,784
Ovary	0,684	0,648	0,683
Pancreatic	0,625	0,689	0,752
Prostate	0,659	0,706	0,717
Skin	0,587	0,607	0,668
Uterus	0,603	0,689	0,68

On the other hand, as it could be seen in Fig. 3, the more rare hotspots we aim to predict the higher the variance of the test ROC AUC on the test set as well as the difference between the train and test ROC AUC. This could be explained by the fact that for the case of rare hotspots there is small number of positive examples in a dataset and its random permutation between the train and test set leads to different results. Moreover, for all cancer types except for the breast cancer difference between the median train and test ROC AUC for the 99.9% labeling type approaches 0.2 ROC AUC and is 2–3 times higher than for the 99.5% and 99% labeling types. Hence, when selecting the best hotspot labeling type it would be reasonable to choose the 99% or 99.5% labeling type according to the highest median test ROC AUC.

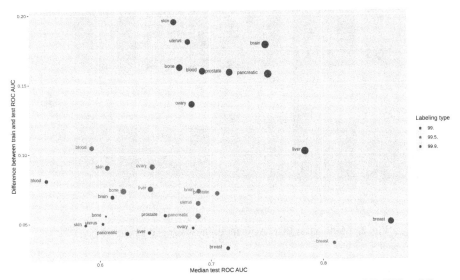

Fig. 3. Dependence of the difference of train and test ROC AUC from test ROC AUC at different labeling types and different standard deviations of test ROC AUC.

3.2 Lift of Precision and Lift of Recall

Next we analyzed the distribution of other quality metrics such as the lift of recall and lift of precision. The results are presented in Fig. 4 and 5 respectively. Here confidence intervals for the mean of these metrics are plotted against different probability quantiles selected as a threshold for model predictions for each cancer and hotspot labeling type. The main conclusion that could be made according to these results is that there is no single labeling type which guarantees the best classification results for all cancer types. However, three groups of cancer types were distinguished: the best labeling type for the blood, brain, liver and pancreatic cancers is 99.9%, for the bone, breast and uterus cancers - 99.5%, for the rest (ovary, prostate, skin cancers) - 99%.

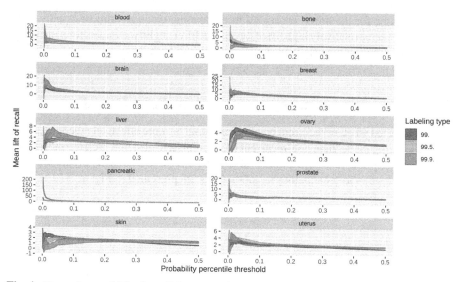

Fig. 4. Dependence of lift of recall from quantile threshold for different aggregation levels (see text for explanation).

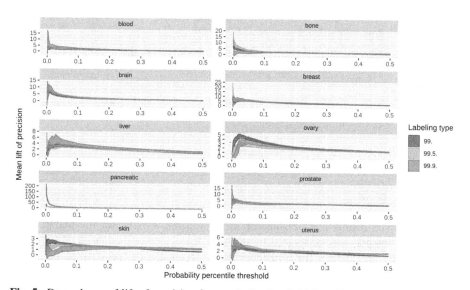

Fig. 5. Dependence of lift of precision from quintile threshold for different aggregation levels (see text for explanation).

When comparing the best labeling type determined with the median test ROC AUC and with the lift of recall/precision, it is the same only for ovary, bone and uterus cancer. As we are mainly interested in selection of minimal number of genome regions with the maximal concentration of hotspots, the final choice of the best labeling type will coincide with the decision according to the lift of recall/precision.

Interestingly, for the breast cancer all three labeling types are almost equally well predicted: they have relatively high lift of recall and differ slightly. Besides, for pancreatic cancer 99.9% labeling type showed significant boost in both lift of precision and lift of recall.

3.3 Prediction of Hotspot Breakpoints Versus Individual Breakpoints

Further, we tested whether recurrent breakpoints could be more effectively recognized by machine learning model than non-recurrent breakpoints and also how well the individual breakpoints are predicted. Distributions of test ROC AUC for hotspots prediction (the best labeling type) and breakpoint prediction tasks are presented in Fig. 6.

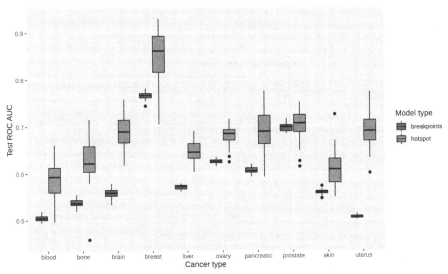

Fig. 6. Distribution of test ROC AUC for the best labeling hotspot profile and all breakpoints prediction.

The observed results can be summarized as follows. Firstly, for the majority of cancer types hotspots are recognized by machine learning models considerably better than individual breakpoints for all cancer types except for the prostate cancer. The quantitative estimate of the difference is given in Fig. 7 and Table 2. The highest ratio of the median test ROC AUC for hotspot prediction model to the median test ROC AUC for breakpoint prediction model is observed for the uterus and brain cancer (1.36 and 1.23 respectively) while for the prostate cancer they are almost equal. For the other cancer types the metric for hotspot model is 9–18% higher than for the breakpoints. Thus, in general, breakpoints are harder to recognize than hotspots using the same genomic features.

Also it could be seen that variance of ROC AUC is significantly lower for breakpoints and this could be a consequence of having a considerably higher number of positive examples for the model.

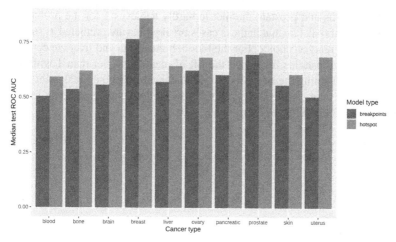

Fig. 7. Median test ROC AUC for best labeling hotspot profile and breakpoints prediction.

Table 2. Median test ROC AUC for best labeling hotspot profile and breakpoints prediction.

Cancer type	Breakpoints median test ROC AUC	Hotspots median test ROC AUC	Ratio of median test ROC AUC for hotspots and breakpoints prediction models
Prostate	0,698	0,706	1,011
Skin	0,559	0,607	1,087
Ovary	0,625	0,684	1,095
Breast	0,766	0,861	1,124
Liver	0,572	0,645	1,128
Pancreatic	0,604	0,689	1,14
Bone	0,537	0,621	1,157
Blood	0,505	0,593	1,175
Brain	0,559	0,689	1,234
Uterus	0,505	0,689	1,363

Secondly, the quality of breakpoint prediction is quite low so that it is a difficult task to predict cancer breakpoints by a machine learning model. For 6 cancer types including skin, liver, bone, blood, brain and uterus cancer the median test ROC AUC does not exceed 0.6. In contrast, the highest value of the metric (0.77) is achieved for breast cancer.

The conclusion is confirmed by the statistics of the lift of recall given in Fig. 8. All cancer types could be divided into 2 groups. For the pancreatic and skin cancer breakpoints are unrecognizable as the lift of recall is very low (almost equal to zero). For the ovary, breast, uterus and prostate the metric hardly achieves 1 for the probability percentile threshold of 0–0.1 so that in these cases breakpoints are predicted as successfully as in the case of a random choice. For the blood, bone, brain and liver cancers there are some probability thresholds for which the lift of recall is higher than 1 with the brain cancer model performing the best. In total, for 6 cancer types the breakpoint prediction model quality does not significantly differ from a random choice and only for 4 cancer types the prediction is slightly better.

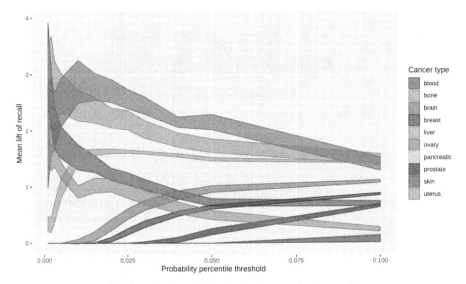

Fig. 8. Lift of recall for breakpoints prediction model

Additionally, it should be noted that for the blood, bone and brain cancer the lift of recall decreases with probability threshold. This could mean that for these cancer types some breakpoints are highly pronounced and could be more easily identified than the rest of breakpoints.

Besides, a set of cancer type models achieving the best performance for the task of breakpoint prediction according to the median ROC AUC (prostate, ovary, breast) differs from a set determined by the lift of recall (blood, bone, brain and liver). This difference outlines the fact that it is very important to choose the right performance metric for a given machine learning task. As the ROC AUC measures a quality of overall examples' ordering produced by the model and the lift of recall measures ordering quality of examples with the highest probabilities, they describe the model performance from different perspectives.

4   Conclusions and Discussion

In this study, using machine learning approach we systematically investigated the effect of different selection criteria for cancer breakpoint predictions for 10 large types of cancer. We built machine-learning models predicting hotspot breakpoints defined by different density thresholds and investigated distributions of the train and test ROC AUC as well as the lift of precision and recall. Almost for all types of cancer the median test ROC AUC increases with an increase of threshold for hotspot selection, though the total quantity of those regions decreases and variance of quality metrics grows up. This fact confirms the fact that the machine learning models better recognize regions with increased density of breakpoint mutations, or regions with recurrent breakpoints, which requires further research. This result could be considered as expected however we empirically found an exception for prostate cancer where median test ROC AUC for hotspots and individual breakpoint do not differ much both being close to 0.70. This suggests that mutagenic processes of individual and recurrent breakpoints in prostate cancer most likely have similar nature. However the effect needs further investigation.

We would like to emphasize that, without actual tests of machine-learning performance on individual breakpoints and breakpoint hotspots, it is not evident, which one of the two will have a higher prediction power. Indeed, breakpoint hotspots are regions enriched with breakpoints, and genomic features of these regions should explain their recurrent formation. On the opposite, rare events are often harder to predict except for the cases when strong predictors of a rare event are available. In the research we posed a question whether the considered genome features identify breakpoints hotspots better than individual breakpoints, and whether individual breakpoints have also distinctive features that influence their formation. The comprehensive analysis of the features is out of scope of the present study is the subject of further systematic research.

The lift of recall and lift of precision signify how many times recall or precision is higher compared to a random choice. Analysis of the distributions of the lift of recall and lift of precision showed that it is impossible to choose one breakpoint density threshold that would lead to the maximum prediction power of models for all types of cancer. Three groups of cancer with similar behavior according to the lift of recall and lift of precision were distinguished. Common properties of cancer breakpoint formation in these three groups of cancer require further investigations.

Selection criteria for the best hotspot labeling threshold based on the median test ROC AUC and the lift of recall and precision coincide only for three types of cancer – ovary, bone, uterus. Moreover, concerning evaluation of prediction power of breakpoints these metrics produce different results. As the ROC AUC and lift of recall measure quality of examples' ordering by the model at different scales (based on all examples and examples with the highest probabilities respectively) we recommend to use the lift of recall and lift of precision metrics to choose the hotspot thresholds.

Comparison of breakpoint predictions and breakpoint hotspots with a chosen selection criterion based on the best machine-learning model showed that the median test AUC is always higher for hotspots rather than for individual breakpoints. We tested additionally several machine learning models such as logistic regression and XGBoost (results are not presented here due to the paper size limitation) and they all show approximately the same relative distribution of ROC AUC, lift of recall and lift of precision across different types of cancer.

Overall the results of our study showed that machine learning model prediction power depends on density threshold for cancer hotspots, and the threshold is different for different types of cancer. Besides, we demonstrated that though individual breakpoints are harder to predict than breakpoint hotspots, individual breakpoints can be predicted to a certain extent, and, moreover, in prostate cancer they are predicted equally well as hotspots. While choosing a selection criterion, the test ROC AUC only is not enough to choose the best model, the lift of recall and lift of precision should be taken into consideration at the level of individual type of cancer.

References

1. Harewood, L., et al.: Hi-C as a tool for precise detection and characterisation of chromosomal rearrangements and copy number variation in human tumours. Genome Biol. **18**, 125 (2017)
2. Nakagawa, H., Fujita, M.: Whole genome sequencing analysis for cancer genomics and precision medicine. Cancer Sci. **109**, 513–522 (2018)
3. Nakagawa, H., Wardell, C.P., Furuta, M., Taniguchi, H., Fujimoto, A.: Cancer whole-genome sequencing: present and future. Oncogene **34**, 5943–5950 (2015)
4. Salk, J.J., Fox, E.J., Loeb, L.A.: Mutational heterogeneity in human cancers: origin and consequences. Annu. Rev. Pathol. **5**, 51–75 (2010)
5. Vogelstein, B., Papadopoulos, N., Velculescu, V.E., Zhou, S., Diaz Jr., L.A., Kinzler, K.W.: Cancer genome landscapes. Science **339**, 1546–1558 (2013)
6. International Cancer Genome Consortium (ICGC). https://icgc.org/
7. Zhang, K., Wang, H.: Cancer genome atlas pan-cancer analysis project. Zhongguo Fei Ai Za Zhi **18**, 219–223 (2015)
8. Cancer Genome Atlas Research, N., et al.: The cancer genome atlas pan-cancer analysis project. Nat. Genet. **45**, 1113–1120 (2013)
9. Pancancer Analysis of Whole Genomes (PCAWG). https://dcc.icgc.org/pcawg
10. The Cancer Genome Atlas (TCGA). https://www.cancer.gov/about-nci/organization/ccg/res earch/structural-genomics/tcga
11. Consortium, I.T.P.-C.A.o.W.G.: Pan-cancer analysis of whole genomes. Nature **578**, 82–93 (2020)
12. Javadekar, S.M., Raghavan, S.C.: Snaps and mends: DNA breaks and chromosomal translocations. FEBS J. **282**, 2627–2645 (2015)
13. Li, Y., et al.: Patterns of somatic structural variation in human cancer genomes. Nature **578**, 112–121 (2020)
14. Polak, P., et al.: Cell-of-origin chromatin organization shapes the mutational landscape of cancer. Nature **518**, 360–364 (2015)
15. Georgakopoulos-Soares, I., Morganella, S., Jain, N., Hemberg, M., Nik-Zainal, S.: Noncanonical secondary structures arising from non-B DNA motifs are determinants of mutagenesis. Genome Res. **28**, 1264–1271 (2018)
16. Cheloshkina, K., Poptsova, M.: Tissue-specific impact of stem-loops and quadruplexes on cancer breakpoints formation. BMC Cancer **19**, 434 (2019)

Integer Linear Programming Formulation for the Unified Duplication-Loss-Coalescence Model

Javad Ansarifar[1], Alexey Markin[1], Paweł Górecki[2], and Oliver Eulenstein[1(✉)]

[1] Department of Computer Science, Iowa State University, Ames, USA
{javad,amarkin,oeulenst}@iastate.edu
[2] Faculty of Mathematics, Informatics and Mechanics, University of Warsaw, Warsaw, Poland
gorecki@mimuw.edu.pl

Abstract. The classical Duplication-Loss-Coalescence parsimony model (DLC-model) is a powerful tool when studying the complex evolutionary scenarios of simultaneous duplication-loss and deep coalescence events in evolutionary histories of gene families. However, inferring such scenarios is an intrinsically difficult problem and, therefore, prohibitive for larger gene families typically occurring in practice. To overcome this stringent limitation, we make the first step by describing a non-trivial and flexible Integer Linear Programming (ILP) formulation for inferring DLC evolutionary scenarios. To make the DLC-model more practical, we then introduce two sensibly constrained versions of the model and describe two respectively modified versions of our ILP formulation reflecting these constraints. Using a simulation study, we showcase that our constrained ILP formulation computes evolutionary scenarios that are substantially larger than the scenarios computable under our original ILP formulation and DLCPar. Further, scenarios computed under our constrained DLC-model are overall remarkably accurate when compared to corresponding scenarios under the original DLC-model.

Keywords: Phylogenetics · Duplications · Losses · Coalescence · Reconciliation

1 Introduction

Reconstructing evolutionary histories of gene families, or gene trees, is of central importance for the understanding of gene and protein function. Gene trees make comparative and investigative studies possible that illuminate relationships between the structure and function among orthologous groups of genes, and are an indispensable tool for assessing the functional diversity and specificness of biological interlinkage for genes within the same family [1,9,11,15,16].

Crucial for understanding evolutionary histories of gene families (gene trees) is contemplating them against a respective species phylogeny; i.e., the evolutionary history of species that host(ed) the genes under consideration. This approach

© Springer Nature Switzerland AG 2020
Z. Cai et al. (Eds.): ISBRA 2020, LNBI 12304, pp. 229–242, 2020.
https://doi.org/10.1007/978-3-030-57821-3_20

is known as *gene tree reconciliation*, and it can directly reveal the most valuable points of interest, such as (i) *gene duplication* events, (ii) *gene loss* events, and (iii) *deep coalescence* or *incomplete lineage sorting* events (which appear as a result of a genetic polymorphism surviving speciation).

Traditional tree reconciliation approaches, while computationally efficient, are rather limited in practice, as they either only account for duplication and loss events, or, on the other hand, only for deep coalescence events [7,12,19]. Beyond the traditional approaches, recently, a robust unified *duplication-loss-coalescence (DLC)* approach has been developed that simultaneously accounts for duplications, losses, and deep coalescence events. In particular, Rasmussen and Kellis [17] originally developed a rigorous statistical model referred to as *DLCoal*. Then a computationally more feasible parsimony framework, which we refer to here as *DLC-model* was developed by Wu et al. [20]. That is, DLC-model is a discrete version of the DLCoal model, and it was shown to be very effective in practice in terms of identification of ortholog/paralog relations and accurate inference of the duplication and loss events. Wu et al. additionally presented an optimized strategy for enumerating possible reconciliation scenarios and a dynamic programming solution to find the optimum reconciliation cost; this algorithm is known as *DLCPar*.

While it has been demonstrated that DLC-model is computationally more feasible when compared to DLCoal, the exact DLCPar algorithm is still only applicable to reconciliation problems involving less than 200 genes. Limiting evolutionary studies to such a small number of genes is highly restrictive in practice, where frequently gene families with thousands of genes and hundreds of host species appear [10]. Further, the DLCPar algorithm is not scalable due to its exponential runtime [3]. Naturally, there is a demand for novel models that are (i) efficiently computable and (ii) comparable to DLCPar in terms of its accuracy.

In this work, we present a non-trivial and flexible *integer linear programming (ILP)* formulation of the DLC-model optimization problem. Then we formulate two novel and constrained DLC-models, and use our ILP formulation to validate these constrained models. That is, our models have smaller solution space and, therefore, are more efficiently computable than the original DLC-model. The validation is performed via a comprehensive simulation study with realistic parameters derived from a real-world dataset. The simulations demonstrate that both our models are applicable to larger datasets than DLCPar. Moreover, one of the models, despite the constraints, almost always provides the same reconciliation cost as the unconstrained algorithms.

Related Work. In recent years, there has been an increased interest in phylogenetic methods involving simultaneous modeling of duplication, loss, and deep coalescence events [6,18]. For example, recently, an approach for *co-estimation* of the gene trees and the respective species tree based on the DLCoal model was presented [5]. Further, Chen et al. [2] presented a parsimony framework for the reconciliation of a gene tree with a species tree by simultaneously modeling

DLC events as well as horizontal gene transfer events. While promising, their approach remains computationally challenging.

Note that to the best of our knowledge, no models were proposed that would be more efficiently computable than DLC-model but be comparable with it in terms of effectiveness.

Our Contribution. We developed a flexible ILP formulation that solves the DLCPar optimization problem. During the development of this formulation, we observed formal issues with the original definition of the DLC-model in [20]. Consequently, in this work, we also present corrected and improved model definitions, which are equivalent to the Wu et al. model. For example, we corrected problems with the definition of a partial order on gene tree nodes, which could otherwise lead to incorrect scoring of deep coalescence events (see Sect. 2 for the full updated model definitions).

Further, the ILP formulation enabled us to test the viability of a *constrained DLC-model*, which we present in this work. In particular, we observed that the advanced time complexity of DLCPar originates from allowing the duplications to appear at any edge of the gene tree, even if there is no direct "evidence" for such occurrences. While this flexibility allows accounting for all feasible DLC scenarios, we show that constraining the duplication locations to those with direct evidence of duplications will enable one to dramatically improve the efficiency of computing optimum reconciliations (without losing the accuracy).

To study the performance of the ILP formulation and test our constrained models, we designed a coherent simulation study with parameters derived from the 16 fungi dataset [17], which became a standard for multi-locus simulations [4,14,20]. We compared the runtimes of the unconstrained ILP (DLCPar-ILP), the constrained ILPs, and the DLCPar algorithm by Wu et al. While we observed that DLCPar was generally faster than DLCPar-ILP there were multiple instances where DLCPar-ILP was able to compute optimum reconciliations, whereas DLCPar failed. Out of 30 instances, when DLCPar failed, DLCPar-ILP was able to provide an optimum in 17 cases. Therefore, we suggest using those two methods as complements of each other. Further, an advantage of using ILPs, is that one can terminate an ILP solver early, but still get a good approximation of the optimum reconciliation cost due to the intricate optimization algorithms used by ILP solvers.

Finally, the constrained ILP models proved to be efficient even on larger datasets with more than 200 genes, where DLCPar and DLCPar-ILP failed. Moreover, we observed that one of our constrained models was accurate in 98.17% of instances.

2 Model Formulation

We use definitions and terminology similar to [20], but modify them for improved clarity and correctness.

A (phylogenetic) tree $T = (V(T), E(T))$ is a rooted binary tree, where $V(T)$ and $E(T)$ denote the set of nodes and the set of directed edges (u, v), respectively.

Leaves of a phylogenetic tree are labeled by species names. By $L(T)$ we denote the set of leaves (labels) and by $I(T)$ the set of internal nodes of T, i.e., $V(T) \setminus L(T)$. Let $r(T)$ denote the root node. By $\dot{V}(T)$ we denote the set $V(T) \setminus \{r(T)\}$. For a node v, $c(v)$ is the set of children of v (note that $c(v)$ is empty if v is a leaf), $p(v)$ is the parent of v, and $e(v)$ denotes the branch $(p(v), v)$. Let $T(v)$ be the (maximal) subtree of T rooted at v. Further, by $\mathsf{clu}(v)$ we denote the species labels below v.

Let $\leq_T$ be the partial order on $V(T)$, such that $u \leq_T v$ if and only if u is on the path between $r(T)$ and v, inclusively. For a non-empty set of nodes $b \subseteq V(T)$, let $lca_T(b)$ be the least common ancestor of b in T.

A *species tree* S represents the relationships among a group of species, while a *gene tree* G depicts the evolutionary history of a set of genes samples from these species. To represent the correspondence between these biological entities, we define a leaf mapping $\mathsf{Le} \colon L(G) \to L(S)$ that labels each leaf of a gene tree with the species, i.e., a leaf from S, from which the gene was sampled. The LCA mapping, $\mathcal{M}$, from gene tree nodes to species tree notes is defined as follows: if g is a leaf node, then $\mathcal{M}(g) := \mathsf{Le}(g)$; if g has two children g' and g'' then $\mathcal{M}(g) := \mathsf{lca}(\mathcal{M}(g'), \mathcal{M}(g''))$.

Definition 2.1 (DLC scenario). Given a gene tree G, a species tree S, and a leaf mapping $\mathsf{Le} \colon L(G) \to L(S)$, the *DLC (reconciliation) scenario* for G, S, and Le is a tuple $\langle \mathcal{M}, \mathcal{L}, \mathcal{O} \rangle$, such that

- $\mathcal{M} \colon V(G) \to V(S)$ denotes a **species map** that maps each node of gene tree to a species node. In this work, species maps are fixed to the LCA mapping.
- $\mathbb{L}$ denotes the **locus set**.
- $\mathcal{L} \colon V(G) \to \mathbb{L}$ is a *surjective* locus map that maps each node of gene tree to a locus,
- For a species node s, let $\mathsf{parent\_loci}(s)$ be the set of loci that yield a new locus in s defined as $\{\mathcal{L}(p(g)) \colon g \in \dot{V}(G), \mathcal{M}(g) = s \text{ and } \mathcal{L}(g) \neq \mathcal{L}(p(g))\}$. Then, $\mathcal{O}$ is a **partial order** on $V(G)$, such that, for every s and every $l \in \mathsf{parent\_loci}(s)$, $\mathcal{O}$ is a total order on the set of nodes $O(s, l) := \{g \colon g \in \dot{V}(G), \mathcal{M}(g) = s \text{ and } \mathcal{L}(p(g)) = l\}$.

Subject to the constraints.

1. For every locus l, the subgraph of the gene tree induced by $\mathcal{L}^{-1}(\{l\})$ is a tree. Moreover, every leaf of such a tree that is also a leaf in G must be uniquely labeled by species.
2. For every $s \in V(S)$, $l \in \mathsf{parent\_loci}(s)$, $g, g' \in O(s, l)$ if $g \leq_G g'$, then $g \leq_\mathcal{O} g'$.
3. A node g is called *bottom* if no child of g maps to $\mathcal{M}(g)$. We say that a node g is *top* (in $\mathcal{M}(g)$) if g is bottom in $\mathcal{M}(p(g))$. Then, $x >_\mathcal{O} y >_\mathcal{O} z$ for every bottom node $x \in O(s, l)$, every non-bottom node $y \in O(s, l)$, and every top node z in s.

The first constraint assures that all gene nodes with the same locus form a connected component; i.e., each locus is created only once. The second constraint incorporates the gene tree's topology in partial order $\mathcal{O}$. Finally, the third constraint guarantees that bottom and top nodes are properly ordered by $\mathcal{O}$.

Inserting Implied Speciation Nodes. For proper embedding a gene tree into a species tree, we require additional degree-two nodes inserted into the gene tree.

Given a gene tree, we define the transformation called *insertion of an implied speciation* as follows. The operation subdivides an edge $(g, g') \in G$ with a new node h, called an *implied speciation*, and sets $\mathcal{M}(h) = p(\mathcal{M}(g'))$ if (i) either $p(\mathcal{M}(g')) > \mathcal{M}(g)$, or (ii) $p(\mathcal{M}(g')) = \mathcal{M}(g)$ and g is not a bottom node of $\mathcal{M}(g)$. Note that h becomes a bottom node after the insertion.

Then, we transform G by a maximal sequence of implied speciation insertions. It is not difficult to see that the resulting gene tree with implied speciation nodes is well defined and unique.

Counting Evolutionary Events. Note that, we first define the species map $\mathcal{M}$, then we transform gene tree by inserting the implied speciation nodes. Next, we define the locus map and partial order $\mathcal{O}$ on the transformed gene tree. Finally, having the DLC scenario, we can define the evolutionary events induced by the scenario.

We start with several definitions. Let s be a node from the species tree. By $\perp(s)$ and $\top(s)$ we denote the sets of all bottom and all top nodes of s, respectively. By $nodes(s)$ we denote the set of gene nodes mapping to s (i.e., $\mathcal{M}^{-1}(\{s\})$. The *internal* nodes of s are defined as $int(s) = nodes(s) \setminus \perp(s)$.

For G, S, Le and $\alpha = \langle \mathcal{M}, \mathcal{L}, \mathcal{O} \rangle$, we have the following evolutionary events at $s \in V(S)$.

- **Duplication:** A non-root gene tree node g is called a *duplication* (at $\mathcal{M}(g)$) if $\mathcal{L}(g) \neq \mathcal{L}(p(g))$. Additionally, we call g the *locus root*. We then say that a duplication happened on edge $(p(g), g)$.
- **Loss:** A locus l is *lost* at s if l is present in s or at the top of s but l is not present at the bottom of s. Formally, l is lost if $l \in \mathcal{L}(\top(s) \cup nodes(s))$ and $l \notin \mathcal{L}(\perp(s))$.
- **ILS at speciation:** Let $C(s, l)$ be the set of all gene lineages (g, g') such that g is a top node at s, whose loci is l, and g' is mapped to s. Then, locus l induces $\max\{|C(s, l)| - 1, 0\}$ (deep) coalescence events at speciation s.
- **ILS at duplication:** For each duplication d, whose parent loci is l, a gene lineage in species s at locus l is *contemporaneous* with d if the lineage starts before and ends after the duplication node d. Let $K(d)$ denote the set of all edges contemporaneous with d. Formally, $K(d) = \{g : g \in O(s, l)$ and $g >_{\mathcal{O}} d >_{\mathcal{O}} p(g)\}$. Then, the duplication d induces $\max\{|K(d)| - 1, 0\}$ (deep) coalescence events.

Problem 1 (DLCParsimony). Given G, S, Le, and real numbers c_D, c_L. and c_{DC}, the reconciliation cost for a DLC scenario $\alpha = (\mathcal{M}, \mathcal{L}, \mathcal{O})$ is

$$R_\alpha := \sum_{s \in V(S)} c_D \cdot nD_\alpha(s) + c_L \cdot nL_\alpha(s) + c_{DC} \cdot (nCS_\alpha(s) + nCD_\alpha(s)),$$

where $nD_\alpha(s)$, is the total number of duplication nodes at s, $nL_\alpha(s)$ is the total number of lost loci at s, and $nCS_\alpha(s)$ is the total number of coalescence events at

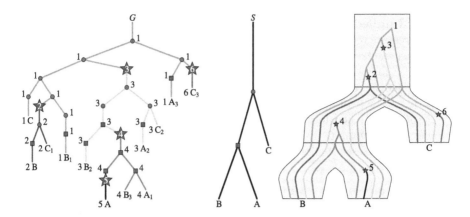

Fig. 1. An example of a DLC scenario with six loci 1 through 6. Stars indicate the duplication events. Left: a gene tree with nodes annotated by locus numbers. Middle: species tree. Right: embedding of the gene tree into the species tree.

speciation s, and $nCD_\alpha(s)$ is the total number coalescence events at duplications mapped to s in the scenario α (Fig. 1).

3   ILP Formulation for DLCParsimony

We now present an Integer Linear Programming (ILP) formulation for solving the DLCParsimony problem. From now on, we refer to this formulation as *DLCPar-ILP*. First, we define global parameters that can be used to constraint the formulation (see constrained models in the next section).

Model Parameters

D_g Binary parameter for each $g \in I(G)$. It is 1, if a duplication event is *allowed* in one of the children of g. In this section $D_g = 1$ for all g, since we do not want to constrain our model.

Next we define the notation that will be used throughout the formulation.

Model Notation

$\mathcal{I}(s)$ Possible order values (indices) of gene nodes within a total ordering of gene nodes induced by $\mathcal{O}$ and restricted to species node s. That is, $\mathcal{I}(s) = \{1, ..., |int(s)|\}$.

$\mathcal{N}$ The maximum possible number of loci; i.e., maximum possible number of duplications plus one. In particular, $\mathcal{N} = 1 + \sum_{g \in I(G)} D_g$. Further, we denote the set $\{1, ..., \mathcal{N}\}$ by $[\mathcal{N}]$.

F_g Indicates the locus index of node g and is defined as $F_g := \sum_{g' \in I(G), g' \leq_{ord} g} D_{g'}$, where $\leq_{ord}$ is some total order on $I(G)$. F_g guarantees that duplication at node g yields a new and distinguished locus F_g in the locus tree.

Now we declare the core variables needed for the ILP formulation.

Decision Variables

x_{uv} A binary variable for edge $(u, v) \in E(G)$. Equals to 1 if v is a duplication; otherwise 0.

y_{gl} Binary variable. 1 if node $g \in V(G)$ is assigned to locus l; otherwise 0.

e_{ls} Binary variable. 1 if locus l is lost at species node/branch s; otherwise 0.

c_{ls} The number of deep coalescence events at a speciation s induced by the locus l.

d_{gl} If g is a duplication and $l = \mathcal{L}(p(g))$, then it denotes the number of corresponding deep coalescence events induced by locus l. Otherwise, $d_{gl} = 0$.

z_{go} Binary variable. 1, if node $g \in V(G)$ is assigned to order $o \in \mathcal{I}(\mathcal{M}(g))$.

w_{gol} Binary variable. 1, if node $g \in V(G)$ is assigned to order o and locus l.

m_{gol} Binary variable. 1, if node g is assigned to order o and locus l and one of children of g is a locus root (i.e., a duplication event happened immediately below g).

Finally, we describe the objective function and the model constraints using the above variables. In particular, the objective function at Eq. 1 minimizes the DLC score. The first term in objective function calculates the total number of duplication events, whereas the second term computes the number of loss events and coalescence events at speciations. The coalescence events at duplications are computed by the last term in the objective function.

Model Constraints

$$\min \zeta = \sum_{e \in E(G)} x_e + \sum_{s \in V(S)} \sum_{l \in [\mathcal{N}]} (e_{ls} + c_{ls})$$

$$+ \sum_{s \in V(S)} \sum_{l \in [\mathcal{N}]} \sum_{g \in int(s)} d_{gl} \qquad (1)$$

s.t.
$$\sum_{e=(g,g') \in E(G)} x_e \leq D_g \qquad g \in V(G) \qquad (2)$$

$$\sum_{g \in \perp(s)} y_{gl} \leq 1 \qquad \forall s \in L(S), l \in [\mathcal{N}] \qquad (3)$$

$$\sum_{l \in [\mathcal{N}]} y_{gl} = 1 \qquad \forall g \in V(G) \qquad (4)$$

$$y_{r(G),1} = 1 \qquad (5)$$

$$F_g x_e \leq \sum_{l \in [\mathcal{N}]} l y_{g'l} \leq F_g x_e + \mathcal{N}(1 - x_e) \qquad \forall e = (g, g') \in E(G) \qquad (6)$$

$$- \mathcal{N} x_{gg'} \leq y_{g'l} - y_{gl} \leq \mathcal{N} x_{gg'} \qquad \forall (g, g') \in E(G), l \in [\mathcal{N}] \qquad (7)$$

$$\sum_{g \in \top(s)} y_{gl} - |V(G)|(e_{ls} + \sum_{g \in \perp(s)} y_{gl}) \leq 0 \quad \forall l \in [\mathcal{N}], s \in V(S) \qquad (8)$$

$$\sum_{g \in \top(s),(g,g') \in E(G),g' \in nodes(s)} y_{gl} - 1 \leq c_{ls} \qquad \forall l \in [\mathcal{N}], s \in V(S) \qquad (9)$$

$$\sum_{o \in \mathcal{I}(s)} z_{go} = 1 \qquad \forall s \in V(S), g \in int(s) \quad (10)$$

$$\sum_{g \in int(s)} z_{go} = 1 \qquad \forall s \in V(S), o \in \mathcal{I}(s) \quad (11)$$

$$\sum_{o' \in \mathcal{I}(s), o' \le o} z_{g'o'} \le 1 - z_{go} \qquad \forall s \in V(S), g, g' \in int(s),$$

$$(g, g') \in E(G), o \in \mathcal{I}(s) \qquad (12)$$

$$2w_{gol} \le y_{gl} + z_{go} \le 1 + w_{gol} \qquad \forall s \in V(S), l \in [\mathcal{N}],$$

$$g \in int(s), o \in \mathcal{I}(s) \qquad (13)$$

$$\sum_{g \in \top(s), (g,g') \in E(G), g' \in nodes(s)} y_{g'l} - 1 \le n_{ls} \qquad \forall l \in [\mathcal{N}], s \in V(S) \qquad (14)$$

$$n_{ls} + \sum_{g' \in int(s) \setminus \{g\}} \sum_{o' < o} (w_{g'o'l} - m_{g'o'l}) \qquad \forall l \in [\mathcal{N}], s \in V(S),$$

$$\le d_{gl} + |\top(s)|(1 - m_{gol}) \qquad o \in \mathcal{I}(s), g \in int(s) \qquad (15)$$

$$2m_{gol} \le w_{gol} + \sum_{e = (g,g') \in E(G)} x_e \le 1 + m_{gol} \qquad \forall s \in V(S), l \in [\mathcal{N}],$$

$$g \in int(s), o \in \mathcal{I}(s) \qquad (16)$$

$$d_{gl}, e_{ls}, c_{ls}.n_{ls} \ge 0 \qquad (17)$$

$$m_{gol}, w_{gol}, x_e, y_{gl}, z_{go} \in \{0, 1\} \qquad (18)$$

In a most parsimonious reconciliation scenario for each internal gene node g only one of its children can be a new locus root [20]. This condition is enforced by inequality 2. Inequality 3 enforces that extant gene nodes mapping to the same extant species must be assigned to different loci. Further, each gene node must be assigned to one locus and it is enforced by Constraint 4. Constraint 5 assigns the original locus (locus 1) to the root of the gene tree. Constraint 6 forces the child gene and its parent to map to different loci if there exists a duplication event between them. Constraint 7 guarantees that if there is no duplication event at gene edge (g, g'), then the locus of g and g' must be the same.

Constraint 8 enforces the correct calculation of loss events. In particular, it ensures that e_{ls} for locus l and species s is 1 if there exists a gene node from $\top(s)$ with locus l, while there is no gene node in $\bot(s)$ with the same locus. Constraint 9 ensures the correct assignment of c_{ls} variables (i.e., the number of coalescence events at speciations). Constraints 10 and 11 jointly assign the partial orders to interior nodes at each species branch. Based on these constraints each order must be assigned to one interior node and each interior node must be assigned to one position in the order. Constraint 12 corresponds to the constraint 2 in Definition 2.1. Constraint 13 ensures proper assignment of the w_{gol} variables. Constraints 14 and 15 should be considered together (note that n_{ls} is an additional variable that joins those two equations; it is required to properly count extra gene lineages at duplications). Those constraints together ensure proper counting of the deep coalescence events at a duplication that happens in one of the children of node g for locus l at species node s. Constraint 16 assures the correct assignment of m_{gol} variables.

3.1   Designing Efficiently Computable Formulations

While the original DLCPar model is very flexible in terms of edges, where duplications can appear, this flexibility contributes substantially to the computational complexity of DLCPar (see the Scalability study for more details). Therefore, in this section, we consider a strategy of restraining the duplication placement only to those edges, where there is *evidence* that a duplication has occurred.

In particular, we call a node $g \in V(G)$ with children g' and g'' an *apparent duplication parent* if $\mathsf{clu}(g') \cap \mathsf{clu}(g'')$ is not empty. That is, there exist extant species, which both child lineages of g sort out to.

We then constraint the DLCPar model in the way that only children of apparent duplication parents can be locus roots. In fact, there are two options for how this constraint can be implemented, which we call ILP-C1 and ILP-C2 that are formalized below.

ILP-C1. Observe that D_g variables defined in the previous section allow us to constrain the locations of gene duplication events easily. That is, we define the ILP-C1 formulation by properly setting the D_g variables: $D_g = 1$ if and only if g is an apparent duplication parent.

ILP-C2. Since apparent duplication parents provide strong evidence of duplications, we define, in addition, a tighter model (ILP-C2). In this model, we require that one of the children of each apparent duplication parent *must* be a duplication. Note that, while this is a strong condition, it allows us to simplify the ILP formulation and reduce the number of variables. That is, we anticipate that ILP-C2 formulation performs fastest in practice.

More precisely, in this model, we "know", where duplications must appear (at least we know the parents of duplications). Therefore, Inequality 2 in DLCPar-ILP should become an equality (which tightens the solution space); further, the m_{gol} variables become redundant, so they can be removed.

3.2   Size of ILP Formulations

We analyze the size of our ILP formulations in terms of their number of variables and constraints. Let n denote the number of nodes in the gene tree and let m denote the number of nodes in the species tree. Further, let k denote the maximum possible number of loci in the gene tree. Note that $k < n$ and k in the ILP-C1 and ILP-C2 models can be expected to be significantly smaller than in the DLCPar-ILP model due to the modified D_g variables.

Then in the DLCPar-ILP and ILP-C1 models, the upper bound on the number of variables is

$$2km + (2k + 1)(n + n^2) = O(k(m + n^2)),$$

and the number of constraints is

$$(3k + 1)n^2 + (k + 2m + 3)n + 4mk + 1 = O(kn^2 + m(n + k)).$$

Finally, the ILP-C2 model has

$$2km + (2k + 1)n + (k + 1)n^2$$

variables, and

$$(k + 1)n^2 + (k^2 + 2m + 3)n + 4mk + 1$$

constraints. Observe, that the ILP-C2 model has fewer variables than the other two models (while asymptotically the same).

3.3 Searching for Multiple Optimal Solutions

The proposed formulations can be extended to detect multiple optimal solutions through an iterative algorithm. At each iteration of that algorithm, our models identify one more alternative optimal solution (if such a solution exists). In particular, for a fixed model, at the first iteration, we solve the original model and save the optimal variables x^*, y^*, and z^* as a part of an optimal solution. To identify a different optimal solution with the same objective value, we add a new constraint such that the ILP model does not repeat identifying previously detected optimal solutions. This constraint is defined as

$$\sum_{e \in E(G)} (x_e - 1)x_e^* + \sum_{g \in V(G)} \sum_{l \in [\mathcal{N}]} (y_{gl} - 1)y_{gl}^* + \sum_{g \in V(G)} \sum_{l \in [\mathcal{N}]} (z_{go} - 1)z_{go}^* \leq -1.$$

We repeat this process as long as the optimal DLC score is the same as the previous iterations.

4   Simulation Study

We present a broad simulation study that (i) compares the computational efficiency and scalability of the developed ILP models with DLCPar and (ii) validates the accuracy of the constrained ILP formulations. Note that we carry out our studies under varied simulation parameters controlling the rate of duplication/loss events as well as the rate of ILS.

Experimental Setup. The process for converting an instance of the DLCParsimony problem to an ILP formulation was implemented in Python 3. Then ILP instances were solved with the Gurobi optimizer version 9.0 [8]. As for DLCPar [20], we used the exact version of the software without heuristic options for a fair comparison. Further, we set the DLCParsimony cost parameters as $c_D = c_L = c_{DC} = 1$. We performed the experiments on a standard workstation with 1.2 GHz (3.6 GHz maximum) CPU.

Simulated Data. We used the standard *SimPhy* simulator [13] to generate the DLCParsimony instances. SimPhy works by first simulating a birth-death species tree and then applying the 2-step DLCoal process by Rasmussen et al. [17] to simulate the multi-locus gene trees. We use the standard simulation parameters derived from the real-world 16 fungi dataset [4,14,17]. In particular, we follow the parameter settings by Molloy and Warnow [14].

To conduct a comprehensive analysis and properly evaluate the proposed constrained DLCPar model, we perform our experiments under various realistic levels of the gene duplication and loss (GDL) and incomplete lineage sorting (ILS). More precisely, we use three different GDL levels: $1e-10$ duplication&loss events per year (low GDL rate), $2e-10$ (moderate GDL rate), and $5e-10$ (high GDL rate). Further, we use two different ILS levels by controlling the tree-wide effective population size; i.e., we use the effective population sizes of $1e7$ and $5e7$ (that correspond to low and moderate ILS levels respectively, according to [14]).

Finally, we simulated DLCParsimony instances with the number of species varying from 5 to 50. That is, overall, we had $3 \times 2 \times 10 = 60$ different parameter settings for DLCParsimony instances. Then to ensure consistency, for each of the 60 parameter combinations, we generated 10 independent DLCParsimony instances. Then we executed DLCPar, DLCPar-ILP, and two constrained ILP models (referred to here as *ILP-C1* and *ILP-C2*) on each of the 600 generated problem instances. Due to a large number of instances and the advanced complexity of the models, we constrained each execution time to 10 min.

4.1   Results and Discussion

Run-Time Comparison. Table 1 shows the breakdown for each algorithm, on how many instances did it fail to complete within 10 min. Further, Fig. 2 demonstrates the scalability ILP-C1 and ILP-C2 algorithms using examples of high-GDL and low-ILS levels. Note that we omitted DLCPar and DLCPar-ILP from the figure as there were multiple instances where those algorithms did not complete (introducing noise).

As expected, we observed that the constrained ILP formulations generally performed faster than both DLCPar and DLCPar-ILP, particularly for instances with more than 50 genes. Overall, ILP-C1 and ILP-C2 were not able to complete within 10 min only on 3 and 2 instances out of 600, respectively. The smallest instance size, where all algorithms failed, contained 50 species and 202 genes. Note, however, that ILP-C1 and ILP-C2 were able to complete on other instances with up to 272 genes (which was the largest number of genes in our study).

Further, we generally observed that DLCPar-ILP failed to complete on more instances than DLCPar (54/600 compared to 30/600), and DLCPar was faster than DLCPar-ILP on average. However, we observed that there were 17 instances, where DLCPar-ILP was able to complete, while DLCPar failed. At the same time, there were 38 instances, where DLCPar was able to complete, while DLCPar-ILP failed. That is, there is no clear domination of one method over the other, and the two methods can be used as complements of each other.

Validating Constrained Models. Given that for the vast majority of instances DLCPar-ILP or DLCPar have completed, we were able to validate the assumptions of the constrained models. That is, we compare the optimum DLC reconciliation score from the constrained models against the overall optimum DLC score (in the unconstrained case). See Table 2 for the results breakdown.

Table 1. Number of instances with running time above 600 s out of 100 instances for each combination

Combination	Population size	GDL	Number of instances			
			DLCPar-ILP	ILP-C1	ILP-C2	DLCPar
1	1e7	1e−10	2	0	0	0
2	1e7	2e−10	9	0	0	1
3	1e7	5e−10	10	1	1	8
4	5e7	1e−10	8	0	0	2
5	5e7	2e−10	9	0	0	3
6	5e7	5e−10	16	2	1	16

Interestingly, we observed that in 98.17% of instances (where we know the optimum unconstrained cost) ILP-C1 provided exactly the same reconciliation cost as the original DLC-model. Moreover, in the 10 instances, where ILP-C1 provided a slightly higher cost, the difference in costs was *at most* 2. On the other hand, ILP-C2, which showed to be faster on average than ILP-C1, provided overestimated reconciliation costs more often. It was exactly correct in 89.9% cases, and the difference in costs in the other 55 cases was at most 8.

That is, overall, ILP-C1 proved to be both very effective and efficient in practice, almost always providing the globally optimum reconciliation cost. Therefore, we suggest the use of this constrained model in practice.

ILP-C2 proved to be faster than ILP-C1 on average, but it gives worse accuracy due to the strength of the constraints. Indeed, ILP-C2 can be very effective in domains with low levels of ILS, since it over-estimated costs significantly less frequently when population size was smaller (see Table 2).

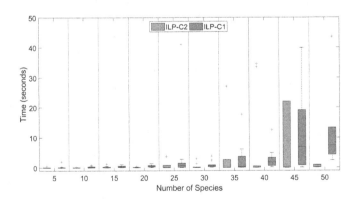

Fig. 2. Computational time comparison for ILP-C1 and ILP-C2 on the example of high-GDL and low-ILS instances.

Table 2. Number of Instances, where ILP-C1 and ILP-C2 score was larger than the DLCPar-ILP/DLCPar score.

Combination	Population size	GDL	Number of Instances	
			ILP-C1	ILP-C2
1	1e7	1e−10	0/98	0/98
2	1e7	2e−10	1/91	1/91
3	1e7	5e−10	2/90	14/90
4	5e7	1e−10	0/92	6/92
5	5e7	2e−10	2/91	14/91
6	5e7	5e−10	5/84	20/84

Acknowledgements. We would like to thank the three anonymous reviewers for their valuable suggestions and comments, and Mukul Bansal for his support in building the foundations of the constrained DLC-model and helpful discussions. AM and OE are supported by the National Science Foundation Grant No. 1617626. PG is supported by the National Science Center grant 2017/27/B/ST6/02720.

References

1. Arvestad, L., Berglund, A.C., Lagergren, J., Sennblad, B.: Gene tree reconstruction and orthology analysis based on an integrated model for duplications and sequence evolution. In: Proceedings of RECOMB 2004, pp. 326–335 (2004)
2. Chan, Y.B., Ranwez, V., Scornavacca, C.: Inferring incomplete lineage sorting, duplications, transfers and losses with reconciliations. J. Theoret. Biol. **432**, 1–13 (2017)
3. Du, H., et al.: Multiple optimal reconciliations under the duplication-loss-coalescence model. IEEE/ACM Trans. Comput. Biol. Bioinform. (2019)
4. Du, P., Hahn, M.W., Nakhleh, L.: Species tree inference under the multispecies coalescent on data with paralogs is accurate. bioRxiv p. 498378 (2019)
5. Du, P., Nakhleh, L.: Species tree and reconciliation estimation under a duplication-loss-coalescence model. In: Proceedings of the 2018 ACM International Conference on Bioinformatics, Computational Biology, and Health Informatics, pp. 376–385 (2018)
6. Du, P., Ogilvie, H.A., Nakhleh, L.: Unifying gene duplication, loss, and coalescence on phylogenetic networks. In: Cai, Z., Skums, P., Li, M. (eds.) Bioinformatics Research and Applications, pp. 40–51. Springer, Cham (2019). https://doi.org/10.1007/978-3-030-20242-2_4
7. Górecki, P., Tiuryn, J.: DLS-trees: a model of evolutionary scenarios. Theor. Comput. Sci. **359**(1–3), 378–399 (2006)
8. Gurobi Optimization, L.: Gurobi optimizer reference manual (2020). http://www.gurobi.com
9. Koonin, E.V.: Orthologs, paralogs, and evolutionary genomics. Annu. Rev. Genet. **39**, 309–338 (2005)
10. Li, H., et al.: Treefam: a curated database of phylogenetic trees of animal gene families. Nucleic Acids Res. **34**(suppl_1), D572–D580 (2006)

11. Lynch, M., Conery, J.S.: The evolutionary fate and consequences of duplicate genes. Science **290**(5494), 1151–1155 (2000)
12. Maddison, W.P.: Gene trees in species trees. Syst. Biol. **46**(3), 523–536 (1997)
13. Mallo, D., de Oliveira Martins, L., Posada, D.: Simphy: phylogenomic simulation of gene, locus, and species trees. Syst. Biol. **65**(2), 334–344 (2015)
14. Molloy, E.K., Warnow, T.: FastMulRFS: Statistically consistent polynomial time species tree estimation under gene duplication. BioRxiv (2019)
15. Ohno, S.: Evolution by Gene Duplication. Springer, Berlin (1970). https://doi.org/10.1007/978-3-642-86659-3
16. Page, R.D.: Maps between trees and cladistic analysis of historical associations among genes, organisms, and areas. Syst. Biol. **43**(1), 58–77 (1994)
17. Rasmussen, M.D., Kellis, M.: Unified modeling of gene duplication, loss, and coalescence using a locus tree. Genome Res. **22**(4), 755–765 (2012)
18. Szöllősi, G.J., Tannier, E., Daubin, V., Boussau, B.: The inference of gene trees with species trees. Syst. Biol. **64**(1), e42–e62 (2014)
19. Wu, T., Zhang, L.: Structural properties of the reconciliation space and their applications in enumerating nearly-optimal reconciliations between a gene tree and a species tree. BMC Bioinform. **12**(S-9), S7 (2011)
20. Wu, Y.C., Rasmussen, M.D., Bansal, M.S., Kellis, M.: Most parsimonious reconciliation in the presence of gene duplication, loss, and deep coalescence using labeled coalescent trees. Genome Res. **24**(3), 475–486 (2014)

In Silico-Guided Discovery of Potential HIV-1 Entry Inhibitors Mimicking bNAb N6: Virtual Screening, Docking, Molecular Dynamics, and Post-Molecular Modeling Analysis

Alexander M. Andrianov[1]([⊠]), Grigory I. Nikolaev[2], Yuri V. Kornoushenko[1], Anna D. Karpenko[2], Ivan P. Bosko[2], and Alexander V. Tuzikov[2]

[1] Institute of Bioorganic Chemistry, National Academy of Sciences of Belarus, Minsk, Republic of Belarus
alexande.andriano@yandex.ru
[2] United Institute of Informatics Problems, National Academy of Sciences of Belarus, Minsk, Republic of Belarus

Abstract. An integrated computational approach to in *silico* drug design was used to identify novel HIV-1 entry inhibitor scaffolds mimicking broadly neutralizing antibody (bNAb) N6 targeting CD4-binding site of the viral gp120 protein. This computer-based approach included (i) generation of pharmacophore models representing 3D-arrangements of chemical functionalities that make bNAb N6 active towards CD4-binding site of gp120, (ii) shape and pharmacophore-based identification of the N6-mimetic candidates by a web-oriented virtual screening platform Pharmit, (iii) molecular docking of the identified compounds with gp120, (iv) optimization of the docked ligand/gp120 complexes using semiempirical quantum chemical method PM7, and (v) molecular dynamics simulations of the docked structures followed by binding free energy calculations. As a result, six hits able to mimic the key interactions of N6 with the Phe-43 cavity of gp120 were selected as the most probable N6-mimetic candidates. The pivotal role in the interaction of these compounds with gp120 is shown to play multiple van der Waals contacts with conserved residues of the hydrophobic Phe-43 cavity critical for the HIV-1 binding to cellular receptor CD4, as well as hydrogen bond with $Asp-368_{gp120}$ that increase the chemical affinity without activating unwanted allosteric effect. According to the data of molecular dynamics, the complexes of the identified molecules with gp120 are energetically stable and show the lower values of binding free energy compared with the HIV-1 entry inhibitors NBD-11021 and DMJ-II-121 used in the calculations as a positive control. Taken together, the findings obtained suggest that these compounds may serve as promising scaffolds for the development of novel, highly potent and broad anti-HIV-1 therapeutics.

Keywords: HIV-1 · Gp120 protein · HIV-1 entry inhibitors · Virtual screening · Molecular docking · Molecular dynamics · Anti-HIV drugs

© Springer Nature Switzerland AG 2020
Z. Cai et al. (Eds.): ISBRA 2020, LNBI 12304, pp. 243–249, 2020.
https://doi.org/10.1007/978-3-030-57821-3_21

1 Introduction

Discovery of potent broadly neutralizing antibodies (bNAbs) isolated from HIV-1 long-term non-progressors gave hope for the possibility of eliciting naturally produced bNAbs through vaccination that may provide a pragmatic way forward. Over 40 bNAbs are currently considered as potential candidates for the development of a globally safe and effective anti-HIV-1 vaccine. These bNAbs target four functionally conserved regions on the trimeric spikes of the HIV-1 envelope (Env) that play a key role in HIV attachment, co-receptor binding, and membrane fusion. In particular, the newly identified antibody from an HIV-infected person, named N6, neutralizes up to 98% of HIV-1 isolates tested, including 16 of 20 strains resistant to other bNAbs that target the CD4-binding site of gp120 [1]. BNAb N6 interacts mostly with relatively conserved regions among worldwide HIV-1 strains and depends less on a variable region V5 of gp120 than its VRC-class predecessors [1]. This unique mode of binding enables N6 to endure Env changes, including N-linked glycosylation of the V5 region which is a major reason of the HIV-1 resistance to other VRC01-like antibodies [1]. In addition, N6 shows wonderful breadth and potency, making this antibody a relevant candidate for further development of both prevention and treatment strategies [1].

Despite significant progress towards the identification of anti-HIV-1 bNAbs and specific modes of their binding to the viral Env, the major challenges in the development of immunogens able to induce potent cross-reactive neutralizing antibodies still remain. Unfortunately, current HIV-1 vaccine candidates are unable to elicit neutralizing antibodies against most circulating virus strains, and thus the induction of a protective antibody response continues to be a major priority for HIV-1 vaccine development. In this context, development of small-molecule HIV-1 entry inhibitors able to show structural and functional mimicry of anti-HIV-1 bNAbs paratopes may be of great interest.

In this work, an integrated computational approach to in *silico* drug design was used to identify novel HIV-1 entry inhibitor scaffolds mimicking structural and pharmacophore features of the bNAb N6 paratope. This computer-based approach included (i) generation of pharmacophore model representing 3D-arrangements of chemical functionalities that make bNAb N6 active towards the CD4-binding site of gp120, (ii) pharmacophore-based identification of the N6-mimetic candidates by a web-oriented virtual screening platform Pharmit (http://pharmit.csb.pitt.edu), (iii) docking of the identified compounds with the molecular target, and (iv) molecular dynamics (MD) simulations of the docked structures followed by binding free energy calculations and selection of the most probable N6 peptidomimetics.

As a result, six hits able to mimic the key interactions of N6 with the Phe-43 cavity of gp120 were selected as the most probable N6-mimetic candidates.

2 Methods

The Pharmit web platform providing an interactive environment for the virtual screening of large compound databases using pharmacophores, molecular shape and energy minimization (http://pharmit.csb.pitt.edu) [2] was applied for generating the pharmacophore model of anti-HIV-1 antibody N6 based on the crystal N6/gp120 complex as the input dataset (Protein Data Bank; code 5TE7; https://www.rcsb.org) [1]. According to the X-ray data [1], the N6 residues Tyr-54^H and Arg-71^H that specifically interact with the hydrophobic Phe-43 cavity of the gp120 CD4 binding site play a key role in the N6 attachment to the HIV-1 gp120 protein. These two N6 residues were therefore used to construct the pharmacophore model of the antibody paratope by the Pharmit platform software [2] combined with the X-ray data [1]. This model was used for virtual screening of the Pharmit chemical databases containing over 1 billion 200 million conformers for ~ 96 million chemical compounds (http://pharmit.csb.pitt.edu) [2] to identify small-molecule drug candidates able to mimic the pivotal interactions of bNAb N6 with gp120. As a result, a set of small-molecule compounds that satisfied the given pharmacophore model and exhibited negative values of binding energies to gp120 was identified. The efficacy of intermolecular interactions between these compounds and the Phe-43 cavity of gp120 was then estimated in terms of the values of binding free energy and dissociation constant using molecular docking and molecular dynamics simulations.

Molecular docking of the designed compounds with gp120 was performed by the QuickVina 2 program [3] in the approximation of rigid receptor and flexible ligands. The HIV-1 inhibitors NBD-11021 [4] and DMJ-II-121 [5] presenting a new generation of the viral entry antagonists were used in the calculations as a positive control. The 3D structure of gp120 was isolated from the crystal N6/gp120 complex (Protein Data Bank; code 5TE7; https://www.rcsb.org) [1]. The 3D structures of NBD-11021 and DMJ-II-121 were taken from the X-ray complexes of these compounds with gp120 (the PDB files 4RZ8 and 4I53) [4, 5]. The gp120 and ligand structures were prepared by adding hydrogen atoms with the Open Babel software (http://openbabel.org/wiki/Main_Page) followed by their optimization in the UFF force field (https://doi.org/10.1021/ja00051a040). The ligands were docked to the crystal gp120 structure [1] using QuickVina 2 [3]. The grid box included the Phe-43 cavity of the gp120 CD4-biding site and was the region of the crystal structure [1] with the following boundary X, Y, Z values: $X \in \{38 \text{ Å}, 63 \text{ Å}\}$, $Y \in \{34 \text{ Å}, 59 \text{ Å}\}$, $Z \in \{55 \text{ Å}, 75 \text{ Å}\}$. The value of "exhaustiveness" parameter defining number of individual sampling "runs" was set to 50 [3].

The classical dynamics of the ligand/gp120 complexes in water was made with the implementation of Amber 16 using the Amber ff10 force field (http://ambermd.org/). The ANTECHAMBER module was employed to set the Gasteiger atomic partial charges (http://ambermd.org/). To prepare the force field parameters, the general AMBER GAFF force field [6] was used. Hydrogen atoms were added to gp120 by the tleap program of the AMBER 16 package (http://ambermd.org/). Initially, the ligand/gp120 complexes were each placed in an octahedron box with periodic boundary conditions. In addition to the ligand/gp120 complex, the box for the MD simulations included TIP3P water [7] as an explicit solvent, Na^+ and Cl^- ions providing overall salt concentration of 0.15 M. After setting up the system, an energy minimization was performed using 500 steps of the steepest descent algorithm followed by 1000 steps of the conjugate-gradient method. The atoms of the complex assembly were then fixed by an additional harmonic potential with the force constant of 1.0 kcal/mol and the system was subject to the equilibration phase. The system equilibration was carried out in three consecutive stages: 1) the system was gradually heated from 0 K to 310 K for 1 ns in NVT ensemble using a Langevin thermostat with a collision frequency of 2.0 ps^{-1} (http://ambermd.org/); 2) pressure equilibration was made for 1 ns at 1.0 bar in NPT ensemble using Berendsen barostat with a 2.0 ps characteristic time (http://ambermd.org/); 3) the constraints on the complex assembly were removed and the system was equilibrated again at 310 K over 2 ns under constant volume conditions. After equilibration was achieved, the MD simulations were carried out for 30 ns in NPT ensemble at temperature T = 310 K and P = 1 bar. Bonds involving hydrogen atoms were constrained using SHAKE algorithm (http://ambermd. org/) to achieve the integration time-step of 2 fs. Long-range electrostatic interactions were calculated using Particle Mesh Ewald (PME) algorithm (http://ambermd.org/). Coulomb interactions and van der Waals interactions were truncated at 10 Å.

3 Results

Based on the findings obtained, six potential peptidomimetics of the cross-reactive neutralizing anti-HIV-1 antibody N6 targeting CD4-binding site of the viral gp120 protein were identified (Fig. 1). Analysis of the complexes of the identified compounds with gp120 reveals a large number of intermolecular interactions involving the residues of gp120 pivotal for the HIV-1 binding to cellular receptor CD4. As an example, Fig. 2 casts shed on the complex of the top-scoring hit with gp120. With the data obtained, this compound forms 7 intermolecular hydrogen bonds (Fig. 2) and 33 van der Waals contacts with the gp120 residues associated with the key hotspots of the CD4-binding site, namely, the Phe-43 cavity, Asp-368$_{gp120}$ and Met-426$_{gp120}$. In particular, the analyzed molecule is involved in the H-bonding with Asp-368$_{gp120}$ mimicking the critical H-bond interaction of this highly conserved gp120 residue with Arg-59$_{CD4}$, as well as in the H-bonding with Met-426$_{gp120}$ which was also defined as a binding hotspot of the CD4-gp120 interface.

Along with hydrogen bonds, the best compound forms multiple van der Waals contacts centered on the residues of gp120 significant for the HIV-1 binding to CD4. According to Fig. 2, the molecule of interest is involved in van der Waals interactions with the gp120 residues Asp-368, Glu-370, Ile-371, Arg-425 and Trp-427, which are highly conserved among various viral isolates and the most important for the HIV-1 attachment to the primary receptor [8].

Thus, inspection of the complex between gp120 and the N6-mimetic candidate with the best scoring function indicates that, in a mechanism similar to that of N6, the identified compound makes hydrogen bonds and van der Waals contacts with the gp120 residues that play a key role in the HIV-1 binding to CD4, resulting in destruction of the critical interactions of gp120 with Phe-43$_{CD4}$ and Arg-59$_{CD4}$. The mechanism of interactions between the other identified molecules and gp120 is close to that described above for the top-ranking compound.

This mechanism is generally provided by intermolecular hydrogen bonds with Asp-368$_{gp120}$, Met-426$_{gp120}$ and multiple van der Waals contacts with the gp120 residues that bind to Phe-43$_{CD4}$.

The data of MD simulations support the results obtained for the docked structures of the N6-mimetic candidates with gp120. These complexes show relative conformational stability within the MD simulations, keep the hydrogen bonds identified in the static

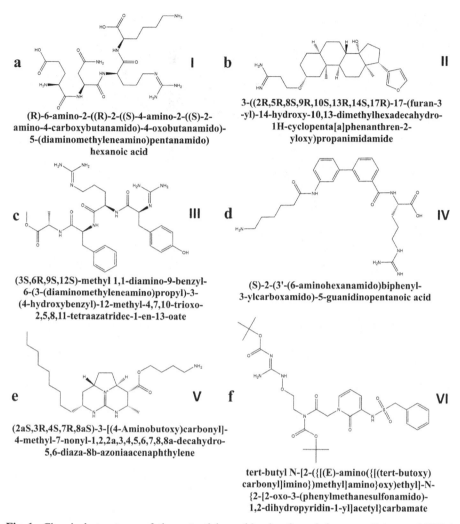

a   I

(R)-6-amino-2-((R)-2-((S)-4-amino-2-((S)-2-
amino-4-carboxybutanamido)-4-oxobutanamido)-
5-(diaminomethyleneamino)pentanamido)
hexanoic acid

b   II

3-((2R,5R,8S,9R,10S,13R,14S,17R)-17-(furan-3
-yl)-14-hydroxy-10,13-dimethylhexadecahydro-
1H-cyclopenta[a]phenanthren-2-
yloxy)propanimidamide

c   III

(3S,6R,9S,12S)-methyl 1,1-diamino-9-benzyl-
6-(3-(diaminomethyleneamino)propyl)-3-
(4-hydroxybenzyl)-12-methyl-4,7,10-trioxo-
2,5,8,11-tetraazatridec-1-en-13-oate

d   IV

(S)-2-(3'-(6-aminohexanamido)biphenyl-
3-ylcarboxamido)-5-guanidinopentanoic acid

e   V

(2aS,3R,4S,7R,8aS)-3-[(4-Aminobutoxy)carbonyl]-
4-methyl-7-nonyl-1,2,2a,3,4,5,6,7,8,8a-decahydro-
5,6-diaza-8b-azoniaacenaphthylene

f   VI

tert-butyl N-[2-({[(E)-amino({[(tert-butoxy)
carbonyl]imino})methyl]amino}oxy)ethyl]-N-
{2-[2-oxo-3-(phenylmethanesulfonamido)-
1,2-dihydropyridin-1-yl]acetyl}carbamate

Fig. 1. Chemical structures of the potential peptidomimetics of the neutralizing anti-HIV-1 antibody N6. Systematic names of these compounds are given.

models and expose the high percentage occupancies of intermolecular H-bonds, in line with the low values of binding free energy predicted for the analyzed molecules based on the MD studies (Table 1). Importantly to note that these values are lower than those calculated using the identical computational protocol for the HIV-1 entry inhibitors NBD-11021 [4] and DMJ-II-121 [5] which were involved in the calculations as a positive control.

Thus, the post-molecular modeling analysis shows that the identified N6-mimetic candidates (Fig. 1) exhibit the similar modes of binding to gp120, resulting in destruction of the critical interactions of Phe-43$_{CD4}$ and Arg-59$_{CD4}$ with the two well-conserved hotspots of the HIV-1 CD4-binding site, namely the Phe-43 cavity and Asp-368$_{gp120}$.

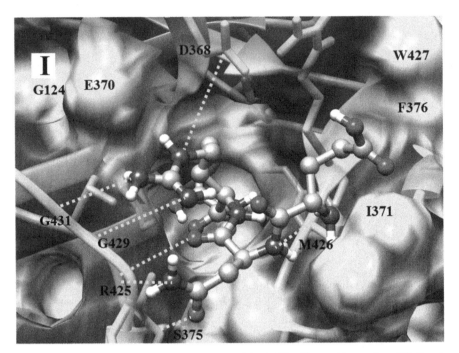

Fig. 2. The docked structure of compound I with gp120. The residues of gp120 forming intermolecular contacts with the ligang are indicated. Hydrogen bonds are shown by dotted lines.

Table 1. Mean values of binding free energy $<\Delta G>$ for the complexes of the antibody N6 mimetics with the HIV-1 gp120 protein and their standard deviations ΔG_{STD}.

Ligand	$<\Delta H>$ kcal/mol	$(\Delta H)_{STD}$ kcal/mol	$<T\Delta S>$ kcal/mol	$(T\Delta S)_{STD}$ kcal/mol	$<\Delta G>$ kcal/mol	ΔG_{STD} kcal/mol
I	−54.87	7.12	−27.66	9.08	−27.20	11.54
II	−41.62	5.92	−17.57	5.54	−24.05	8.11
III	−48.09	5.25	−26.87	8.09	−21.23	9.64
IV	−43.01	10.54	−24.78	7.72	−18.24	13.07
V	−39.22	4.53	−24.64	8.95	−14.59	10.03
VI	−39.59	5.75	−25.20	8.82	−14.38	10.53
DMJ-II-121	−38.81	3.83	−26.42	8.94	−12.39	9.72
NBD-11021	−32.17	6.02	−19.87	6.09	−12.30	8.56

$<\Delta H>$ and $<T\Delta S>$ are the mean values of enthalpic and entropic components of free energy respectively; $(\Delta H)_{STD}$ and $(T\Delta S)_{STD}$ are standard deviations corresponding to these values.

These binding modes are mainly provided by numerous van der Waals interactions with the gp120 residues that are the major contributors to the gp120-CD4 interaction, and the hydrogen bond with Asp-368$_{gp120}$ that is associated with increasing the binding affinity without triggering the undesirable allosteric signal [5] is also highly important.

4 Conclusion

Taken together, the data obtained suggest that the identified N6-mimetic candidates (Fig. 1) may serve as promising scaffolds for the development of novel, highly potent and broad anti-HIV-1 therapeutics.

Acknowledgments. This research was funded by grant from the Belarusian Republican Foundation for Fundamental Research (project X18КИ-002).

References

1. Huang, J., et al.: Identification of a CD4-binding-site antibody to HIV that evolved near-pan neutralization breadth. Immunity. **45**, 1108–1121 (2016). https://doi.org/10.1016/j.immuni. 2016.10.027
2. Sunseri, J., Koes D.R.: Pharmit: interactive exploration of chemical space. Nucl. Acids Res. **44** (Web Server issue), W442–W448 (2016). https://doi.org/10.1093/nar/gkw287
3. Handoko, S.D., Ouyang, X., Su, C.T.T., Kwoh, C.K., Ong, Y.S.: QuickVina: accelerating AutoDock Vina using gradient-based heuristics for global optimization. IEEE/ACM Trans. Comput. Biol. Bioinform. **9**, 1266–1272 (2012). https://doi.org/10.1109/tcbb.2012.82
4. Curreli, F., et al.: Structure-based design of a small molecule CD4-antagonist with broad spectrum anti-HIV-1 activity. J. Med. Chem. **58**, 6909–6927 (2015). https://doi.org/10.1021/acs.jmedchem.5b00709
5. Courter, J.R., et al.: Structure-based design, synthesis and validation of CD4-mimetic small molecule inhibitors of HIV-1 entry: Conversion of a viral entry agonist to an antagonist. Acc. Chem. Res. **47**, 1228–1237 (2014). https://doi.org/10.1021/ar4002735
6. Wang, J., Wolf, R.M., Caldwell, J.W., Kollman, P.A., Case, D.A.: Development and testing of a general Amber force field. J. Comput. Chem. **25**, 1157–1174 (2004). https://doi.org/10.1002/jcc.20035
7. Jorgensen, W.L., Chandrasekhar, J., Madura, J.D., Impey, R.W., Klein, M.L.: Comparison of simple potential functions for simulating liquid water. J. Chem. Phys. **79**, 926–935 (1983). https://doi.org/10.1063/1.445869
8. Kwong, P.D., Wyatt, R., Robinson, J., Sweet, R.W., Sodroski, J., Hendrickson, W.A.: Structure of an HIV gp120 envelope glycoprotein in complex with the CD4 receptor and a neutralizing human antibody. Nature **393**, 648–659 (1998). https://doi.org/10.1038/31405

Learning Structural Genetic Information via Graph Neural Embedding

Yuan Xie[1]([✉]), Yulong Pei[2], Yun Lu[3], Haixu Tang[1], and Yuan Zhou[1,4]

[1] Indiana University Bloomington, Bloomington, IN, USA
xieyuan@iu.edu, hatang@indiana.edu
[2] Eindhoven University of Technology, Eindhoven, The Netherlands
y.pei.1@tue.nl
[3] School of Software and Microelectronics, Peking University, Beijing, China
luyun-anastasia@pku.edu.cn
[4] University of Illinois Urbana-Champaign, Urbana, IL, USA
yuanz@illinois.edu

Abstract. Learning continuous vector representations of genes has been proved to be conducive for many bioinformatics tasks as it can incorporate information of various sources including gene interactions and gene-disease interactions. However, most of the existing approaches, following a paradigm stemmed from the natural language processing community, treat the embedding context in a flat fashion such as a sequence, and tend to overlook the fact that proteins are more likely to function together. In this study, we propose an unsupervised gene embedding algorithm which utilizes graph convolutional network to learn structural information of genes from their neighborhoods in genetic interaction networks. We also propose a neighborhood sampling strategy to generate training samples. Our approach does not assume conditional independence of the node neighborhood and focuses on learning structural information. We compare our method against state-of-the-art baselines and experimental results demonstrate the effectiveness of our approach.

Keywords: Gene embedding · Graph convolutional network · Protein-protein interaction · Essential gene identification

1 Introduction

Like words to natural language processing (NLP), genes are the fundamental building blocks of molecular biological systems. It has been proved that treating genes as atomic units may not be the best solution in many scenarios [3, 6, 35], whereas high dimensional vector representation of genes learned through unsupervised training could be used as features to improve performances of many downstream tasks including protein-protein interaction prediction and essential gene identification. Such representations capture properties of genes based on their biological properties and allow them to generalize across tasks.

Z. Cai et al. (Eds.): ISBRA 2020, LNBI 12304, pp. 250–261, 2020.
https://doi.org/10.1007/978-3-030-57821-3_22

Most of the current approaches for gene embedding are based on the Skip-gram model [23,24] introduced for the NLP community which utilizes the co-occurrence statistics from sequential context of words. In particular, given a sequence of words, Skip-gram model tries to maximize the probability of one word given another using a simple neural network model whose weights are obtained as learned embeddings for each word. Du et al. [6] transferred this concept to the domain of bioinformatics and successfully learned meaningful embeddings for genes using co-expression data. Furthermore, Dai et al. [3] applied the Node2Vec algorithm for gene network embedding which generates sequential contexts for each gene via biased random walks and then utilizes the Skip-gram model to the generated contexts to learn embeddings for each gene. However, all of the aforementioned approaches tend to overlook the fact that proteins rarely act alone and tend to team up as "molecular machines" characterized by intricate physicochemical dynamic connections to perform complex biological functions [4]. Specifically, the Skip-gram model based approaches assume the conditional independence of neighborhood genes and treat the gene interactions in a flat and pair-wise fashion, ignoring the rich structural information displayed by complex protein-protein interaction networks.

Recent years have witnessed the success of neural networks, including convolutional neural networks, recurrent neural networks, and graph convolutional network (GCN) on many applications [8,32]. Graph neural networks, in particular [5,12,15,19,27,33], have been proved to be particularly useful for capturing structural information on graphs. It extends the notion of convolutional neural networks on grid-like input to irregular input such as graphs and provides the flexibility to incorporate both node features and topological features. Several studies have been conducted to utilize GCN for incorporating structural information into word or gene embeddings. In particular, Vashishth et al. [31] employed GCN to efficiently incorporate syntactic as well as semantic information into pre-trained word embeddings and Li et al. [20] proposed a GCN based approach to learn node embeddings from heterogeneous networks made by genes and diseases for disease gene prioritization. On the other hand, most of the previous approaches which are based on the Skip-gram architecture treat the nodes in a pair-wise fashion and fail to utilize the structural nature of genetic interaction networks. To overcome this drawback and by recognizing the advantages of GCN, we make the following contributions in this paper.

- We propose GeneGCN to incorporate structural information of genetic interaction networks into continuous vector representations of genes in an unsupervised fashion. We also propose a graph neighborhood sampling strategy to generate training samples for each gene.
- We train our model using the genetic information mined from the stringDB [28] database and evaluate the learned embeddings on two downstream tasks including protein-protein interaction prediction and essential gene identification. For each task, we compare our model against state-of-the-art baselines to demonstrate the effectiveness of our approach.

The rest of the paper is organized as follows. In Sect. 2, we review related works. In Sect. 3, we describe our approach in detail. Section 4 presents the empirical results and Sect. 5 concludes the paper.

2 Related Work

2.1 Unsupervised Learning for Distributional Representations

Learning continuous representations for words has been widely recognized to be effective for many NLP tasks and many models have been proposed to incorporate the contextual information into word embeddings. In particular, Mikolov et al. [23,24] proposed the popular neural network based Skip-gram model to learn word embeddings in an unsupervised fashion. Based on this, Grover et al. [10] proposed the Node2Vec model to generate sequential contexts which are then fed to the Skip-gram model to learn node embeddings. Du et al. [6] and Dai et al. [3] applied the two aforementioned models to protein-protein interaction networks to learn vector representations for genes. But all of these approaches treat the words or nodes in a pair-wise fashion and tend to overlook the structural properties of the input networks.

2.2 Graph Convolutional Networks

By extending the notion of neural network from grid-like input to irregular graphs, GCN has achieved remarkable success in many fields [26], including word and gene embedding. Li et al. [20] employed GCN to learn gene-disease associations from heterogeneous network made by genes and diseases. Vashishth et al. [31] utilized GCN trained on large-scale unlabeled data to incorporate both syntactic and semantic information into word embeddings. These studies together inspired our work.

2.3 Protein-Protein Interaction

As indicated in [4], proteins within a tissue rarely act alone and tend to team up to undertake complex biological functions together through multimodal physicochemical dynamic connections. Some studies [1,13] proved that groups of genes linked by molecular interactions are more likely to have correlated gene profiles than genes chosen at random and some sub-networks of the entire protein-protein interaction network could have high statistical significance for the identification of certain diseases [14]. Furthermore, recent studies [3,6] have proved that learning gene embeddings from protein-protein interaction networks in an unsupervised fashion is indeed conducive for downstream bioinformatics tasks.

3 Methodology

In this section, we first introduce the background information for GCN, then describe the sampling strategy to generate training data and finally present our GeneGCN.

3.1 Background: Graph Convolutional Networks

Inspired by deep convolutional neural networks on grid-like input, GCN learns node embeddings in a graph by conducting convolutional operations over the neighborhood of nodes. In general, to generate new embedding of a node given the current learned embeddings in the network, GCN collects embeddings of the neighboring nodes, multiply the embeddings with trainable weights, and then merge them using some aggregate function and this procedure could be repeated for K hops which is usually termed the depth of graph convolution.

Formally, given a graph $G = \{\mathcal{V}, \mathcal{E}, \mathcal{X}\}$ where $\mathcal{V}$ denotes the set of nodes, $\mathcal{E}$ the set of edges and $\mathcal{X}$ the input node features, GCN generates embedding for node v on the $(k+1)$-th layer as follows.

$$\mathbf{h}_v^{k+1} = f\left(\sum_{u \in \mathcal{N}(v)} W^k \mathbf{h}_u^k + b^k\right) \tag{1}$$

where k denotes the depth of the graph convolution, $\mathbf{h}^k \in \mathbb{R}^d$ the node representation at the k-th layer, $W^k \in \mathbb{R}^{d \times d}$ and $b^k \in \mathbb{R}^d$ the trainable parameters of the k-th layer, and $\mathcal{N}_v$ the neighborhood of node v. f could be an activation function or an aggregation function and d is the input dimension (embedding dimension). The graph convolution process is illustrated in Fig. 1.

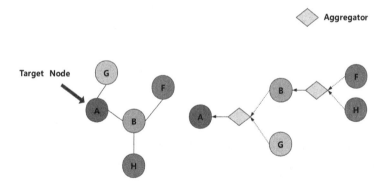

Fig. 1. The graph convolution process.

3.2 GeneGCN

Similar to the learning scheme proposed in [31], we learn gene embeddings in an unsupervised fashion. Specifically, given the GCN embedding ($\mathbf{h}_t$) of a target gene (g_t) and its target embedding v_{g_t}, we use its neighbors in the input network to predict the target gene. Formally, we seek to maximize the following objective.

$$L = \sum_{t=1}^{|V|} \log Pr(g_t|\mathcal{N}(g_t)) \tag{2}$$

where g_t is the target gene, $\mathcal{N}(g_t)$ the neighborhood of g_t in the input network and V the set of all the distinct genes in the input network. The probability $Pr(g_t|\mathcal{N}(g_t))$ is calculated as follows.

$$Pr(g_t|\mathcal{N}(g_t)) = \frac{\exp(v_{g_t}^T \mathbf{h}_t)}{\sum_{i=1}^{|V|} \exp(v_{g_i}^T \mathbf{h}_t)} \tag{3}$$

where $\mathbf{h}_t$ is the GCN representation of the target gene and v_{g_t} is its target embedding. Using Eq. 3, L can be further reduced to

$$L = \sum_{t=1}^{|V|} \left(v_{g_t}^T \mathbf{h}_t - \log \sum_{i=1}^{|V|} \exp(v_{g_i}^T \mathbf{h}_t) \right) \tag{4}$$

Unlike Node2Vec [10], our GeneGCN assumes no conditional independence of the nodes in a target gene's neighborhood and aims to learn the structural information from the input network. The embedding process of GeneGCN is illustrated in Fig. 2.

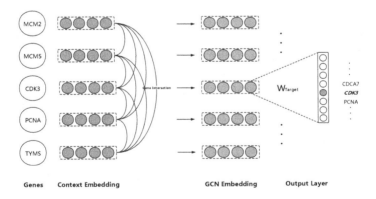

Fig. 2. An example of the embedding process of GeneGCN.

3.3   Graph Neighborhood Sampling

In this section, we propose an efficient neighborhood sampling strategy to generate training samples for our GeneGCN. Unlike Node2Vec which generates samples for each node, we employ an edge-based sampling strategy. In particular, we first sort the neighborhood for each node according to the edge-associated weights (e.g., interaction confidence) and then keep only the top n nodes as the considered neighborhood. Next, for each edge (u, v) in the input graph, we first use u as the target node (gene) and then uniformly sample m nodes without replacement from its considered neighborhood. Finally, we add v to the sampled neighborhood to create the sampled neighborhood $\mathcal{N}_s^T(u)$. Here n and m are the hyperparameters to control the neighborhood size and we use multiple values of m to create training samples of different sizes. We repeat the procedure for v to generate the training examples associated with v. The detailed algorithm is described in Algorithm 1. The proposed sampling strategy allows different sizes of neighborhood to be considered and assumes no conditional independence of the neighborhood.

Algorithm 1: Edge-based Graph Neighborhood Sampling

---

input : Graph $\mathcal{G}(\mathcal{V}, \mathcal{E}, \mathcal{W})$; considered neighborhood size n; set of sample sizes S; times to repeat the sampling process Z; random sampling without replacement function: SAMPLE(X, y).

output: List of tuples L: $[(g_t, \mathcal{N}_s^T(g_t))]$

1 For each node $v \in \mathcal{V}$, sort its neighborhood according to the connection weights. Keep only the top n nodes for each neighborhood to obtain $\mathcal{N}_s(v)$;

2 **for** $i \leftarrow 1$ **to** Z **do**

3   **for** $k \in S$ **do**

4   **for** $(u, v) \in \mathcal{E}$ **do**

5             $\mathcal{N}_s^T(u) \leftarrow$ SAMPLE($\mathcal{N}_s(u)$ - $\{v\}$, k) $\cup \{v\}$

6             $L \leftarrow L \cup \{(u, \mathcal{N}_s^T(u))\}$

7             $\mathcal{N}_s^T(v) \leftarrow$ SAMPLE($\mathcal{N}_s(v)$ - $\{u\}$, k) $\cup \{u\}$

8             $L \leftarrow L \cup \{(v, \mathcal{N}_s^T(v))\}$

9   **end**

10   **end**

11 **end**

12 **return** L

---

4   Experimental Results

In order to demonstrate the effectiveness of our approach, we evaluate our genetic network embedding approach on two tasks: protein-protein interaction prediction and essential gene identification. We first describe the datasets that we use in the experiments, then detail the evaluation process and finally present the empirical results.

4.1   Datasets

We use the stringDB [28] as our major source of network embedding data. The STRING database collects, scores and integrates all publicly available sources of protein–protein interaction information, and complements these with computational predictions to achieve a comprehensive and objective global network, including direct (physical) as well as indirect (functional) interactions. For the first task which is protein-protein interaction prediction, we use only the co-expression data in stringDB as did in [6] for network embedding (we exclude edges whose connection confidence is lower than 0.05) and for the second which is essential gene identification, we use the combined scores of protein-protein interactions as the network input.

We followed the procedures in [18] to generate the gene-gene interaction prediction data based on the shared Gene Ontology (GO) annotations which is obtained using R package "org.Hs.eg.db". We download the GO structure file from[1] and we define the gene pairs that share GO annotations as the positive set of functional associations as did in [6]. Specifically, we choose the GO category "Biological Process" with experimental evidence: IDA (inferred from direct assay), IMP (inferred from mutant phenotype), IPI (inferred from protein interaction), IGI (inferred from genetic interaction), and TAS (traceable author statement). We also exclude the highly over-represented GO terms to minimize generalized annotations as did in [6]. This creates a positive set of 270,704 pairs involving 5369 genes. We collect all gene-pairs that do not share any GO term or their children GO terms as negative set of functional associations, which results in a total of 40,879,714 gene pairs involving 12,521 human genes.

For the second task, which is predicting human essential genes, we download the data from the supplementary files of [11] which are extracted from the DEG database[2]. It contains 12015 genes among which 1516 are essential genes and 10499 are non-essential genes.

4.2   Task 1: Protein-Protein Interaction Prediction

In our first task, we evaluate the effectiveness of our approach by predicting protein-protein interactions. Given a pair of genes, we employ the same gene-gene interaction predictor neural network (GGIPNN) proposed in [6] to predict whether there is any association between these two genes. We first use GeneGCN to learn gene embeddings from the co-expression network and then use the learned embeddings as the input of GGIPNN. We randomly sample negative pairs with equal number of positive pairs to avoid the impact of the imbalanced labels distribution and we divide the data into training, validation and testing with a ratio of 7 : 1 : 2 as did in [6]. We use pairs that both two genes appear only in the training set for training; we use pairs that both two genes appear only in the validation set for validation; we use pairs that both two genes appear only in the testing set for testing. We also remove genes that do not appear in the stringDB data. This ends up with a training set that has 156,221 pairs (involving 8722 genes), validation set that has 3416 pairs (1054 genes) and testing set that has 13,022 pairs (2370 genes).

---

[1] http://geneontology.org/ontology/go.obo.

[2] http://tubic.tju.edu.cn/deg/.

For this task, we compare our approach against three baselines: random initialization, Gene2Vec [6] and Node2Vec [10]. We use the F1 score, Matthews Correlation Coefficient (MCC) and Area Under ROC Curve (AUC) as the performance comparison metrics. For the hyperparameters, we use 50000 for the considered neighborhood size n, 10 for the repetition times Z and [2, 5, 10] for the sample sizes S. We use 100 for the embedding dimension and we fix all the learned embeddings during training. For each algorithm, we repeat the experiments for 5 random runs and the results are reported in Table 1.

Table 1. Performance comparison with baselines on the protein-protein interaction prediction task.

Algorithm	F1	MCC	AUC
Random initialization	0.364 ± 0.172	0.008 ± 0.012	0.504 ± 0.002
Gene2Vec [6]	0.628 ± 0.035	0.331 ± 0.012	0.721 ± 0.011
Node2Vec [3]	0.662 ± 0.030	0.349 ± 0.012	0.734 ± 0.010
GeneGCN (Our Approach)	**0.685** ± 0.02	**0.370** ± 0.006	**0.753** ± 0.003

From the table, we can see that GeneGCN outperforms all the baseline models on all the metrics. Besides, we observe that our approach could greatly reduce variance compared to the baseline models.

4.3   Task 2: Essential Gene Identification

In this section, we further evaluate GeneGCN on the task of predicting essential genes. Genes can be deemed as essential if they play indispensable role in cell viability and fertility [34]. Since there is a mismatch between the gene set of stringDB and the gene pairs we obtained from the DEG database, we only consider the genes that are available in both sources and this results in a total number of 6193 genes 1207 of which are positive genes and 4986 are negative genes. In order to make a fair comparison, we use support vector machine (SVM) as the downstream classifier as did in [3]. We conduct five fold cross validation and for each fold, we randomly down sample the negative data to obtain 1 : 1 ratio with the positive data as did in [3]. We use the same set of hyperparameters that we used for Task 1 except that we use 15 for Z. For this task, we compare our GeneGCN against the λ-interval Z curve method [11], Gene2Vec, and Node2Vec. We also use the F1 score, MCC and AUC as the comparison metrics and repeat the experiments for each algorithm for 5 random runs. We report the results in Table 2.

From the table we observe similar behavior as we do in Task 1 that GeneGCN outperforms all the baselines on all the metrics and generally achieves a smaller variance.

From both two tasks, we see that our approach could indeed mine additional structural information from the input network, which is reflected by the improved performance over the baselines.

Table 2. Performance comparison with baselines on the essential gene identification task

Algorithm	F1	MCC	AUC
Z curve [11]	0.783 ± 0.011	0.568 ± 0.016	0.834 ± 0.006
Gene2Vec [6]	0.805 ± 0.009	0.669 ± 0.015	0.935 ± 0.005
Node2Vec [3]	0.865 ± 0.006	0.731 ± 0.012	0.933 ± 0.005
GeneGCN (Our Approach)	**0.883** ± 0.007	**0.770** ± 0.013	**0.949** ± 0.002

4.4   Embedding Visualization by t-SNE

In this section, we seek to visualize and interpret the gene embeddings learned by GeneGCN. We use the gene embeddings learned in the first task and we employ the t-Distributed Stochastic Neighbor Embedding (t-SNE) [21] which maps high dimensional data to 2 or 3 dimensional vectors. As recommended in [6], we first use the principal component analysis (PCA) to reduce the learned embedding dimension to 50 and then apply the multi-core version of Barnes-Hut t-SNE [30] to speed up the mapping process. For the hyperparameters, we use a value of 30 for the perplexity, 200 for the learning rate and 100,000 for number of iterations. The mapping results are plotted in Fig. 3.

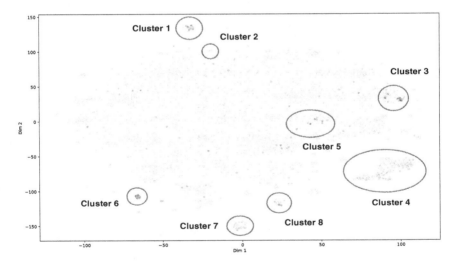

Fig. 3. Gene embedding map generated using t-SNE. "Dim 1" and "Dim 2" are the two embedding dimensions. Blue: leucine-rich repeat (LRR) genes [16]; pink: cytochromes P450 (CYP) genes [9]; green: homebox genes [2]; orange: ankryin repeat genes [22]; cyan: ZNF genes; brown: kinesin genes [17]; red: receptor tyrosine kinase genes [25]; purple: RhoGAP gens [29]. (Color figure online)

From the figure we can see that most of the genes form a single cloud while several small groups of genes scatter around it. In order to further illustrate the embedding map, we color the genes that belong to the several well known gene families and these genes display strong trend of clustering. In particular, in cluster 4 which is the island on the lower right corner, we can see an abundance of the ZNF genes which are colored in cyan [7]. Similarly, the kinesin genes (brown) [17] and the ankryin repeat genes (orange) [22] are mapped around cluster 6 and cluster 7 respectively. We also spot clusters of closely related genes within the single cloud, such as the leucine-rich repeat containing genes (blue, cluster 1) [16] and the homebox genes (green, cluster 3) [2]. The clustering behavior of these gene families demonstrates the effectiveness of our approach.

5   Conclusion

In this paper, we propose a novel approach to learn continuous vector representations for genes by using graph convolutional neural network. We introduce an edge-based graph neighborhood sampling algorithm to generate context graphs for each node from the input network. We use the graph convolutional neural network to predict each target gene given its sampled context graphs. Our approach focuses on learning the structural information contained in the input networks and does not assume the conditional independence of neighborhood nodes as some of the baseline models did. We evaluate our approach on two tasks which are predicting protein-protein interaction from gene co-expression data and human essential gene identification. Experimental results prove the effectiveness of our approach as our model outperforms all the baselines on all the metrics and achieves smaller variances. We further visualize the learned embeddings using t-SNE which maps the high dimensional embeddings to a 2-dimensional space where we can identify clearly clustering patterns. Our approach is more suitable for the structural nature of gene interactions and provides a new perspective for many bioinformatics applications.

References

1. Ayati, M., Erten, S., Chance, M.R., Koyutürk, M.: Mobas: identification of disease-associated protein subnetworks using modularity-based scoring. EURASIP J. Bioinform. Syst. Biol. **2015**(1), 7 (2015)
2. Bürglin, T.R., Affolter, M.: Homeodomain proteins: an update. Chromosoma **125**(3), 497–521 (2015). https://doi.org/10.1007/s00412-015-0543-8
3. Dai, W., Chang, Q., Peng, W., Zhong, J., Li, Y.: Identifying human essential genes by network embedding protein-protein interaction network. In: Cai, Z., Skums, P., Li, M. (eds.) ISBRA 2019. LNCS, vol. 11490, pp. 127–137. Springer, Cham (2019). https://doi.org/10.1007/978-3-030-20242-2_11
4. De Las Rivas, J., Fontanillo, C.: Protein-protein interactions essentials: key concepts to building and analyzing interactome networks. PLoS Comput. Biol. **6**(6), e1000807 (2010)

5. Defferrard, M., Bresson, X., Vandergheynst, P.: Convolutional neural networks on graphs with fast localized spectral filtering. In: Advances in Neural Information Processing Systems, pp. 3844–3852 (2016)

6. Du, J., Jia, P., Dai, Y., Tao, C., Zhao, Z., Zhi, D.: Gene2vec: distributed representation of genes based on co-expression. BMC Genom. **20**(1), 82 (2019)

7. Fernandez-Zapico, M.E., et al.: A functional family-wide screening of SP/KLF proteins identifies a subset of suppressors of KRAS-mediated cell growth. Biochem. J. **435**(2), 529–537 (2011)

8. Gilmer, J., Schoenholz, S.S., Riley, P.F., Vinyals, O., Dahl, G.E.: Neural message passing for quantum chemistry. In: Proceedings of the 34th International Conference on Machine Learning, vol. 70, pp. 1263–1272. JMLR. org (2017)

9. Gonzalez, F.J., Gelboin, H.V.: Human cytochromes P450: evolution and cDNA-directed expression. Environ. Health Perspect. **98**, 81–85 (1992)

10. Grover, A., Leskovec, J.: node2vec: Scalable feature learning for networks. In: Proceedings of the 22nd ACM SIGKDD International Conference on Knowledge Discovery and Data Mining, pp. 855–864 (2016)

11. Guo, F.B., et al.: Accurate prediction of human essential genes using only nucleotide composition and association information. Bioinformatics **33**(12), 1758–1764 (2017)

12. Hamilton, W., Ying, Z., Leskovec, J.: Inductive representation learning on large graphs. In: Advances in Neural Information Processing Systems, pp. 1024–1034 (2017)

13. Ideker, T., Ozier, O., Schwikowski, B., Siegel, A.F.: Discovering regulatory and signalling circuits in molecular interaction networks. Bioinformatics **18**(suppl_1), S233–S240 (2002)

14. Ivanov, A.A., Khuri, F.R., Fu, H.: Targeting protein-protein interactions as an anticancer strategy. Trends Pharmacol. Sci. **34**(7), 393–400 (2013)

15. Kipf, T.N., Welling, M.: Semi-supervised classification with graph convolutional networks. arXiv preprint arXiv:1609.02907 (2016)

16. Kobe, B., Deisenhofer, J.: The leucine-rich repeat: a versatile binding motif. Trends Biochem. Sci. **19**(10), 415–421 (1994)

17. Lawrence, C.J., et al.: A standardized kinesin nomenclature. J. Cell Biol. **167**(1), 19–22 (2004)

18. Lee, I., Blom, U.M., Wang, P.I., Shim, J.E., Marcotte, E.M.: Prioritizing candidate disease genes by network-based boosting of genome-wide association data. Genome Res. **21**(7), 1109–1121 (2011)

19. Lee, J., Lee, I., Kang, J.: Self-attention graph pooling. arXiv preprint arXiv:1904.08082 (2019)

20. Li, Y., Kuwahara, H., Yang, P., Song, L., Gao, X.: PGCN: disease gene prioritization by disease and gene embedding through graph convolutional neural networks. bioRxiv p. 532226 (2019)

21. van der Maaten, L., Hinton, G.: Visualizing data using t-SNE. J. Mach. Learn. Res. **9**(Nov), 2579–2605 (2008)

22. Michaely, P., Tomchick, D.R., Machius, M., Anderson, R.G.: Crystal structure of a 12 ANK repeat stack from human ankyrinR. EMBO J. **21**(23), 6387–6396 (2002)

23. Mikolov, T., Chen, K., Corrado, G., Dean, J.: Efficient estimation of word representations in vector space. arXiv preprint arXiv:1301.3781 (2013)

24. Mikolov, T., Sutskever, I., Chen, K., Corrado, G.S., Dean, J.: Distributed representations of words and phrases and their compositionality. In: Advances in Neural Information Processing Systems, pp. 3111–3119 (2013)

25. Robinson, D.R., Wu, Y.M., Lin, S.F.: The protein tyrosine kinase family of the human genome. Oncogene **19**(49), 5548–5557 (2000)
26. Sanchez-Lengeling, B., Wei, J.N., Lee, B.K., Gerkin, R.C., Aspuru-Guzik, A., Wiltschko, A.B.: Machine learning for scent: Learning generalizable perceptual representations of small molecules. arXiv preprint arXiv:1910.10685 (2019)
27. Scarselli, F., Gori, M., Tsoi, A.C., Hagenbuchner, M., Monfardini, G.: The graph neural network model. IEEE Trans. Neural Netw. **20**(1), 61–80 (2008)
28. Szklarczyk, D., et al.: String v11: protein-protein association networks with increased coverage, supporting functional discovery in genome-wide experimental datasets. Nucleic Acids Res. **47**(D1), D607–D613 (2018)
29. Tcherkezian, J., Lamarche-Vane, N.: Current knowledge of the large RhoGAP family of proteins. Biol. Cell **99**(2), 67–86 (2007)
30. Van Der Maaten, L.: Accelerating t-SNE using tree-based algorithms. J. Mach. Learn. Res. **15**(1), 3221–3245 (2014)
31. Vashishth, S., Bhandari, M., Yadav, P., Rai, P., Bhattacharyya, C., Talukdar, P.: Incorporating syntactic and semantic information in word embeddings using graph convolutional networks. In: Proceedings of the 57th Conference of the Association for Computational Linguistics, pp. 3308–3318. Association for Computational Linguistics, Florence, July 2019. https://www.aclweb.org/anthology/P19-1320
32. Xie, Y., Liu, B., Liu, Q., Wang, Z., Zhou, Y., Peng, J.: Off-policy evaluation and learning from logged bandit feedback: error reduction via surrogate policy. In: 7th International Conference on Learning Representations, ICLR 2019 (2019)
33. Ying, Z., You, J., Morris, C., Ren, X., Hamilton, W., Leskovec, J.: Hierarchical graph representation learning with differentiable pooling. In: Advances in Neural Information Processing Systems, pp. 4800–4810 (2018)
34. Zhang, R., Lin, Y.: Deg 5.0, a database of essential genes in both prokaryotes and eukaryotes. Nucleic Acids Res. **37**(suppl_1), D455–D458 (2009)
35. Zou, Q., Xing, P., Wei, L., Liu, B.: Gene2vec: gene subsequence embedding for prediction of mammalian n6-methyladenosine sites from mRNA. RNA **25**(2), 205–218 (2019)

The Cross-Interpretation of QSAR Toxicological Models

Oleg Tinkov[1]($\boxtimes$) , Pavel Polishchuk[2] , Veniamin Grigorev[3] ,
and Yuri Porozov[4,5,6]

[1] Military Institute of the Ministry of Defense, 3300 Tiraspol, Transdniestria, Moldova
`oleg.tinkov.chem@mail.ru`
[2] Institute of Molecular and Translational Medicine, Faculty of Medicine and Dentistry, Palacký University and University Hospital in Olomouc, 779 00 Olomouc, Czech Republic
[3] Institute of Physiologically Active Compounds, Russian Academy of Sciences, 142432 Moscow Oblast, Chernogolovka, Russia
[4] Laboratory of Bioinformatics, I.M. Sechenov First Moscow State Medical University (Sechenov University), 119991 Moscow, Russia
[5] Department of Food Biotechnology and Engineering, ITMO University, 197101 Saint-Petersburg, Russia
[6] Department of Computational Biology, Sirius University of Science and Technology, 354340 Sochi, Russia

Abstract. The investigation of influence of the molecular structure of different organic compounds on acute, developmental toxicity, mutagenicity has been carried out with the usage of 2D simplex representation of molecular structure and Support Vector Machine (SVM), Random Forest (RF), Gradient Boosting Machine (GBM), Partial Least Squares (PLS). Suitable QSAR (Quantitative Structure - Activity Relationships) models were obtained. The study was focused on QSAR model interpretation. The aim of the study was to develop a set of structural fragments that steadily increase various types of toxicity. The interpretation allowed to detail the molecular environment of known toxicophors and to propose new fragments.

Keywords: Toxicity · Simplex descriptors · Modeling · Interpretation

1 Introduction

Integrated toxicity assessment is an important task in drug development. Experimental testing of compounds on different types of toxicity being costly is also criticized on ethical reasons. This is the key reason why Quantitative Structure–Activity Relationship (QSAR) modeling continues to be a viable approach to reduce the amount of efforts and cost of experimental toxicity assessments [1].

To address this challenge, many QSAR studies have been conducted and reported for different toxicity endpoints [2]. Toxicity is a complex phenomenon which includes action of chemicals through different biochemical mechanisms. Such complexity hampers the process of QSAR modeling. For example, a detailed analysis of 150 QSAR models led

© Springer Nature Switzerland AG 2020
Z. Cai et al. (Eds.): ISBRA 2020, LNBI 12304, pp. 262–273, 2020.
https://doi.org/10.1007/978-3-030-57821-3_23

to the conclusion that many available models have only limited usefulness due to their poor or modest statistical quality or because they were obtained from limited data sets [3]. Nevertheless, several works with reasonable results appeared in the past years [4–6]. Today the trend in QSAR modeling is associated with the study of large sets of toxicants using different machine learning methods [7, 8].

The attention of the researchers is directed to the complex prediction of the toxicity of chemical compounds. Many programs and web- resources allow simultaneously predicting acute, reproductive toxicity for different organisms, mutagenicity [9–11].

Unfortunately, for most existing QSAR models there is no structural interpretation. Organisation for Economic Co-operation and Development (OECD) principles include the fifth requirement for mechanistic interpretation of QSAR models used for regulatory purposes [12]. This information could provide for the formation of new hypotheses as to mechanisms of chemical toxic action and allows to carry out the molecular design [13].

Thus, the aim of this study was to compare the results of interpretation of QSAR models to identify common mechanisms of toxic action for different organisms and identify fragments that persistently increase different types of toxicity.

Several articles have been published recently on new approaches for the interpretation of QSAR models that allow us to estimate the contributions of structural fragments from QSAR models built by any machine learning methods [14, 15].

Many of these approaches utilize the idea of matched molecular pair (MMP) analysis [16] to calculate atom or fragment contributions.

Following this trend, we have applied Universal Approach for Structural Interpretation [17] to search for structural fragments that steadily increase different toxicity endpoints:

1. 96 h fathead minnow LC_{50};
2. 48 h Daphnia magna LC_{50};
3. 48 h Tetrahymena pyriformis IGC_{50};
4. Oral rat LD_{50};
5. Bioaccumulation factor;
6. Developmental toxicity;
7. Ames mutagenicity.

In Fig. 1, we briefly recapitulate our earlier proposed Universal Approach for Structural Interpretation.

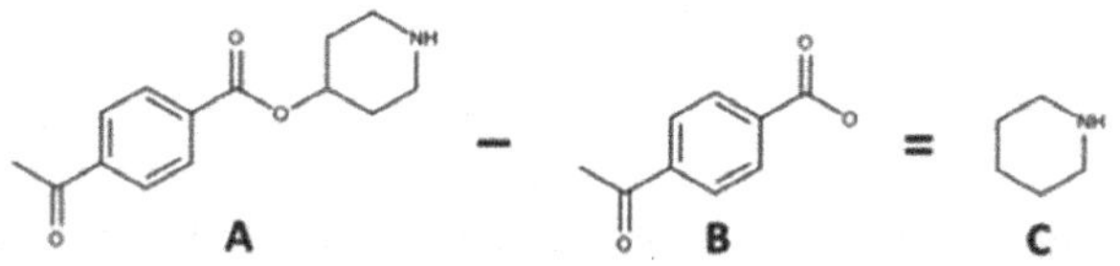

Interpretation	$Activity_{pred}(A)$	$Activity_{pred}(B)$	Contribution(C)
Structural	$f(A) = x$	$f(B) = y$	$W(C) = x - y$

Fig. 1. Schemes of structural interpretation of QSAR models.

If we have a compound **A** consisting of two fragments **B** and **C** then the contribution of the fragment **C** can be calculated as the difference between predicted activity values for the initial compound **A** and the counter-fragment **B** (obtained by removal of the fragment **C** from the molecule **A**). Thus we calculate overall contribution of the fragment **C** in units of a studied activity.

This simple procedure can be used for the interpretation of both regression and binary classification QSAR models based on any combination of machine learning methods and descriptors. In the case of regression models, predicted numerical values are used for the calculation of fragment contributions, and thus, the contributions have the same units as the investigated property value and reflect the change in the value of the investigated property with the addition of certain fragments. In the case of binary classification models, predicted probabilities of belonging to the active class of compounds are used. Thus, the fragment contributions are probabilities to change class upon addition of those fragments. The developed approach for structural interpretation can estimate contributions of scaffolds and linkers as well as contributions of single substituents. After removal of the linker or scaffold, the remaining structure will consist of two or more disconnected fragments. This creates a certain limitation, since not all descriptors can be calculated for such multifragment structures, which can be chemically not meaningful. However, simplex descriptors and fingerprints can handle such structures perfectly. Therefore, in this study we used simplex descriptors [18] because they provide great flexibility and opportunity to analyze the contributions of any fragments.

2   Materials and Methods

2.1   Data Sets

All data sets were obtained from the Toxicity Estimation Software Tool (T.E.S.T.), version 4.2, provided by the U.S. Environmental Protection Agency [11]. Structures were checked for errors [19] and duplicates which were removed. The distribution of chemical compounds after removing duplicates is given in Table 1.

All modeling steps including descriptor calculation, model development and validation, molecule fragmentation and calculation of fragment contributions were performed by means of open-source SPCI software [20, 21].

2.2   Simplex Representation of Molecular Structure (SiRMS)

The main concept of SiRMS approach is that any molecule can be represented as a system of different simplexes (tetratomic fragments with fixed composition and topological structure). At the 2D level, the connectivity of atoms in simplex, atom type and bond nature (single, double, triple, aromatic) are taken into consideration (Fig. 2). Atoms were differentiated not only by their atom types but also by other physico-chemical characteristics, such as partial charge, lipophilicity, refraction and the ability for an atom to be a hydrogen bond donor or acceptor. The usage of sundry variants of differentiation of simplex vertexes (atoms) represents the principal feature of the proposed approach.

In this study, we used simplex descriptors labeled by partial atomic charge, lipophilicity, refractivity, and H-bonding. All of these parameters were calculated using the

Table 1. Sets of investigation compounds.

Toxicity endpoints	Brief description	Number of substances in the set
96 h, *fathead minnow*, LC_{50}	Concentration of the test chemical in water in mg/L that causes 50% of fathead minnow to die after 96 h	803
48 h, *Daphnia magna*, LC_{50}	Concentration of the test chemical in water in mg/L that causes 50% of Daphnia magna to die after 48 h	335
48 h, *Tetrahymena pyriformis*, IGC_{50}	Concentration of the test chemical in water in mg/L that causes 50% growth inhibition to Tetrahymena pyriformis after 48 h	1780
Oral, rat, LD_{50}	Amount of chemical in mg/kg body weight that causes 50% of rats to die after oral ingestion	7205
BCF	Ratio of the chemical concentration in fish as a result of absorption via the respiratory surface to that in water at steady state	676
Developmental toxicity	Whether or not a chemical causes developmental toxicity effects to humans or animals	285
Ames mutagenicity	A compound is positive for mutagenicity if it induces revertant colony growth in any strain of *Salmonella typhimurium*	5718

ChemAxon cxcalc software tool [22]. When we carried out structural interpretation, we removed the descriptors of all four groups mentioned above for each selected fragment. The structures of compounds were standardized using ChemAxon Standardizer [23].

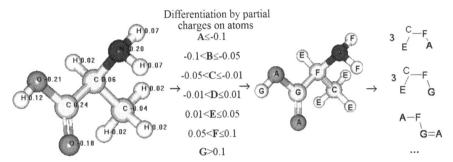

Fig. 2. Simplex representation of molecular structure

2.3 Model Building

The Python sklearn package was used for modeling. Support Vector Machine (SVM) with radial basis function (RBF) kernel, Random Forest (RF), and gradient boosting method (GBM) models were built to overcome the classification problem. For regression problems, SVM, RF, GBM, and partial least squares (PLS) techniques were used. Consensus predictions were produced by averaging the predictions of regression models or by choosing the major voted class among classification models. The performance of all models was assessed by fivefold cross-validation. The tuning parameters of the models were optimized by a grid search.

2.4 Fragmentation of Molecules

Exhaustive fragmentation was applied to generate fragments from compounds of the data set. Fragments were enumerated using a SMARTS pattern [#6+0;!$(*=,#[!#6])]!@!=!#[*] matching bonds which can be cleaved during fragmentation.

3   Results and Discussion

The statistical characteristics of individual models, their consensus predictions are given in Table 2 and Table 3. Predictive performance of SVM, RF and GBM models was reasonable whereas PLS model had low predictivity. Therefore, consensus prediction was obtained by averaging of predictions of SVM, RF and GBM models.

Table 3 also shows the statistical characteristics that have been obtained for these sets by other researchers in the program T.E.S.T. v.4.2 [11]. Predictive ability of the developed QSAR models was comparable to the models developed by other researchers with use of other modeling approaches (Table 2 and Table 3).

The fragment contributions calculated from individual models were in good agreement (R Pearson = 0.69–0.94). Therefore, further analysis was focused only on a discussion of interpretation results of the consensus model.

First, structural interpretation of the obtained QSAR models was carried out using for 75 molecular fragments. These toxicophors were described in details in our previous

Table 2. Predictive performance of regression QSAR models.

Toxicity endpoints	SiRMS			T.E.S.T.		
	Model	R^2	RMSE	Model	R^2	RMSE
96 h, *fathead minnow*, LC_{50}	GBM	0.64	0.87	Hierarchical	0.71	0.80
	RF	0.66	0.85	Single model	0.70	0.80
	SVM	0.59	0.94	FDA	0.63	0.92
	PLS	0.45	1.09	Group contribution	0.69	0.81
				Nearest neighbor	0.67	0.88
	Consensus	0.67	0.84	Consensus	0.73	0.77
48 h, *Daphnia magna*, LC_{50}	GBM	0.55	1.14	Hierarchical	0.70	0.98
	RF	0.58	1.1	Single Model	0.70	0.99
	SVM	0.58	1.1	FDA	0.57	1.19
	PLS	0.39	1.33	Group contribution	0.67	0.80
				Nearest neighbor	0.73	0.98
	Consensus	0.61	1.06	Consensus	0.74	0.91
48 h, *Tetrahymena pyriformis*, IGC_{50}	GBM	0.79	0.48	Hierarchical	0.72	0.54
	RF	0.77	0.51	FDA	0.75	0.49
	SVM	0.74	0.54	Group contribution	0.68	0.58
	PLS	0.67	0.60	Nearest neighbor	0.60	0.64
	Consensus	0.81	0.46	Consensus	0.76	0.48
Oral, rat, LD_{50}	GBM	0.57	0.62	Hierarchical	0.58	0.65
	RF	0.61	0.59	FDA	0.56	0.65
	SVM	0.56	0.63	Nearest neighbor	0.56	0.66
	PLS	0.44	0.71			
	Consensus	0.62	0.59	Consensus	0.63	0.60
BCF	GBM	0.74	0.70	Hierarchical	0.73	0.71
	RF	0.75	0.68	Single model	0.74	0.68
	SVM	0.61	0.86	FDA	0.71	0.75
	PLS	0.39	1.07	Group contribution	0.68	0.76
				Nearest neighbor	0.61	0.88
	Consensus	0.73	0.71	Consensus	0.76	0.66

publications [20, 24, 25]. According to literature analysis they are associated with high acute and reproductive toxicity.

Interpretation analysis involved only those molecular fragments that were found in 5 or more compounds of the training set, which, in our opinion, allowed us to focus on fragments that stably affect the studied type of toxicity and to a certain extent avoid

Table 3. Predictive performance of classification QSAR models.

Toxicity endpoints	SiRMS				T.E.S.T.			
	Model	BA	SEN	SP	Model	CON	SEN	SP
Developmental toxicity	GBM	0.71	0.89	0.52	Hierarchical	0.72	0.83	0.47
	RF	0.70	0.93	0.47	Single model	0.73	0.85	0.44
	SVM	0.66	0.98	0.34	FDA	0.72	0.78	0.59
					Nearest neighbor	0.80	0.84	0.67
	Consensus	0.70	0.94	0.46	Consensus	0.79	0.90	0.53
Ames mutagenicity	GBM	0.81	0.83	0.78	Hierarchical	0.76	0.78	0.75
	RF	0.82	0.84	0.80	FDA	0.78	0.77	0.79
	SVM	0.79	0.80	0.79	Nearest neighbor	0.77	0.78	0.75
	Consensus	0.82	0.84	0.80	Consensus	0.79	0.79	0.79

where: BA– balanced accuracy, SEN –sensitivity, SP –specificity, CON–Concordance

the influence of random factors, for example, errors in experimental data or predicted toxicity values and fragment contributions. Then we compared fragments for pairs of endpoints.

Analyzing the calculated contributions of fragments, we can note the following:

1) fragment contributions for aquatic toxicity (*Tetrahymena pyriformis, Daphnia magna, Fathead minnow*) were well correlated.
2) aliphatic and aromatic fragments substituted with chlorine or oxygen had much greater contributions to aquatic toxicity endpoints rather than to the acute oral toxicity on rats. At the same time phosphorous containing fragments had comparable contributions to both endpoints.
3) nitro groups and fragment containing them and CCl_3 groups had high contributions to mutagenicity and also contributed to aquatic toxicity.
4) nitrosamine, nitro-groups, acyl halides had high contributions to mutagenicity and moderate or low to acute oral toxicity on rats.
5) in some cases small fragments (like methyl, ethyl) had considerably different contributions. We assume that this was caused by the context of those fragments rather than their toxicity themselves.
6) for *Tetrahymena pyriformis* and bioconcentration factor (BCF) there was a clear difference. While aromatic fragments had comparable contributions, the aliphatic fragments had very high contributions for *Tetrahymena pyriformis* but low for BCF.

Summarizing the obtained relationships and previously proposed mechanisms of toxic action [26, 27] we can assume that aquatic organisms have similar mechanisms of toxic action. Inhibition of acetyl cholinesterase is the most important. Also in this case, an important role is played by oxidative phosphorylation in mitochondria.

Probably, the toxicity for *Tetrahymena pyriformis* is not due to the accumulation of a toxicant in the organism, but is realized through mentioned mechanisms of toxic action within a short period of time.

Alkylating agents are also highly significant for mutagenicity as opposed to acute toxicity for rats.

Table 4. Molecular fragments which steadily improve acute aquatic toxicity

Fragment, «A» is the place of attachment of a fragment to the rest of the molecule	SMARTS	Representative structures
4,5-dichlorobenzene	Clc1c([*:1])c([*:1])c([*:1])cc1Cl	*Fathead minnow,*-lg(LC$_{50}$)=5.4 *Daphnia magna,*-lg(LC$_{50}$)=6.0 *T. pyriformis,*-lg(IGC$_{50}$)=5.2
3-phenoxy	O([*:1])c1cc([*:1])ccc1	*Daphnia magna,*-lg(LC$_{50}$)=5.6 *Fathead minnow,*-lg(LC$_{50}$)=4.3 *T. pyriformis,*-lg(IGC$_{50}$)=3.9

(continued)

Table 4. (*continued*)

4-chlorobenzene A	Clc1ccc([*:1])cc1	 *Daphnia magna,*-lg(LC50)=5.7 *Fathead minnow,*-lg(LC50)=5.0 *T. pyriformis,*-lg(IGC50)=4.5
4-nitrobenzene A	[O-][N+](=O)c1cc([*:1])c([*:1])cc1	 *Daphnia magna,*-lg(LC50)=4.0 *T. pyriformis,*-lg(IGC50)=3.6 *Fathead minnow,*-lg(LC50)=4.7

(*continued*)

Table 4. (*continued*)

esters of phthalic acid	O=C(O([*:1]))c1cc ccc1C(=O)O([*:1])	
		Daphnia magna,-lg(LC$_{50}$)= 4.9 *Fathead minnow,*-lg(LC$_{50}$)= 5.4 *T. pyriformis,*-lg(IGC$_{50}$)=4.6

Besides, from the above fragment sets, those fragments that occur in at least three sets with high calculated fragment contributions to the corresponding activity or property were determined. Thus, we have identified fragments that have a higher contribution to toxicity to aquatic organisms than known toxicophors, such as carbamates and phosphoryl (Table 4). The main toxicophors for aquatic organisms are described in publications [28–30] and included as "structural alerts" (Endpoint "Acute Aquatic Toxicity") in the expert system OCHEM (https://ochem.eu//alerts/show.do?render-mode=full). Unfortunately, toxicophors proposed by other researchers have a rather general structure. For example, toxicophore "Aryl halide" does not allow us to understand the number of halogens in the benzene ring and their location relative to other substituents in the molecule. The analysis made it possible to define more precisely the toxicophors "Mononitroaromatics", "Aromatic alcohols", "Aryl halide", which are proposed in the OCHEM [31]. These new molecular fragments can be used as structural alerts for virtual screening of potentially hazardous organic compounds.

4 Conclusion

The proposed QSAR/QSPR models are expected to be useful for the prediction ecological endpoints used in the regulatory assessment of chemicals.

The structural interpretation was performed for the QSAR/QSPR models obtained using simplex descriptors. It allowed us to analyze the relationship between molecular fragments for different types of toxicity and suggest similar mechanisms of toxicity for different organisms. In this investigation, we have identified a set of structural fragments that increase the acute aquatic toxicity. The information can be used directly for fragment-based drug design or to establish structure–toxicity relationship trends and uncover possible mechanism(s) of toxic action.

The task of the next stage of research is to identify common fragments that persistently increase other types of toxicity, as well as to detail the molecular surroundings of other known toxicophors.

References

1. Zhu, H., et al.: Combinatorial QSAR modeling of chemical toxicants tested against Tetrahymena pyriformis. J. Chem. Inf. Model. **48**(4), 766–784 (2008)
2. Zhang, L., et al.: Applications of machine learning methods in drug toxicity prediction. Curr. Top. Med. Chem. **18**(12), 987–997 (2018)
3. Devillers, J., Devillers, H.: Prediction of acute mammalian toxicity from QSARs and interspecies correlations. SAR QSAR Environ. Res. **20**(5–6), 467–500 (2009)
4. Lagunin, A., et al.: ROSC-Pred: web-service for rodent organ-specific carcinogenicity prediction. Bioinformatics **34**(4), 710–712 (2018)
5. Stolbov, L., et al.: AntiHIV-Pred: web-resource for in silico prediction of anti-HIV/AIDS activity. Bioinformatics **36**(3), 978–979 (2020)
6. Khan, K., et al.: QSAR modeling of Daphnia magna and fish toxicities of biocides using 2D descriptors. Chemosphere **229**, 8–17 (2019)
7. Li, X., et al.: In silico prediction of chemical acute oral toxicity using multi-classification methods. J. Chem. Inf. Model. **54**(4), 1061–1069 (2014)
8. Lagunin, A., et al.: QSAR modelling of rat acute toxicity on the basis of PASS prediction. Mol. Inform. **30**(2–3), 241–250 (2011)
9. Web-based platform OCHEM. https://ochem.eu. Accessed 21 Apr 2020
10. Web-based platform Way2Drug. http://www.way2drug.com. Accessed 21 Apr 2020
11. Toxicity Estimation Software Tool. https://www.epa.gov/chemical-research/toxicity-estimation-software-tool-test. Accessed 21 Apr 2020
12. Guidance Document on the Validation of (Quantitative) Structure-Activity Relationship [(Q)SAR] Models. https://www.oecd.org/env/guidance-document-on-the-validation-of-quantitative-structure-activity-relationship-q-sar-models-9789264085442-en.htm. Accessed 21 Apr 2020
13. Polishchuk, P.: Interpretation of quantitative structure-activity relationship models: past, present, and future. J. Chem. Inf. Model. **57**(11), 2618–2639 (2017)
14. Webb, S.J., Hanser, T., Howlin, B., Krause, P., Vessey, J.D.: Feature combination networks for the interpretation of statistical machine learning models: application to Ames mutagenicity. J. Cheminform. **6**(1), 1–21 (2014). https://doi.org/10.1186/1758-2946-6-8
15. Sushko, Y., et al.: Prediction-driven matched molecular pairs to interpret QSARs and aid the molecular optimization process. J. Cheminform. **6**(1), 48 (2014)
16. Leach, A.G., et al.: Matched molecular pairs as a guide in the optimization of pharmaceutical properties; a study of aqueous solubility, plasma protein binding and oral exposure. J. Med. Chem. **49**(23), 6672–6682 (2006)
17. Polishchuk, P.G., et al.: Universal approach for structural interpretation of QSAR/QSPR models. Mol. Inform. **32**(9–10), 843–853 (2013)
18. Kuz'min, V.E., Artemenko, A.G., Muratov, E.N.: Hierarchical QSAR technology based on the Simplex representation of molecular structure. J. Comput. Aided Mol. Des. **22**(6–7), 403–421 (2008). https://doi.org/10.1007/s10822-008-9179-6
19. Software Marvin, ChemAxon. http://www.chemaxon.com. Accessed 21 Apr 2020
20. Polishchuk, P., et al.: Structural and physicochemical interpretation (SPCI) of QSAR models and its comparison with matched molecular pair analysis. J. Chem. Inf. Model. **56**(8), 1455–1469 (2016)

21. Polishchuk P.G.: SPCI - Tool for mining structure-property relationships from chemical datasets. https://github.com/DrrDom/spci. Accessed 21 Apr 2020

22. Software cxcalc, version 5.4; Chemaxon. https://chemaxon.com/marvin-archive/5_2_0/marvin/help/applications/calc.html. Accessed 21 Apr 2020

23. Software Standardizer, version 5.4; ChemAxon. https://chemaxon.com/presentation/introducing-the-standardizer-gui. Accessed 21 Apr 2020

24. Tin'kov, O.V., et al.: Analysis and prediction of the reproductive toxicity of organic compounds of different classes using 2D simplex representations of molecular structure. Pharm. Chem. J. **47**(8), 426–432 (2013)

25. Tinkov, O.V., et al.: QSAR investigation of acute toxicity of organic acids and their derivatives upon intraperitoneal injection in mice. Pharm. Chem. J. **49**(2), 104–110 (2015)

26. Kienzler, A., et al.: Mode of action (MOA) assignment classifications for ecotoxicology: an evaluation of approaches. Environ. Sci. Technol. **51**(17), 10203–10211 (2017)

27. Zhang, X., et al.: Discrimination of excess toxicity from narcotic effect: comparison of toxicity of class-based organic chemicals to Daphnia magna and Tetrahymena pyriformis. Chemosphere **93**(2), 397–407 (2013)

28. Verhaar, H.J.M., Leeuwen, C.J.V., Hermens, J.L.M.: Classifying environmental pollutants part 1: structural activity relationship for prediction of aquatic toxicity. Chemosphere **25**, 471–491 (1992)

29. Hermens, J.L.: Electrophiles and acute toxicity to fish. Environ. Health Perspect. **87**, 219–225 (1990)

30. von der Ohe, P., et al.: Structural alerts-a new classification model to discriminate excess toxicity from narcotic effect levels of organic compounds in the acute daphnid assay. Chem. Res. Toxicol. **18**(3), 536–555 (2005)

31. Sushko, I., et al.: Online chemical modeling environment (OCHEM): web platform for data storage, model development and publishing of chemical information. J. Comput. Aided Mol. Des. **25**(6), 533–554 (2011). https://doi.org/10.1007/s10822-011-9440-2

A New Network-Based Tool to Analyse Competing Endogenous RNAs

Selcen Ari Yuka$^{(\boxtimes)}$ and Alper Yilmaz

Chemical and Metalurgical Faculty, Bioengineering Department,
Yildiz Technical University, Esenler/Istanbul, Turkey
selcenay@yildiz.edu.tr

Abstract. Interactions between microRNA targets are defined as competing endogenous RNAs. After discovery of the repressive activity of microRNAs with different mechanisms, various experimental or computational approaches have been developed to understand the relationships among their targets. We developed a package ceRNAnetsim that provides network-based computational method as considering the expressions and interaction factors of microRNAs and their targets. By using ceRNA targets that have similar expression value as trigger on a relatively small network with 4 microRNAs and 20 gene targets, the perturbation efficiency of these ceRNAs on the network has been shown to be significantly different. However, the change was observed in the time (or iteration) to gaining steady-state of nodes on the network. So, we have provided the package which defines a user-friendly method for understanding complex ceRNA relationships, simulating the fluctuating behaviors of ceRNAs, clarifying the mechanisms of regulation and defining potentially important ceRNA elements. The ceRNAnetsim package can be found in Bioconductor software packages.

Keywords: Competing endogenous RNA · microRNA · Network modelling

1 Motivation

MicroRNAs are the family of the non-coding RNA group that plays an important role in the regulation of transcripts by post-transcription mechanisms [4,17]. This family shows activity on free gene transcripts in the cell, by mechanisms of degradation or repression of translation [3]. Following the determination that microRNAs have different mRNA transcript targets, it has been understood that the regulations of microRNA:targets is a rich and complex interaction network [20,21]. The state of change due to expression changes between different targets of a microRNA was explained by two mRNAs and one microRNA by study of Ala et al. [1]. Briefly, the ceRNA hypothesis is explained by the change in the expression of one of the mRNAs in the balanced/steady state (i.e. closely expressed genes that are targeted by a common microRNA), and the amount of other free

© Springer Nature Switzerland AG 2020
Z. Cai et al. (Eds.): ISBRA 2020, LNBI 12304, pp. 274–281, 2020.
https://doi.org/10.1007/978-3-030-57821-3_24

target mRNAs in the cell as a result of the change in the repression interest of the microRNA. Following this, genes acting as competing endogenous RNA have been identified in different studies [2,16,18]. Initial studies to understand interactions between microRNA and target genes include algorithms based on the detection of genes that are regulated in the same direction [7,19]. In these studies correlation based analysis with expression datasets identified mRNA genes showing comparable trends in microRNA based regulation. Similarly, List et al. have developed an R package [15] that allows to discovers and displays gene pairs with correlation coefficients using the sparse partial correlation method from inputs of gene expression and microRNA expression matrices. The method provided an understandable and user-friendly method for determining potential ceRNAs by analyzing their interactions on genome-wide [13].

Determining the regulation of potential ceRNA behavior will provide a good insight into studies that focus on using microRNA or mRNA targets as potential therapeutics. For this purpose, computational approaches have been developed for the expression relationships between microRNA:target genes. These approaches have been implemented by taking into account a small number of microRNAs:targets interaction content with few factors that are important in interactions [5,6,8,9]. However, considering the multi-layered and complex interactions of the ceRNAs, it needs to be handled in genome-wide network while incorporating more detailed interaction parameters [20].

2   Methods

2.1   Context of ceRNAnetsim

We have prepared an R package that processes the microRNA:target dataset into a network object and simulates microRNA:target regulation on this network via user-provided iteration. It can simulate the regulation of the ceRNAs on the network as function of microRNA, target gene expression and numerical coefficients that are important in interaction (such as affinity, degradation activity).

In the interaction of a balanced microRNA:target gene, the proportional distribution of the microRNAs to their targets are assigned as variables in edge data of network object. At this stage, if the user specify numeric values that are important in interactions in edge data (optionally), can add them as coefficients for the proportional distribution of microRNAs. Initially, the expression of microRNAs is distributed by taking into account the ratio of the target expression to the total target expression.

This situation is characterized as a steady or balanced state. When the user changes the expression of a network element (node) which will define its disturbing activity, the network found at steady state identifies this change as a trigger. This causes steady-state proportional distribution to spread to primary neighbors due to a perturbed node, followed by the changes in the edge data of the primary neighbors of the affected neighbors.

The number of iterations required for the system to become steady-state is a user-defined argument however, our package provides function which assist user to select appropriate iteration number. It creates the graphical output of the number of effected nodes in the system for each iteration, when trigger and expression change arguments are provided. So, user can visually pick the appropriate number of iterations for simulation.

On the other hand, *find_node_perturbation* function calculates overall effect of each node when perturbed. This function gives the user an idea about the effectiveness of the nodes used as triggers under the same expression change and iteration conditions. In other words, it provides the importance of network nodes to be determined by the coefficient of perturbation effectiveness. Additionally, there are additional functions in the ceRNAnetsim package such as displaying simulation results, calculating the perturbation efficiency of individual elements.

2.2 Simulation Evaluation in Small Dataset

We used networks that can be considered small compared to biological networks, in order to demonstrate capabilities of our approach. For this, we simulated the midsamp dataset containing interactions between 20 genes and 4 microRNAs and parameters (seed, binding energy and binding region on the target) converted into numerical expressions that may be important in interactions. We randomly selected two nodes; one central and another with lower centrality (Gene17 circled with black, Gene5 circled with red in Fig. 2E, respectively).

The *find_iteration* function was used to determine the number of iterations needed to reach steady-state in simulations. Consequently, we performed simulations using the Gene17 and Gene5 nodes as triggers with 3 fold upregulation and 10 iterations. To evaluate how perturbation activity will be affected by interaction parameters, we simulated with and without interaction parameters on the same node (Gene17).

3   Results and Discussion

We have devised a new approach to microRNA-mediated regulation of mRNAs that can take into account target gene and microRNA expressions and factors that are important in microRNA:target gene interactions (Workflow shown at Fig. 2A, briefly). Our approach can integrate many factors affecting repression or distribution of microRNA targets. For instance, seed location is known to affect repression whereas seed type and binding energy are known to affect distribution of microRNA over its targets [6,10,12]. The network-based approach has the potential to provide explanations on how genes that do not interact directly can be associated.

The ceRNAnetsim package contains functions for visualizing changes in network after set number of iterations. The user can obtain the network data of the steady state by using the *simulate()* function in dataset, as well as monitoring the perturbation spread by creating the network image for each iteration or a specific iteration with the visualisation functions (i.e. *simulate_vis()* or *vis_graph()*). The outputs of the visualisation functions in the first 4 iterations of the Gene17 perturbation are given at Fig. 1. Based on this, it has been shown that the simulation model can be used to spread effect of perturbation (at Fig. 1B-C) following the perturbation of steady state with a change (at Fig. 1A) in microRNA:target interactions, and to follow the re-regulation behavior (at Fig. 1D) of competing gene targets by indirect interactions. We found that ceRNA re-regulations display fluctuating trends, supporting previous literature information.

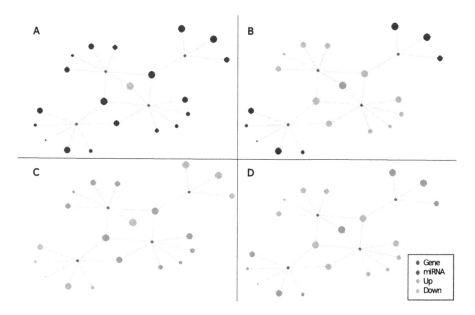

Fig. 1. Simulation visualisation of first four iterations when perturbation triggered by Gene17 in small sample dataset. (**A**) First response of network: upregulation of Gene17, (**B**) and (**C**) spreading of perturbation on network, (**D**) Fluctuating trend of perturbation. (Color figure online)

We have shown that all ceRNA nodes are affected (Fig. 2B) when simulation is triggered without taking into account the interaction parameters (see Fig. 2E, node circled in red with expression value of 5000), in a small sample network. However, different perturbation efficiency (Fig. 2C) was obtained by using a gene at another location that exhibits approximately the same expression (see Fig. 2E, node circled in black with expression value of 6000) as the trigger under the same conditions. However, when additional conditions (node circled in black in Fig. 2E, additional parameters are seed type and binding energy) are taken into account lower perturbation efficiency is calculated (see Fig. 2D). We observed that the time to reach steady-state in networks with fewer nodes is shorter and ceRNA behaviors are easier to understand.

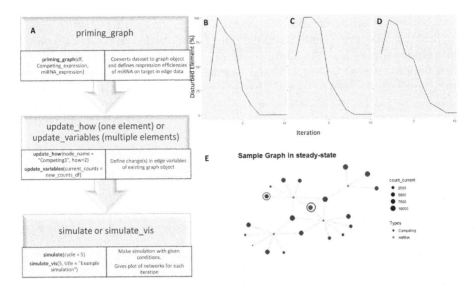

Fig. 2. (A) General workflow and essential functions in ceRNAnetsim, (B) perturbation efficiency of node circled with red and (C) black in sample graph, (D) perturbation efficiency of node marked with black circle (in considering interaction parameters), (E) Sample graph at steady-state expression of ceRNA node circled in red is 5000, node circled in black is 6000. (Color figure online)

The fact that not every network element has the same perturbation activity in a given microRNA:target network has raised an important question. Can the importance of all network elements in terms of perturbation be found with our model? Simply, the function *find_node_perturbation* scans each node under the same conditions in order to determine perturbation efficiency, giving the number of affected elements and the percentage of mean expression change of nodes in the network. Aforementioned function can identify the importance of existing genes as regulators, independent of network topological features.

Perturbation efficiencies with (seeds and binding energies) and without interaction parameters were compared in Table 1. There have been changes in the top six elements (0.25 of the total number of nodes) that are significant for perturbation in the network. For instance, the overall perturbation effect of Gene6 is reduced when we take into account seed and binding energy interaction parameters, while it is a significant element when interaction parameters are not taken into account. On the other hand, Gene9 has more perturbation efficiency with interaction parameters. Although Gene9 has almost half the expression of Gene6 expression in dataset (given as 6000 and 10000, respectively) Gene6 could not show effective perturbation due to the fact that Gene9 has higher seed interaction parameter value.

Table 1. Top six nodes which have highest perturbation efficiency in network

Without interaction parameters		With interaction parameters	
Node name	Perturbation efficiency	Node name	Perturbation efficiency
Mir4	8,25	Mir4	17,53
Mir3	4,47	Mir3	3,87
Mir2	2,89	Mir2	2,65
Gene6	2,04	Gene14	1,67
Gene14	1,33	Gene11	1,59
Gene11	1,07	Gene9	1,06

4 Conclusion

microRNAs are responsible for the regulation of many functions such as cell division, tissue differentiation, metabolism, signaling and immunity. Especially with the introduction of the ceRNA hypothesis, microRNA:target regulation has been a crucial topic for the diagnosis and treatment of different diseases [11,14]. microRNA:target interactions were explored by experimental methods or integrative analysis of microRNA and gene expressions. However, an appropriate computational method has not been developed which encompasses the whole microRNA:target network.

Graph-based approach to microRNA:target interactions is advantageous since complex calculations on edges and nodes can be performed easily. We provided the simulation of regulations by calculating microRNA activity as a function of these variables via network edge variables, in case the important parameters of microRNA:target interactions are given in numerical expressions. So, this package may be a new tool in the clarification of complex microRNA:ceRNA interactions, and it can identify critical nodes on the network.

In the case of simulating large data sets (i.e the network object contains too many edges and nodes), calculations take longer time despite parallel processing. We plan to speed up simulations and critical node detection so that large networks can be analyzed easily.

In summary, ceRNAnetsim package allows studying ceRNA networks with user-friendly and easy to understand functions. Our package is flexible so users can plug in factors regulating ceRNA effects. Although microRNAs play role as one of the key regulators in gene regulation, many other cellular molecules also have effect on gene regulation and our package is extendable to any type biological (molecular) elements affecting microRNAs, such as circular RNAs or non-coding RNAs.

References

1. Ala, U., et al.: Integrated transcriptional and competitive endogenous RNA networks are cross-regulated in permissive molecular environments. Proc. Natl. Acad. Sci. **110**(18), 7154–7159 (2013). https://doi.org/10.1073/pnas.1222509110
2. Arvey, A., Larsson, E., Sander, C., Leslie, C.S., Marks, D.S.: Target mRNA abundance dilutes microRNA and siRNA activity. Mol. Syst. Biol. **6** (2010). https://doi.org/10.1038/msb.2010.24
3. Bartel, D.P.: MicroRNAs. Cell **116**(2), 281–297 (2004). https://doi.org/10.1016/S0092-8674(04)00045-5, https://linkinghub.elsevier.com/retrieve/pii/S0092867404000455
4. Bartel, D.P.: MicroRNAs: target recognition and regulatory functions. Cell **136**(2), 215–233 (2009). https://doi.org/10.1016/j.cell.2009.01.002, https://linkinghub.elsevier.com/retrieve/pii/S0092867409000087
5. Bosson, A., Zamudio, J., Sharp, P.: Endogenous miRNA and target concentrations determine susceptibility to potential ceRNA competition. Mol. Cell **56**(3), 347–359 (2014). https://doi.org/10.1016/j.molcel.2014.09.018, https://linkinghub.elsevier.com/retrieve/pii/S1097276514007503
6. Breda, J., Rzepiela, A.J., Gumienny, R., van Nimwegen, E., Zavolan, M.: Quantifying the strength of miRNA-target interactions. Methods **85**, 90–99 (2015). https://doi.org/10.1016/j.ymeth.2015.04.012, https://linkinghub.elsevier.com/retrieve/pii/S1046202315001589
7. Davis, J.A., Saunders, S.J., Mann, M., Backofen, R.: Combinatorial ensemble miRNA target prediction of co-regulation networks with non-prediction data. Nucleic Acids Res. **45**(15), 8745–8757 (2017)
8. Denzler, R., Agarwal, V., Stefano, J., Bartel, D., Stoffel, M.: Assessing the ceRNA hypothesis with quantitative measurements of miRNA and target abundance. Mol. Cell **54**(5), 766–776 (2014). https://doi.org/10.1016/j.molcel.2014.03.045, https://linkinghub.elsevier.com/retrieve/pii/S1097276514002780
9. Denzler, R., McGeary, S., Title, A., Agarwal, V., Bartel, D., Stoffel, M.: Impact of MicroRNA levels, target-site complementarity, and cooperativity on competing endogenous RNA-regulated gene expression. Mol. Cell **64**(3), 565–579 (2016). https://doi.org/10.1016/j.molcel.2016.09.027, https://linkinghub.elsevier.com/retrieve/pii/S1097276516305767
10. Hausser, J., Syed, A.P., Bilen, B., Zavolan, M.: Analysis of CDS-located miRNA target sites suggests that they can effectively inhibit translation. Genome Res. **23**(4), 604–615 (2013). https://doi.org/10.1101/gr.139758.112, http://genome.cshlp.org/cgi/doi/10.1101/gr.139758.112
11. Hayes, J., Peruzzi, P.P., Lawler, S.: MicroRNAs in cancer: biomarkers, functions and therapy. Trends Mol. Med. **20**(8), 460–469 (2014)
12. Helwak, A., Kudla, G., Dudnakova, T., Tollervey, D.: Mapping the human miRNA interactome by CLASH reveals frequent noncanonical binding. Cell **153**(3), 654–665 (2013). https://doi.org/10.1016/j.cell.2013.03.043, https://linkinghub.elsevier.com/retrieve/pii/S009286741300439X
13. List, M., Dehghani Amirabad, A., Kostka, D., Schulz, M.H.: Large-scale inference of competing endogenous RNA networks with sparse partial correlation. Bioinformatics **35**(14), i596–i604 (2019)
14. Lu, M., et al.: An analysis of human microRNA and disease associations. PloS One **3**(10) (2008)

15. Markus List, M.S.: SPONGE (2017). https://doi.org/10.18129/B9.bioc.SPONGE, https://bioconductor.org/packages/SPONGE
16. Poliseno, L., Salmena, L., Zhang, J., Carver, B., Haveman, W.J., Pandolfi, P.P.: A coding-independent function of gene and pseudogene mRNAs regulates tumour biology. Nature **465**(7301), 1033–1038 (2010)
17. Rüegger, S., Großhans, H.: MicroRNA turnover: when, how, and why. Trends Biochem. Sci. **37**(10), 436–446 (2012). https://doi.org/10.1016/j.tibs.2012.07.002, https://linkinghub.elsevier.com/retrieve/pii/S096800041200103X
18. Salmena, L., Poliseno, L., Tay, Y., Kats, L., Pandolfi, P.: A ceRNA hypothesis: the rosetta stone of a hidden RNA language? Cell **146**(3), 353–358 (2011). https://doi.org/10.1016/j.cell.2011.07.014, https://linkinghub.elsevier.com/retrieve/pii/S0092867411008129
19. Seo, J., Jin, D., Choi, C.H., Lee, H.: Integration of microRNA, mRNA, and protein expression data for the identification of cancer-related microRNAs. PloS One **12**(1) (2017)
20. Smillie, C.L., Sirey, T., Ponting, C.P.: Complexities of post-transcriptional regulation and the modeling of cerna crosstalk. Critical Rev. Biochem. Mol. Biol. **53**(3), 231–245 (2018)
21. Tay, Y., Rinn, J., Pandolfi, P.P.: The multilayered complexity of ceRNA crosstalk and competition. Nature **505**(7483), 344–352 (2014). https://doi.org/10.1038/nature12986, http://www.nature.com/articles/nature12986

Deep Ensemble Models for 16S Ribosomal Gene Classification

Heta P. Desai[(⊠)], Anuja P. Parameshwaran, Rajshekhar Sunderraman[(⊠)], and Michael Weeks

Department of Computer Science, Georgia State University, Atlanta, GA 30302, USA
{hdesai1,aparameshwaran1}@student.gsu.edu,
{raj,mweeks}@cs.gsu.edu

Abstract. In bioinformatics analysis, the correct identification of an unknown sequence by subsequent matching with a known sequence is a crucial and critical initial step. One of the constantly evolving open and challenging areas of research is understanding the adaptation of microbiome communities derived from different environment as well as human gut. The critical component of such studies is to analyze 16s rRNA gene sequence and classify it to a corresponding taxonomy. Thus far recent literature discusses such sequence classification tasks being solved using many algorithms such as early methods of k-mer frequency matching, and assembly-based clustering or advanced methods of machine learning algorithms– for instance, random forests, naïve Bayesian techniques, and recently deep learning architectures. Our previous work focused on a comprehensive study of 16s rRNA gene classification by implementing simplistic singular neural models of Recurrent Neural Networks (RNNs) and Convolutional Neural Networks (CNNs). The outcome of this study demonstrated very promising classification results for family, genus and species taxonomic levels, prompting an immediate investigation into deep ensemble models for problem at hand. In this study, we attempt to classify 16s rRNA gene using deep ensemble models along with a hybrid model that emulates an ensemble in its early convolutional layers followed by a recurrent layer.

Keywords: 16S rRNA gene · Bacterial classification · RNNs · CNNs · Deep learning in genomics · Ensemble deep models

1 Introduction

In the early millennia, the first ever human genome was successfully sequenced. Ever since, a plethora of sequences including that of microbes, archaea and plants, have been sequenced and publicly made available for various genomic studies. In more recent decades, progressive trend in emerging next generation sequencing technologies have been seen, which vastly enhanced accuracy and rapidness of not only the whole genome shotgun sequencing (WGS) but also targeted gene sequencing or amplicon sequencing (AS) [1]. This phenomenon is noticeable in many areas of bioinformatics, especially in metagenomics. Metagenomics focuses on studying the composition of environmental

© Springer Nature Switzerland AG 2020
Z. Cai et al. (Eds.): ISBRA 2020, LNBI 12304, pp. 282–290, 2020.
https://doi.org/10.1007/978-3-030-57821-3_25

and human gut samples for abundance and identification of microbiome community and its chronological comparisons [2]. Metagenomics studies are crucial due to their applications in various fields such as ecology, biomedicine, environmental sciences, and microbiology. They are also important for studying gut microbiota for its role in maintaining healthy weight, blood sugar, cholesterol and immune system [3–6]. One of the most commonly used markers to correctly identify the composition of a microbiome community is 16S ribosomal ribonucleotide acid (rRNA) gene sequence [7]. In every cell of prokaryotic organisms, 16S rRNA gene is part of 30S subunit [7, 8]. This 30S subunit together with 50S subunit makes 70S ribosome –a site of protein synthesis [7, 8]. Because 16S rRNA gene is present in all bacteria and archaea, it serves as an identification card or a biological marker to study the presence of a species/taxa in biological samples. The sequence of 16S rRNA consists of nine hypervariable regions wrapped in between highly conserved regions. These hypervariable (V1–V9) regions make 16S rRNA gene to be rendered as a biological marker [9]. 16S rRNA gene sequencing is preferred due to it having low sequencing cost per Gigabyte, not requiring laboratory cell culture [10, 11] and requiring relatively low input DNA at the beginning [7]. On the contrary to popular belief that metagenomics and 16S rRNA are similar, metagenomics differs from 16s rRNA gene study on an important instance; while 16S rRNA gene study is an examination of relationship among different taxa based on a single gene, metagenomics is a study of all translated genes (entire translated genome) of all microbiomes in a sample [2]. While 16S rRNA gene study allows one to identify underlying taxa composition, it has limitation when the taxa composition of two different samples is predicted to be exactly the same or when two species have a very high sequence identity of >99.5% such as Streptococcus mitis and Streptococcus pneumoniae [7]. In this case, metagenomics whole genome shot gun sequencing may provide with a much deeper resolution of abundance as it sequences all translated genes of all present species including that of low fraction taxa, virus, and fungi. Figure 1 depicts an overview knowledge graph of 16S rRNA motivation and classification techniques.

Some of the basic techniques applied for classification of bacterial taxonomy are based on alignment, assembly [12], machine learning, and more recently deep learning. Many bioinformatics applications involve finding sequence similarity and correctly mapping sequences to sequences in known databases. Finding sequence similarity and correct sequence labeling require sequences to be mapped to databases with known sequence taxonomy known as reference genomes. Metagenomic sequences or 16s rRNA gene sequences are thus mapped to reference genomes using alignment algorithms such as Basic Local Alignment Search Tool (BLAST) to classify and measure abundance of taxa; for example, mothur and kraken known to perform read based sequence matching [13, 14]. Second widely utilized technique is assembly based in which first sequence is assembled into entire genome or large contigs and then gene curation is performed by matching predicted genes from contigs to known database. In either case, sequence matching requires some bioinformatics sequence manipulations and analysis. However, in machine learning or deep learning-based techniques, sequence reads or k-mers from sequences can be directly tested on previously trained models, reducing analysis duration. Some of the known machine learning based techniques such as naïve Bayesian,

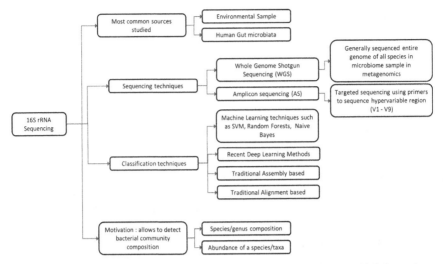

Fig. 1. Overview of 16S rRNA sequencing application and motivation in bioinformatics.

hierarchical clustering, random forests, and support vector machines, also have shown comparable results to aforementioned classifying techniques [15–17].

The recent advances of various affordable sequencing technologies coupled with the advancements of fast hardware (general-purpose graphic processing units (GPGPUs)), categorical big datasets, open source libraries and improved algorithms have enabled researchers, and scientists to develop multi-disciplinary studies [18]. This hardware acceleration aided in refinement of very powerful deep learning architectures for image and text classification; these discoveries then resulted in the rise of deep learning applications in medical imaging and genomics [12, 19]. Thus far, only a few studies have been published including ours that studies direct classification of 16S rRNA using deep learning architectures. Fiannaca et al. implemented a CNN and deep belief network (DBN) based classifiers for both targeted sequencing and whole genome sequencing taxonomy classification [12]. More recently, Busia et al. published a study with deep neural network (DNN) classifier that looked at various length sequences to note the performance [20]. Our published study's main goal was to compare performances of deep learning architectures especially of RNNs such as LSTM, BiLSTM with CNNs for 16S rRNA classification task [21].

2 Methods

Method development focuses on dataset preprocessing, and proposed deep learning models for 16S rRNA classification task. For all proposed models, input dataset is exactly the same, and tested on same training and validation data split. The overall goal of this study is to be able to create a model that can take raw reads with minimum pre-quality check and trimming requirements. This work implements architecturally four different models, three ensemble models and a hybrid model. The ensemble models average three different deep models, while hybrid model consists of both convolution and recurrent layers. The hybrid model, however, emulates the Multi-Filter model in Fig. 3a of published study [21] for its early convolutional layers with one striking difference: variable length of kernels in Multi-Filter model versus the same kernel sizes in the parallel convolutional branches of the hybrid model. For ensemble models, there are three different intrinsic models involved in making three different combination of models. Next two sections further discuss dataset and implemented models.

2.1 Dataset

The dataset used in this study remained same as previously published study [21]. This manually curated dataset is obtained from Genomic-based 16s ribosomal RNA Database (GRD) [22]. 16s rRNA gene or rDNA sequence length is approximately 1500 base pairs long; however, some of the bacteria can have multiple copies of 16s rRNA gene, hence input sequences from this dataset varies in length from 65 to 2900. Input files are same as [21], consists of two raw files; one containing tab delaminated fasta header with its corresponding bacterial taxonomy and other is fasta header tag with a fasta sequence containing all of the sequences in database. This study also focuses family, genus and species taxa levels as opposed to phylum, class, and order that are known to have >99% classification accuracies. Number of classes at each taxonomic level were 272, 840 and 2456 for family, genus, and species respectively. Approximately ~13,000 sequences were used for training the model and ~3500 for validation, which is 80%–20% split for training vs validation dataset. Preprocessing of sequences for input sequences is exactly as first published study, for further details please refer to [21]. The main focus of this study is to demonstrate the effectiveness of ensemble and hybrid models in achieving better classification accuracies compared to simpler deep models.

2.2 Deep Learning Approach

As discussed in introduction, deep learning models are on the rise with many applications in medical and biological fields. Architectures presented in this study are driven from previous study's results. In study [21], we observe a trend where recurrent models, Bidirectional LSTM and LSTM, outdo convolutional models. The outcome in this study [21] shows singular BiLSTM achieving highest accuracies for genus and species taxa; whereas, LSTM achieved the best accuracy for family taxa. The run time of BiLSTM for ~13,500, 100-character long sequences in training was much higher than of simple LSTM and simple CNN. Hence, in this study, the proposed model architectures are explored to grasp whether proposed models can achieve comparable accuracies as BiLSTM.

One type of proposed model is an ensemble model. Ensemble models have multiple classification algorithms incorporated, allowing them to perform better upon completion as oppose to an individual model [23]. Generally, ensemble model is able to improve accuracy if there is a good amount of variety in model architectures that makes up an ensemble model. In this study four different models – model 1, model 2, model 3 and model 4 – are developed. Model 1–3 are an averaging ensemble models, which are made up using combinations of four intrinsic sub-models: 1) a simple CNN model with two convolutional layers, 2) multi-filter CNN [21], 3) a hybrid model with two convolutional layers followed by a LSTM layer, lastly, 4) a simple two layers LSTM model. Specifically, model 1 – CNN-MultiFilterCNN-LSTM, consists aforementioned sub-models 1, 2 and 4; model 2 – CNN-CNN-LSTM, consists of two sub-models 1 and one sub-model 4; while, model 3 – CNN-hybrid-LSTM, consists of sub-model 1, 3, and 4. These three ensemble models average the output weights of its intrinsic sub-models. However, model 4 is a hybrid model, which is a single model that imitates the multi-filter CNN architecture from [21] in its earlier convolutional layers followed by a recurrent LSTM layer before the softmax classifier. Figure 2a illustrates the ensemble model particularly showcasing the model 3, while Fig. 2b illustrates model 4.

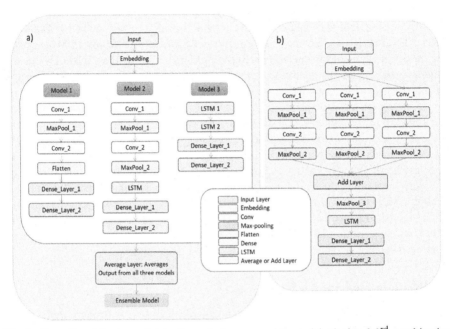

Fig. 2. Overall architecture of data flow of a) an ensemble model (depicted 3rd combination CNN-hybrid-LSTM), and b) hybrid model used for the classification task at hand.

Model 4 draws its architecture inspiration from the sequence to sequence deep model, which is staple model used for machine translation tasks deployed in many Natural Language Processing (NLP) such as speech recognition, language translation, and Computer Vision (CV) applications like video captioning [24, 25]. Sequence to sequence models are broadly made up with one model that acts as an encoder and another that decodes the output of an encoder. Model 4, however, does not have an encoder-decoder arrangement; it is a singular model that incorporates the convolutional and recurrent layers within its instance.

3 Results

Neural networks are known for their ability to learn very complex underlying patterns from large dataset; however, at the same time, their performance heavily relies on initial training weights as well as balanced un-bias training data. Due to such initial conditions, neural networks are susceptible to high variance, and ensemble models are one of the ways to reduce this variance by combining prediction accuracies of different models. Compared to previous studies, performance of the ensemble models and hybrid model aligns with accuracy greater than 85% for family and genus taxa [12, 20]. For species level, however, these models didn't surpass 70% accuracy achieved in our previous study [21].

Figure 3a, b and c shows loss and accuracy curves for family, genus and species taxa respectively. These figures only show model 3 (averaging ensemble) and model 4 (hybrid) curves since they achieved the highest accuracies. As described in Table 1 below, the highest validation accuracies for family and species taxa are 92.22% and 67.95%, achieved with hybrid model. However, at genus level, the highest validation accuracy achieved was 85.98% with CNN-hybrid-LSTM model and second highest validation accuracy of 85.94% with hybrid model. Even though, model 3 and model 4 outperformed previously obtained classification results for family and genus taxa, both models failed to outperform at species level, but stayed within 3% percentile range. All of the models in this study have comparable outcomes within 1–2% accuracies obtained for all taxa amongst each other, unlike our previously explored simplistic single models [21]. This agrees to the notion that ensemble and/or hybrid models tend to achieve better performance predictions than any singular model.

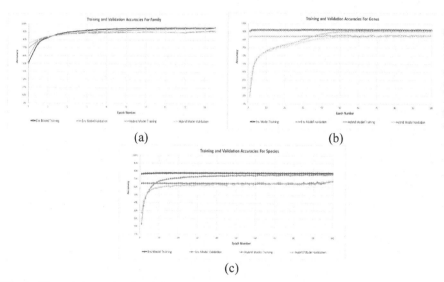

(a)   (b)

(c)

Fig. 3. The training and validation(testing) accuracies of classification for family (a), genus (b) and species (c) level including both hybrid model and ensemble model. For ensemble model, accuracies shown here are from the best performing CNN-hybrid-LSTM model.

Table 1. Final accuracies and losses of all four models for each family, genus and species taxonomic levels. Accuracies highlighted in bold are the highest classification accuracy achieved within each level.

Model Info		Family		Genus		Species	
No.	Name	Val_Loss	Val_Acc	Val_Loss	Val_Acc	Val_Loss	Val_Acc
1	CNN-MF-CNN-LSTM Ensemble Model	0.5760	90.20%	1.2330	85.76%	2.6654	66.86%
2	CNN-CNN-LSTM Ensemble Model	0.7239	91.33%	1.1342	85.80%	2.2033	67.15%
3*	CNN-hybrid-LSTM Ensemble Model	0.4670	91.60%	1.0007	**85.98%**	2.0057	67.39%
4*	Hybrid model	0.5231	**92.22%**	0.9988	85.94%	1.9226	**67.95%**

All models are trained using the same hyperparameters for all taxa classification at hand, except for epochs and non-linearity function. For family taxa classification, all four models are run with 20 epochs; whereas, for genus and species taxa, the models ran for 100 epochs until we saw no further improvements on the outcome loss and accuracy curves. For all LSTM layers, the number of hidden states is set to 500. For all models, the batch size used is 128, with 'adadelta' optimizer, learning rate applied is 0.01, and momentum of 0.0. Further hyperparameter optimization is an open avenue for ongoing improvisation.

4   Conclusion and Future Work

After studying various different deep learning architectures, it is determined that higher accuracies at species taxa level requires further refinement of 1) cleaning and pre-processing of 2456 classes to ensure at least thirty to forty sequences per class is maintained in species, 2) using larger than 100 bp length sequences (this is applicable to improve accuracies of other two taxa as well), and 3) developing a probabilistic model on top of a deep learning model. In this study, input reads are hundred base pair long, in other words, input is hundred characters long string; however, the model can easily be adapted for longer or shorter read lengths. Ongoing experiments are being developed to test different read length effects on outcome. With deep learning architectures such as recurrent neural networks, longer strings may provide a finer representation of features in recognizing underlying patterns. These taxonomical data consist of a hierarchical relationship which cannot be used to find abundance of sequences in a sample, but it can certainly be used for sequence classification tasks. Next steps involve of using such information along with higher dimension input feature vector of different sequence regions to further improve accuracies at Family, Genus and Species taxa levels.

References

1. Achtman, M.: A phylogenetic perspective on molecular epidemiology. In: Molecular Medical Microbiology, pp. 485–509. Academic Press (2002)

2. Ghosh, A., Mehta, A., Khan, A.M.: Metagenomic analysis and its applications. Encycl. Bioinf. Comput. Biol. **3**, 184–193 (2019)
3. Qin, J., et al.: A metagenome-wide association study of gut microbiota in type 2 diabetes. Nat. **490**(7418), 55–60 (2012)
4. Turnbaugh, P.J., Ley, R.E., Mahowald, M.A., Magrini, V., Mardis, E.R., Gordon, J.I.: An obesity-associated gut microbiome with increased capacity for energy harvest. Nature **444**(7122), 1027 (2006)
5. Turnbaugh, P.J., et al.: A core gut microbiome in obese and lean twins. Nature **457**(7228), 480–484 (2009)
6. Karlsson, F.H., et al.: Symptomatic atherosclerosis is associated with an altered gut metagenome. Nat. Commun. **3**(1), 1–8 (2012)
7. Janda, J.M., Abbott, S.L.: 16S rRNA gene sequencing for bacterial identification in the diagnostic laboratory: pluses, perils, and pitfalls. J. Clin. Microbiol. **45**(9), 2761–2764 (2007)
8. Berg, J.M., Tymoczko, J.L., Stryer, L.: Biochemistry. Freeman, New York (2002)
9. Chakravorty, S., Helb, D., Burday, M., Connell, N., Alland, D.: A detailed analysis of 16S ribosomal RNA gene segments for the diagnosis of pathogenic bacteria. J. Microbiol. Methods **69**(2), 330–339 (2007)
10. Woo, P.C.Y., Lau, S.K.P., Teng, J.L.L., Tse, H., Yuen, K.Y.: Then and now: use of 16S rDNA gene sequencing for bacterial identification and discovery of novel bacteria in clinical microbiology laboratories. Clin. Microbiol. Infect. **14**(10), 908–934 (2008)
11. Woese, C.R.: Bacterial evolution. Microbiol. Rev. **51**(2), 221 (1987)
12. Fiannaca, A., et al.: Deep learning models for bacteria taxonomic classification of metagenomic data. BMC Bioinform. **19**(7), 198 (2018)
13. Schloss, P.D., et al.: Introducing mothur: open-source, platform-independent, community-supported software for describing and comparing microbial communities. Appl. Environ. Microbiol. **75**(23), 7537–7541 (2009)
14. Wood, D.E., Salzberg, S.L.: Kraken: ultrafast metagenomic sequence classification using exact alignments. Genome Biol. **15**(3), R46 (2014)
15. Wang, Q., Garrity, G.M., Tiedje, J.M., Cole, J.R.: Naive Bayesian classifier for rapid assignment of rRNA sequences into the new bacterial taxonomy. Appl. Environ. Microbiol. **73**(16), 5261–5267 (2007)
16. La Rosa, M., Fiannaca, A., Rizzo, R., Urso, A.: Probabilistic topic modeling for the analysis and classification of genomic sequences. BMC Bioinform. **16**(6), S2 (2015)
17. Zhang, A.B., Sikes, D.S., Muster, C., Li, S.Q.: Inferring species membership using DNA sequences with back-propagation neural networks. Syst. Biol. **57**(2), 202–215 (2008)
18. LeCun, Y.: 1.1 Deep learning hardware: past, present, and future. In: 2019 IEEE International Solid-State Circuits Conference-(ISSCC), pp. 12–19. IEEE, February 2019
19. Park, Y., Kellis, M.: Deep learning for regulatory genomics. Nat. Biotech. **33**(8), 825 (2015)
20. Busia, A., et al.: A deep learning approach to pattern recognition for short DNA sequences. bioRxiv, 353474 (2019)
21. Desai, H.P., Parameshwaran, A.P., Sunderraman, R., Weeks, M.: Comparative study using neural networks for 16S ribosomal gene classification. J. Comput. Biol. **27**(2), 248–258 (2020)
22. Laboratory for integrated bioinformatics, center for integrative medical sciences, RIKEN. GRD - Genomic-based 16S ribosomal RNA Database (2015)
23. Rokach, L.: Ensemble-based classifiers. Artif. Intell. Rev. **33**(1–2), 1–39 (2010). https://doi.org/10.1007/s10462-009-9124-7

24. Prabhavalkar, R., Rao, K., Sainath, T.N., Li, B., Johnson, L., Jaitly, N.: A comparison of sequence-to-sequence models for speech recognition. In: Interspeech, pp. 939–943, August 2017
25. Venugopalan, S., Rohrbach, M., Donahue, J., Mooney, R., Darrell, T., Saenko, K.: Sequence to sequence-video to text. In: Proceedings of the IEEE International Conference on Computer Vision, pp. 4534–4542 (2015)

Search for Tandem Repeats in the First Chromosome from the Rice Genome

Eugene V. Korotkov[1,2(✉)], Anastasya M. Kamionskaya[1], and Maria A. Korotkova[2]

[1] Institute of Bioengineering, Research Center of Biotechnology of the Russian Academy of Sciences, Bld.2, 33 Leninsky Ave., 119071 Moscow, Russia
genekorotkov@gmail.com, rifampicin@yandex.ru
[2] National Research Nuclear University MEPhI (Moscow Engineering Physics Institute), 31 Kashirskoye Shosse, 115409 Moscow, Russia
bioinf@rambler.ru

Abstract. Using the RPWM method, we searched for tandem repeats of 2 to 50 nucleotides long in the rice genome. We compared the effectiveness of the RPWM method with Mreps, T-reks, Tandem Repeat Finder and ATR Hunter. About 70% of the tandem repeats found could not be found by other algorithms. The correlation of dispersed repeats and transposons with tandem repeats was studied in this work. We assumed that some of the dispersed repeats and transposons originated from tandem repeats

Keywords: Tandem repeats · Dynamic programming · Rice genome

1 Introduction

The search for tandem repeats is an important task in studying the genomes of various organisms. The interest in tandem repeats emanated from the fact that their excessive or insufficient number in some regions of the genome leads to a wide range of human diseases. These diseases include Fragile-X syndrome, Huntington's disease, Friedreich's ataxia, and certain forms of cancer [1].

Two classes of mathematical methods are currently used to search for periodicity in nucleotide sequences. The first class of methods includes spectral approaches. These methods make it possible to perfectly find the periodicity in the nucleotide sequence, even when individual tandem repeats are significantly different from each other [2–4]. However, spectral methods cannot find tandem repeats in the presence of insertions or deletions of nucleotides, which is a very serious drawback and severely limits the use of these methods.

The second class of methods includes algorithms that use dynamic programming and can find tandem periods in the presence of insertions and deletions of nucleotides [5]. These methods use pairwise repetition comparison. Therefore, they cannot detect tandem periods when the similarity is statistically insignificant between any two repeats [4]. Methods based on dynamic programming include algorithms and programs such as TRF [5], Mreps [6], TRStalker [7], ATRHunter [8], T-REKS [9], IMEX [10], CRISPRs

© Springer Nature Switzerland AG 2020
Z. Cai et al. (Eds.): ISBRA 2020, LNBI 12304, pp. 291–295, 2020.
https://doi.org/10.1007/978-3-030-57821-3_26

[11], SWAN [12]. Therefore, the question remains unresolved whether all tandem repeats can be found by the developed mathematical methods and algorithms. It is convenient to introduce a measure of the evolutionary distance between two separate repeats in the form of the average number of nucleotide substitutions per nucleotide x accumulated by them relative to each other. It is shown below that all developed methods can detect tandem repeats for $x < 2.0$. However, DNA regions may exist where tandem repeats have accumulated a significant number of DNA base replacements and $x > 2.0$. In this case, they will be skipped by all previously developed mathematical methods and algorithms.

Previously, based on the generation of random position-weight matrices (RPWM), a new mathematical method was developed to search for tandem repeats [13]. The RPWM method allows the detection of tandem repeats with a significant number of insertions or deletions. In this paper, we wanted to solve three problems. Firstly, we wanted to compare the effectiveness of searching for tandem repeats by the RPWM method and some other popular programs. Secondly, we wanted to find all tandem repeats, including tandem repeats for which $x > 1.25$. These calculations were carried out in the range of period lengths from 2 to 50 nucleotides. Third, we wanted to study the correlation between the tandem repeats we discovered and the dispersed repeats, which were discovered earlier in the rice genome in [14]. Our results showed that the mathematical method we developed allowed us to find many tandem repeats in the rice genome that have not been previously identified. The RPWM method [13] allows this to be done for $0.0 < x < 3.2$. A comparative analysis showed that other developed algorithms can find tandem repeats for x in the range from 0.0. up to 1.25.

2 Methods and Algorithms

To conduct a comparative analysis of the effectiveness of various methods of searching for tandem repeats, we created an artificial nucleotide sequence S with a specific value of x. Here x, as above, is the average number of nucleotide substitutions between any two repeats in the sequence S. The sequence length was 3000 nucleotides, the number of repeats was 100, the length of one repeat was 30 nucleotides. In sequence S, 50 insertions and 50 deletions of nucleotides were made at random positions. For each value of x in the interval from 0.0 to 4.0 with a step equal to 0.25, we generated a set of $Q(x)$ sequences of S. Each set $Q(x)$ contained 100 sequences of S. The sequences in the set $Q(x)$ differed in the initial 30 nucleotide period. Thereafter, we analyzed each $Q(x)$ set with the programs Mreps [6], T-reks [9], Tandem Repeat Finder [5], and ATR Hunter [8]. These programs produced a file with various tandem repeats. In the output file, we counted the number of repeats of 30 nucleotides in length. This calculation was made for all programs for the set $Q(x)$ for different x. We obtained similar data for the RPWM method [13]. This algorithm calculated a quantitative measure for the found tandem repeats in the form of the similarity function $F_{max}(x)$ (see paragraph 2.2). We calculated $F_{max}(x)$ for $Q(10.0)$ and in this case, the sequences were considered as random. The set $Q(10.0)$, as written above, contains 100 sequences of S. Therefore, we got 100 values of $F_{max}(10.0)$. We calculated the average value $\overline{F_{max}(10.0)}$ and $D(F_{max}(10.0))$. Then, we calculated:

$$Z(x) = \frac{F_{max}(x) - \overline{F_{max}(10.0)}}{\sqrt{D(F_{max}(10.0))}} \tag{1}$$

The RPWM algorithm for tandem repeats search was described in detail in [13]. Dispersed repeat coordinates were obtained from [14] from the site https://github.com/oushujun/EDTA. We believed that there is overlap between these genome regions and the regions earlier found with tandem repeats, if the length of the intersecting region was more than 80%. This 80% was taken from the minimum length of the two compared regions.

3 Results and Discussion

In this work, we did not analyze tandem repeats having lengths of 3 nucleotides. This is because these lengths of periods are inherent in the triplet periodicity of the genes [15]. Therefore, a filter was applied that eliminated such tandem repeats without insertions or deletions in the analyzed window with a length of 600 bases. We calculated the information decomposition for a period of length 3 [15]. Triplet periodicity was found in this work only if tandem repeats with length equal to 3 nucleotides contained insertions or deletions of nucleotides and therefore was not screened out by a preliminary filter.

Table 1 shows that Mreps [6], T-reks [9], Tandem Repeat Finder [5], and ATR Hunter [8] found approximately half of the dispersed repeats for x from 0.5 to 0.75. For x greater than 1.25, the ability to correctly find tandem repeats for these methods is completely absent. At the same time, it can be seen from Table 2 that the RPWM method [13] revealed statistically significant tandem repeats up to $x = 3.0$. $Z(3.0) = 5.1$ is a statistically significant value (approximately 1 false positive per million S sequences). This shows that the ability of the method [13] to search for tandem repeats is approximately 3 times higher than that of all previously developed algorithms. The same result was obtained earlier when we compared RPWM with T-reks [13] only.

Table 1. Average percentage of found tandem repeats in test nucleotide sequences from the set Q(x).

X	Mreps	T_REKS	TR Finder	ATR Hunter
0	85	89	100	52
0,25	0	22	100	52
0,5	0	16	99,6	28
0,75	0	6	18,3	16
1,0	0	0	0	2
1,25	0	0	0	0

Then, we studied the complete first chromosome from the rice genome by the RPWM method. The sequence was obtained from https://plants.ensembl.org/info/website/ftp/index.html. A total of 8277 regions with tandem periods of various lengths were found.

In Table 3, we divided all dispersed repeats found in the rice genome in [14] into separate classes of tandem repeats. A total of 33 classes of dispersed repeats (including transposons) were obtained in [14].

Table 2. $Z(x)$ for the previously developed tandem repeat search method [13].

x	0	0,5	1,0	1,5	2,0	2,25	2,5	2,75	3,0	3,5	4,0
Z	114	96	67	43	23	19	12.5	7,0	5,1	3,8	2,8

For each class of dispersed repeats, we calculated the number of intersections with the tandem repeats found. We also estimated the statistical significance of intersections between dispersed and tandem repeats. For this, we used 100 artificial chromosomes, where the positions of each class of dispersed repeats were randomly mixed without changing their length. Here, Z shows the number of deviations from the average, for the number of intersections in the number of standard deviations. Table 3 shows that some classes of dispersed repeats have a strong correlation with the tandem repeats which we found. This was observed for transposons as well.

Table 3. The intersection of the found tandem repeats with dispersed repeats where $Z > 0.0$ [14].

№	Repeat name	Number of repeats	The number of intersections with the found tandem repeats	Expected number of intersections	Z
1	Anona/Helitron	4173	605	326	15,45
2	Anona/MULE	7121	1038	402	31,72
3	NE/unknown	2188	274	164	8,59
4	R/Gypsy	4233	1256	990	8,45
5	Anona/hAT	2585	282	134	1279
6	R/Copia	1412	273	198	5,33
7	Aauto/CACTA	400	154	74	9,30
8	Anona/CACTA	1100	237	60	22,85
9	R/unknown	132	26	10	5,06
10	Anona/CACTG	389	122	26	18,83
11	Aauto/CACTG	512	330	124	18,50
12	Centro/tandem	74	183	20	36,45

For example, such a correlation is visible for Helitron transposons (No. 1), Mutator-like transposable elements (Mule, No. 2), retrotransposon Gypsy (No. 4), and many other dispersed repeats. More than 70% of the identified tandem repeats cannot be detected by previously developed methods [13]. In general, the results show that many families of dispersed repeats and transposons either contain tandem repeats themselves, or are integrated into parts of the genome that contains tandem repeats. This preferential embedding can be explained by the fact that some tandem repeats in genome are well recognized during transposition. It can be assumed that part of the dispersed repeats and transposons could have evolved from tandem repeats.

Acknowledgement. The work was partly supported by the grant RFBR N.20-016-00057A.

References

1. Richard, G.F., Kerrest, A., Dujon, B.: Comparative genomics and molecular dynamics of DNA repeats in eukaryotes. Microbiol. Mol. Biol. Rev. **72**, 686–727 (2008). https://doi.org/10.1128/MMBR.00011-08
2. Lobzin, V.V., Chechetkin, V.R.: Order and correlations in genomic DNA sequences. The Spectral Approach. Uspekhi Fizicheskih Nauk. **170**, 57–68 (2000)
3. Kravatskaya, G.I., Kravatsky, Y.V., Chechetkin, V.R., Tumanyan, V.G.: Coexistence of different base periodicities in prokaryotic genomes as related to DNA curvature, supercoiling, and transcription. Genomics **98**, 223–231 (2011). https://doi.org/10.1016/j.ygeno.2011.06.006
4. Korotkov, E.V., Korotkova, M.A., Kudryashov, N.A.: Information decomposition method to analyze symbolical sequences. Phys. Lett. Sect. A Gen. At. Solid State Phys. **312**, 198–210 (2003)
5. Benson, G.: Tandem repeats finder: a program to analyze DNA sequences. Nucleic Acids Res. **27**, 573–580 (1999)
6. Kolpakov, R., Bana, G., Kucherov, G.: mreps: efficient and flexible detection of tandem repeats in DNA. Nucleic Acids Res. **31**, 3672–3678 (2003)
7. Pellegrini, M., Renda, M.E., Vecchio, A.: TRStalker: an efficient heuristic for finding fuzzy tandem repeats. Bioinform. (Oxford, England) **26**, i358–i366 (2010). https://doi.org/10.1093/bioinformatics/btq209
8. Wexler, Y., Yakhini, Z., Kashi, Y., Geiger, D.: Finding approximate tandem repeats in genomic sequences. J. Comput. Biol. A J. Comput. Mol. Cell Biol. **12**, 928–942 (2005). https://doi.org/10.1089/cmb.2005.12.928
9. Jorda, J., Kajava, A.V.: T-REKS: identification of Tandem REpeats in sequences with a K-meanS based algorithm. Bioinform. (Oxford, England) **25**, 2632–2638 (2009)
10. Mudunuri, S.B., Kumar, P., Rao, A.A., Pallamsetty, S., Nagarajaram, H.A.: G-IMEx: a comprehensive software tool for detection of microsatellites from genome sequences. Bioinform. **5**, 221–223 (2010)
11. Grissa, I., Vergnaud, G., Pourcel, C.: CRISPRFinder: a web tool to identify clustered regularly interspaced short palindromic repeats. Nucleic Acids Res. **35**, W52–W57 (2007). https://doi.org/10.1093/nar/gkm360
12. Boeva, V., Regnier, M., Papatsenko, D., Makeev, V.: Short fuzzy tandem repeats in genomic sequences, identification, and possible role in regulation of gene expression. Bioinform. (Oxford, England) **22**, 676–684 (2006). https://doi.org/10.1093/bioinformatics/btk032
13. Korotkov, E.V., Korotkova, M.A.: Search for regions with periodicity using the random position weight matrices in the C. elegans genome. Int. J. Data Min. Bioinform. **18**(4), 331–354 (2017). https://doi.org/10.1504/IJDMB.2017.088141
14. Ou, S., et al.: Benchmarking transposable element annotation methods for creation of a streamlined, comprehensive pipeline. Genome Biol. **20**, 275 (2019). https://doi.org/10.1186/s13059-019-1905-y
15. Frenkel, F.E., Korotkov, E.V.: Classification analysis of triplet periodicity in protein-coding regions of genes. Gene **421**, 52–60 (2008). https://doi.org/10.1016/j.gene.2008.06.012

Deep Learning Approach with Rotate-Shift Invariant Input to Predict Protein Homodimer Structure

Anna Hadarovich[1,2(✉)], Alexander Kalinouski[1], and Alexander V. Tuzikov[1,2]

[1] United Institute of Informatics Problems, National Academy of Sciences, 220012 Minsk, Belarus
ahadarovich@gmail.com
[2] Belarusian State University, 220030 Minsk, Belarus

Abstract. The ability to predict protein complexes is important for applications in drug design and generating models of high accuracy in the cell. Recently deep learning techniques showed a significant success in protein structure prediction, but a protein docking problem is unsolved yet. We developed a two-staged approach which consists of deep convolutional neural network to predict protein contact map for homodimers and optimization procedure based on gradient descent to build the homodimer structure from the contact map. Neural network uses the distance map calculated as all pairwise Euclidian distances between CB atoms of protein 3D structure as input, which is invariant to rotation and translation. The network has a large receptive filed to capture patterns in contacts between residues. The suggested approach could be generalized to heterodimers because it does not depend on symmetry features inherent in homodimers. The presented algorithm could be also used for scoring protein homodimers models in docking.

Keywords: Protein docking · Protein homodimers · Deep learning

1 Introduction

Structural characterization of protein-protein interactions can significantly boost the development in structural biology and has important applications in drug design. Interaction between proteins is characterized by the 3D structure of their complex. The problem of finding the three-dimensional structure of a complex formed by proteins is called a protein docking. Despite substantial number of docking approaches there is no universal algorithm which could predict the structure of protein complex with high accuracy [1]. Experimental techniques remain the only alternative but consume much time and resources. Computational methods can be helpful in fast and reliable modeling of protein complexes, using as input the three-dimensional structure of individual proteins, their sequences, functional properties and other information available. In this case, the region of binding, or the interface of the protein complex, is of the greatest importance. This region contains amino acids which are necessary for the formation of the complex to perform inherent functionality.

© Springer Nature Switzerland AG 2020
Z. Cai et al. (Eds.): ISBRA 2020, LNBI 12304, pp. 296–303, 2020.
https://doi.org/10.1007/978-3-030-57821-3_27

Methods to model the protein complexes can be divided into two classes. The first class includes methods based on templates. It uses an assumption that structure to be modelled (called target) has to be close to the complex with similar properties (called template). Similarity criteria could be based on sequence identity, functional annotations and etc. [2–9]. The second class of methods is called free modeling (ab initio docking), when the structure of the complex is predicted by searching for a relative orientation of the proteins with minimum binding energy. Since in this case the binding energy is unknown, various functions are used to predict its value for a given orientation of the proteins. These functions can be built on the basis of statistical potentials, physico-chemical properties, geometric complementarity, etc. [10–16]. Recently Deep Learning became very popular [17–19] providing new opportunities for structural biology [20]. It has pushed community to create algorithms to solve protein docking problem by deep learning [21, 22]. Inspired by the success of Deep Mind team, we developed an algorithm which exploits deep neural network to predict the structure of protein homodimer complexes.

2 Materials and Methods

The prediction algorithm comprises two stages. In the first stage deep fully convolutional neural network [23] with only delated operations (without subsampling) [24] and residual blocks was built to predict contact map (contacts between interfacial residues in homodimers) based on protein distance map (distances between residues inside the one chain of homodimer) [23, 25–27]. Interfacial residues were defined as residues at opposite chains of homodimer having less than 8Å between CB atoms. The developed architecture consists of the repetition of blocks shown in Fig. 1.

Binary cross-entropy was used as a loss function, where y_i denotes an existence of the contact (that is, y is equal to 1 if the contact exists, 0 otherwise), $p(y_i)$ stands for probability of the contact prediction, N is the number of CB atoms (CA in case of GLY) [5]:

$$H_p(q) = -\frac{1}{N} \sum\nolimits_{i=1}^{N} y_i \times log(p(y_i)) + (1 - y_i) \times log(1 - p(y_i)) \tag{1}$$

Except for well-known neural networks operations like convolution and maxpooling, which are common operation for CNN, here is some specificity which distinguishes the architecture developed.

Best results were obtained with the following parameters:

The number of epochs (iterations in the learning process, including the feeding of all examples from the training set to the network and the verification of the quality of training on the control set) – 500;
Number of iterations per one epoch – 200;
Learning rate – 0,0001;
Optimization – Adam method [28].

In the second stage an optimization procedure based on gradient descent algorithm was used to build the homodimer structure. This procedure allows to build the model

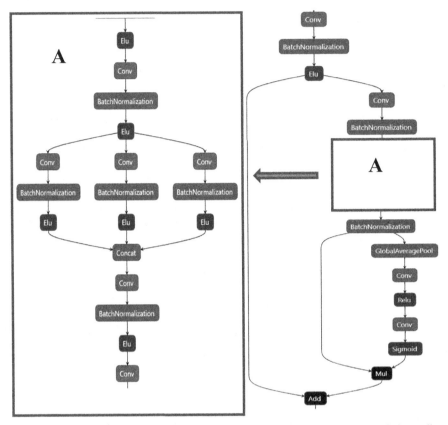

Fig. 1. Architecture of the block of deep neural network to predict the contact map of a homodimer.

of protein homodimer structure provided that the contact map was predicted with high accuracy in the previous stage. It requires a predicted contact map and 3D structure of the protein forming the homodimer as input. Let us consider a binary matrix A = [a(i, j)] of shape (N × N), where N is the number of residues in a single protein of the homodimer. The matrix elements were calculated as follows [1]:

$$a(i,j) = \begin{cases} 1, \textit{if } d(i,j) < 8\text{Å} \\ 0, \textit{else} \end{cases} \tag{2}$$

As in the case of the standard approach of free modeling of a dimeric (consisting of two proteins) complex, one protein structure is fixed in 3D space. Then the problem is formulated as second protein position prediction relative to the first one. It can be unambiguously identified by 6 parameters responsible for 6 degrees of freedom: 3 parameters specify a rotation and 3 parameters determine a shift.

We assume at the beginning that the initial position of the second protein structure coincides with the location of the first one, i.e., all 6 parameters are 0.

For each iteration one needs to calculate:

- Distance matrix between CB atoms of the first and second protein structures.
- Loss function values.
- Derivative of the loss function.
- Parameters responsible for rotation and shift to minimize the loss function in the direction of the calculated derivative.
- Transformation matrix for the new atomic coordinates.
- New coordinates of the second protein structure from the resulting matrix.

At the last stage, 6 parameters are transformed into an affine transformation matrix, which is then applied to the structure of the second protein to obtain its coordinates. The provided output is a matrix of affine transformation T of size 4×4 [4]:

$$T = \begin{pmatrix} r_{11} & r_{12} & r_{13} & t_1 \\ r_{21} & r_{22} & r_{23} & t_2 \\ r_{31} & r_{32} & r_{33} & t_3 \\ 0 & 0 & 0 & 1 \end{pmatrix} \tag{3}$$

Here elements r_{ij} denote rotation parameters, and t_i shift parameters respectively.

After the second stage additional procedure for penetration removal at interface was introduced. It is an iterative process of changing translation values in the obtained transformation matrix for the homodimer with 2Å step until penetration between chains of the homodimer was removed.

3 Datasets

Homodimers dataset was retrieved from biounits files from Protein Data Bank (PDB) [29]. Filtration procedure was applied to the obtained data: files with erroneous information were excluded, biounits files with number of chains different from two were filtered out. Additionally, BSA (Buried Surface Area) was calculated with FreeSASA library [30] to ensure selected complexes have interfaces of reasonable size [31]. Final dataset comprised about 10000 structures. Obtained dataset was split into training and test dataset in proportion 4:1 accordingly.

4 Results and Discussion

For more than half of the complexes it was possible to predict most of the interface residues with an accuracy of more than 50%, out of which the quarter of the complexes reached the accuracy of more than 70%. The predicted interface contacts greatly reduce the search area for a three-dimensional structure, since the binding region is known. This allows to impose restrictions on the search area and significantly reduce the number of possible orientations of protein structures in the complex. For many docking algorithms, the accuracy of the structure prediction increases with provided information about potential interface.

An example of predicted interface in comparison with the real one is shown in Fig. 2. Here the obtained interface can be used as constraint for protein docking or as input data for the second stage of the algorithm.

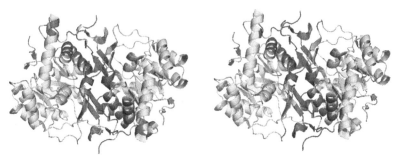

Fig. 2. Example of the interface prediction for homodimer 3w1x (left panel) in comparison with the real interface (right panel). Two protein chains of the homodimer are shown in cyan and green. (Color figure online)

For nearly 30% of homodimers the given algorithm has provided predictions of complex structure of good quality which means RMSD between CB atoms of real and predicted structures was less than 10Å. An example of input and output data for the first stage of the algorithm is presented in Fig. 3. Middle and right columns present the images of the real and predicted contact maps which are hardly distinguishable.

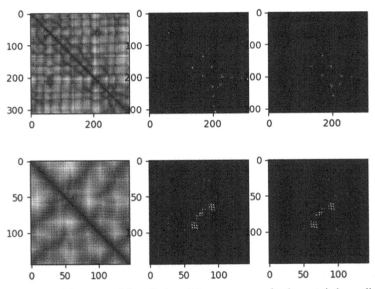

Fig. 3. Examples of the successful prediction of the contact map for the protein homodimer 3w1x (upper row) and 1jzk (lower row). Left column: distance map between CB atoms of the first protein in the homodimer used as input data for deep neural network. Middle column: contact map (ground truth) for the protein homodimer. Right column: predicted contact map for the protein homodimer.

The example of the second stage of the prediction algorithm is shown in Fig. 4, where the predicted position of the second protein in homodimer is compared with the real position. Figure 4 shows results produced after additional procedure for penetration removal described in Materials and methods.

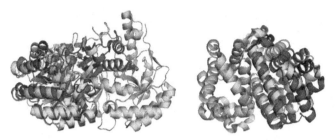

Fig. 4. Example of the predicted protein homodimer structures. Two chains of protein homodimers are shown in green and cyan, position of the structures for second protein (cyan chain) predicted by the algorithm are in red. RMSD between superposition of cyan and red chains is equal to 3Å for 3w1x homodimer (left panel) and 4.3Å for 1jzk homodimer (right panel). (Color figure online)

It should be noted that another type of input – distance maps between the CA atoms – could also be used as input for the neural network. This approach was tested and resulted in less accuracy both in predicting the contact map and in restoring the structure of the complex at the second stage of the algorithm. The rationale behind it that CA atoms belong to the main chain of the protein, and the distance between the corresponding CA atoms of the two proteins is greater than in the case of CB atoms located on the side chains of the protein.

We intend to continue the development of the algorithm in two directions. The first one is to generalize the described approach to heterodimers and to explore other architectures of deep neural networks to increase accuracy of predictions. The second one is to generalize to multimeric homocomplexes having more than two identical proteins.

References

1. Vakser, I.A.: Protein-protein docking: from interaction to interactome. Biophys. J. **107**, 1785–1793 (2014). https://doi.org/10.1016/j.bpj.2014.08.033
2. Mukherjee, S., Zhang, Y.: Protein-protein complex structure predictions by multimeric threading and template recombination. Structure **19**, 955–966 (2011). https://doi.org/10.1016/j.str.2011.04.006
3. Lu, L., Lu, H., Skolnick, J.: MULTIPROSPECTOR: an algorithm for the prediction of protein-protein interactions by multimeric threading. Proteins **49**, 350–364 (2002). https://doi.org/10.1002/prot.10222
4. Baspinar, A., Cukuroglu, E., Nussinov, R., Keskin, O., Gursoy, A.: PRISM: a web server and repository for prediction of protein-protein interactions and modeling their 3D complexes. Nucleic Acids Res. **42**, W285–W289 (2014). https://doi.org/10.1093/nar/gku397
5. Källberg, M., et al.: Template-based protein structure modeling using the RaptorX web server. Nat. Protoc. **7**, 1511–1522 (2012). https://doi.org/10.1038/nprot.2012.085

6. Sinha, R., Kundrotas, P.J., Vakser, I.A.: Docking by structural similarity at protein-protein interfaces. Proteins Struct. Funct. Bioinforma. **78**, 3235–3241 (2010). https://doi.org/10.1002/prot.22812

7. Kundrotas, P.J., Zhu, Z., Janin, J., Vakser, I.A.: Templates are available to model nearly all complexes of structurally characterized proteins. Proc. Natl. Acad. Sci. U. S. A **109**, 9438–9441 (2012). https://doi.org/10.1073/pnas.1200678109

8. Negroni, J., Mosca, R., Aloy, P.: Assessing the applicability of template-based protein docking in the twilight zone. Structure **22**, 1356–1362 (2014). https://doi.org/10.1016/j.str.2014.07.009

9. Vakser, I.A.: Low-resolution structural modeling of protein interactome. Curr. Opin. Struct. Biol. **23**, 198–205 (2013). https://doi.org/10.1016/j.sbi.2012.12.003

10. Pierce, B.G., Hourai, Y., Weng, Z.: Accelerating protein docking in ZDOCK using an advanced 3D convolution library. **6**, e24657 (2011). https://doi.org/10.1371/journal.pone.0024657

11. Pierce, B., Weng, Z.: ZRANK: reranking protein docking predictions with an optimized energy function. Proteins Struct. Funct. Genet. **67**, 1078–1086 (2007). https://doi.org/10.1002/prot.21373

12. Zacharias, M.: Protein-protein docking with a reduced protein model accounting for side-chain flexibility. Protein Sci. **12**, 1271–1282 (2003). https://doi.org/10.1110/ps.0239303

13. Gabb, H.A., Jackson, R.M., Sternberg, M.J.E.: Modelling protein docking using shape complementarity, electrostatics and biochemical information. J. Mol. Biol. **272**, 106–120 (1997). https://doi.org/10.1006/jmbi.1997.1203

14. Neveu, E., Ritchie, D.W., Popov, P., Grudinin, S.: PEPSI-Dock: a detailed data-driven protein-protein interaction potential accelerated by polar Fourier correlation. Bioinformatics **32**, i693–i701 (2016). https://doi.org/10.1093/bioinformatics/btw443

15. Kastritis, P.L., Bonvin, A.M.J.J.: Are scoring functions in protein-protein docking ready to predict interactomes? Clues from a novel binding affinity benchmark. J. Proteome Res. **9**, 2216–2225 (2010). https://doi.org/10.1021/pr9009854

16. Chen, R., Li, L., Weng, Z.: ZDOCK: an initial-stage protein-docking algorithm. Proteins Struct. Funct. Genet. **52**, 80–87 (2003). https://doi.org/10.1002/prot.10389

17. Senior, A.W., et al.: Improved protein structure prediction using potentials from deep learning. Nature **577**, 706–710 (2020). https://doi.org/10.1038/s41586-019-1923-7

18. Billings, W.M., Hedelius, B., Millecam, T., Wingate, D., Corte, D.D.: ProSPr: democratized implementation of alphafold protein distance prediction network. bioRxiv. 830273 (2019). https://doi.org/10.1101/830273

19. Kryshtafovych, A., Schwede, T., Topf, M., Fidelis, K., Moult, J.: Critical assessment of methods of protein structure prediction (CASP)—Round XIII. Proteins Struct. Funct. Bioinforma. **87**, 1011–1020 (2019). https://doi.org/10.1002/prot.25823

20. Lecun, Y., Bengio, Y., Hinton, G.: Deep learning. Nature **521**, 436–444 (2015). https://doi.org/10.1038/nature14539

21. Balci, A.T., Gumeli, C., Hakouz, A., Yuret, D., Keskin, O., Gursoy, A.: DeepInterface: protein-protein interface validation using 3D Convolutional Neural Networks. bioRxiv. 617506 (2019). https://doi.org/10.1101/617506

22. Derevyanko, G., Lamoureux, G.: Protein-protein docking using learned three-dimensional representations. bioRxiv. 738690 (2019). https://doi.org/10.1101/738690

23. Long, J., Shelhamer, E., Darrell, T.: Fully convolutional networks for semantic segmentation. In: Proceedings of the IEEE Computer Society Conference on Computer Vision and Pattern Recognition, pp. 3431–3440. IEEE Computer Society (2015). https://doi.org/10.1109/CVPR.2015.7298965

24. Yu, F., Koltun, V.: Multi-scale context aggregation by dilated convolutions. In: 4th International Conference on Learning Representations, ICLR 2016 - Conference Track Proceedings. International Conference on Learning Representations, ICLR (2015)

25. Fu, J., et al.: Dual attention network for scene segmentation. In: Proceedings of the IEEE Computer Society Conference on Computer Vision and Pattern Recognition, 2019-June, pp. 3141–3149 (2018)

26. He, K., Zhang, X., Ren, S., Sun, J.: Deep residual learning for image recognition. In: Proceedings of the IEEE Computer Society Conference on Computer Vision and Pattern Recognition, pp. 770–778. IEEE Computer Society (2016). https://doi.org/10.1109/CVPR.2016.90

27. Chen, L.C., Papandreou, G., Kokkinos, I., Murphy, K., Yuille, A.L.: DeepLab: semantic image segmentation with deep convolutional nets, atrous convolution, and fully connected CRFs. IEEE Trans. Pattern Anal. Mach. Intell. **40**, 834–848 (2018). https://doi.org/10.1109/TPAMI.2017.2699184

28. Kingma, D.P., Ba, J.L.: Adam: a method for stochastic optimization. In: 3rd International Conference on Learning Representations, ICLR 2015 - Conference Track Proceedings. International Conference on Learning Representations, ICLR (2015)

29. Berman, H.M.: The protein data bank: a historical perspective. Acta Crystallogr. Sect. A: Found. Crystallogr. **64**, 88–95 (2008). https://doi.org/10.1107/S0108767307035623

30. Mitternacht, S.: FreeSASA: an open source C library for solvent accessible surface area calculations. F1000Research. **5**, 189 (2016). https://doi.org/10.12688/f1000research.7931.1

31. Janin, J., Bahadur, R.P., Chakrabarti, P.: Protein-protein interaction and quaternary structure. Q. Rev. Biophys. **41**, 133–180 (2008). https://doi.org/10.1017/S0033583508004708

Development of a Neural Network-Based Approach for Prediction of Potential HIV-1 Entry Inhibitors Using Deep Learning and Molecular Modeling Methods

Grigory I. Nikolaev[1], Nikita A. Shuldov[2], Arseny I. Anischenko[2],
Alexander V. Tuzikov[1], and Alexander M. Andrianov[3(✉)]

[1] United Institute of Informatics Problems, National Academy of Sciences of Belarus, Minsk,
Republic of Belarus
[2] Belarusian State University, Minsk, Republic of Belarus
[3] Institute of Bioorganic Chemistry, National Academy of Sciences of Belarus, Minsk,
Republic of Belarus
alexande.andriano@yandex.ru

Abstract. A generative adversarial autoencoder for the rational design of potential HIV-1 entry inhibitors able to block the region of the viral envelope protein gp120 critical for the virus binding to cellular receptor CD4 was developed using deep learning methods. In doing so, the following studies were carried out: (i) the architecture of the neural network was constructed; (ii) a virtual compound library of potential anti-HIV-1 agents for training the neural network was formed; (iii) molecular docking of all compounds from this library with gp120 was made and calculations of the values of binding free energy were performed; (iv) molecular fingerprints for chemical compounds from the training dataset were generated; (v) training the neural network was implemented followed by estimation of the learning outcomes and work of the autoencoder. The validation of the neural network on a wide range of compounds from the ZINC database was carried out. The use of the neural network in combination with virtual screening of chemical databases was shown to form a productive platform for identifying the basic structures promising for the design of novel antiviral drugs that inhibit the early stages of HIV infection.

Keywords: HIV-1 · Gp120 protein · HIV-1 entry inhibitors · Virtual screening · Molecular docking · Molecular dynamics · Anti-HIV drugs

1 Introduction

Modern methods of computer-aided drug design significantly expand the capabilities of the pharmaceutical industry, vastly reducing the time and costs required for developing novel therapeutic agents. Despite the efficacy of computer methods in drug design is currently universally recognized, the development of new mathematical approaches and the availability of powerful and cheap computing resources promote their continuous

© Springer Nature Switzerland AG 2020
Z. Cai et al. (Eds.): ISBRA 2020, LNBI 12304, pp. 304–311, 2020.
https://doi.org/10.1007/978-3-030-57821-3_28

improvement. Among these approaches, machine learning methods and, in particular, deep learning technique, which offers great potential for further progress in this research area, occupy an important place. To date, computer-aided design of potential drugs using machine learning methods is one of the most important and rapidly developing areas of chemoinformatics [1]. Unlike physical models based on explicit physical equations, such as quantum chemistry or molecular dynamics modeling, machine learning approaches use pattern recognition algorithms to determine the mathematical relationships between empirical observations of small molecules and their extrapolation to predict chemical, biological and physical properties of new compounds. In addition, compared to physical models, machine learning methods are more efficient and can easily scale to big data sets. One of the benefits of using machine learning to design drugs is to help researchers understand and use the relationships between chemical structures and their biological activity [2]. Modern machine learning methods can be used to model the quantitative structure-activity relationship (QSAR) or quantitative structure-property relations (QSPR) and develop intelligent tools that can accurately predict the effect of chemical modifications of a compound on its biological activity, pharmacokinetic and toxicological characteristics [1]. In this context, the use of machine learning methods for computer design of potential drugs is of great scientific and practical importance [3].

The goal of this study was to develop a generative adversarial autoencoder for the rational design of potential HIV-1 entry inhibitors able to block the region of the viral envelope gp120 protein critically important for the virus attachment to cellular receptor CD4.

In doing so, the following studies were carried out: (i) architecture of the generative adversarial autoencoder was constructed; (ii) a virtual compound library of potential anti-HIV-1 agents for training the neural network was formed using the concept of click chemistry; (iii) molecular docking of all compounds from this library with gp120 and calculation of binding free energy values were performed; (iv) molecular fingerprints for chemical compounds from the training dataset were generated; (v) training the neural network was implemented followed by evaluation of the learning outcomes and work of the autoencoder.

2 Methods

The structure of the developed adversarial autoencoder is based on the model of the basic neural network that was designed to generate chemical compounds with anticancer properties [4]. The model consists of two neural networks including autoencoder and discriminator that work during training in a competitive mode (Fig. 1). The developed autoencoder is a seven-layer neural network with input and output layers, a latent layer, and 4 fully connected layers (Fig. 1). Fingerprints of chemical compounds are fed to the input layer, the data of which pass through two fully connected layers (encoder) and fall on the latent layer where a numerical estimate of the binding energy to the molecular target is added to the result obtained. Next, fingerprints pass through 2 fully connected layers (decoder) and go to the output, which, like the input, is a fingerprint vector. A network operating in this mode reduces the number of neurons entering the latent layer containing compressed information about the vector fed to the input of the network followed by its expansion at the output. The latent layer consists of 3 neurons two of which receive values from the encoder, and the third neuron receives the value of the binding free energy. In the autoencoder generative mode, random numbers are fed to the latent layer containing the most important information about the object and then pass through a decoder generating fingerprints of molecules with the desired properties. For the generation of such molecules, it is important that the data entering the latent layer after passing through the encoder have a normal distribution, to which the random number generator and discriminator are trained. To ensure this condition, in the course of competitive training of the encoder and discriminator, it was ensured that the encoder was able to encode data with a normal distribution on the latent layer, and that the discriminator distinguishes the standard normal distribution (generated data) from the distribution obtained on the latent layer.

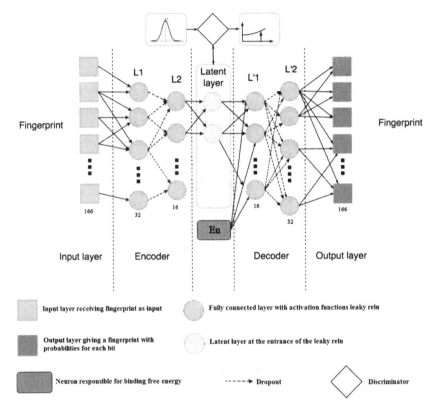

Fig. 1. The architecture of a generative adversarial autoencoder for prediction of potential anti-HIV-1 agents targeting the viral envelope gp120 protein. The encoder consists of two consecutive layers L1 and L2 with 32 and 16 neurons, respectively. The decoder includes two layers L'1 and L'2 containing 16 and 32 neurons, respectively. The latent layer consists of 3 neurons.

The training dataset was generated using the AutoClickChem program [5]. In the first stage of the study, virtual screening of the Drug-Like subset of the ZINC15 database

[6] was performed to form two molecular libraries of small modular units with the functional groups involved in the reaction of azide-alkyne cycloaddition. To do this, the DataWarrior program [7] providing data visualization and analysis was employed. Using DataWarrior, small aromatic molecules with molecular mass < 250 Da containing azide or alkyne groups were selected from ZINC15 and placed into library 1. In the same way, all low-molecular compounds (molecular mass < 250 Da) with the above functional groups were collected in library 2. The choice of aromatic molecules as modular units for the design of CD4-mimetic candidates is due to the fact that their aromatic moieties can mimic the key interactions of Phe-43$_{CD4}$ with the Phe-43 cavity of gp120 that dominate the HIV-1 binding to CD4. According to the X-ray data [8], the benzene ring of Phe-43$_{CD4}$ is buried into the Phe-43 cavity of gp120, resulting in the blockade of the gp120 residues critically important for viral adsorption to CD4$^+$ cells. In addition, novel HIV-1 entry antagonists, such as NBD-11021and NBD-14010, show the similar interaction modes to bind this hydrophobic pocket of gp120 [9, 10]. As a result of screening of the ZINC15 database, a total of 1388 and 3769 compounds were included into libraries 1 and 2, respectively. These small modular units were then used as reactants to mimic the click-reaction of azide-alkyne cycloaddition by AutoClickChem [5], resulting in 1 655 301 hybrid molecules. 120 000 compounds out of the designed molecules that fully satisfied the Lipinski's "rule of five" [11] were included into the training dataset.

Molecular docking of these drug-like molecules with gp120 was performed by the QuickVina 2 program [12] in the approximation of rigid receptor and flexible ligands. The 3D structure of gp120 was isolated from the crystal complex of this glycoprotein with CD4 and antibody 17b [8]. The gp120 and ligand structures were prepared by adding hydrogen atoms with the OpenBabel software followed by their optimization in the UFF force field (http://openbabel.org/wiki/Main_Page). The ligands were docked to the crystal gp120 structure [8] using QuickVina 2 [12]. The grid box included the Phe-43 cavity of the gp120 CD4-biding site and was the region of the crystal structure [8] with the following boundary X, Y, Z values: $X \in (24 \text{ Å}; 34 \text{ Å}$, $Y \in (-15 \text{ Å}; -5 \text{ Å})$, $Z \in (78 \text{ Å}; 88 \text{ Å})$. The value of "exhaustiveness" parameter defining number of individual sampling "runs" was set to 50 [12].

Molecular fingerprints of the compounds from the dataset were calculated using the open-source chemoinformatics software package RDKit (https://www.rdkit.org/). A three-stage iterative procedure was used for training and validation of the developed autoencoder, namely: (i) training the discriminator to distinguish a given normal distribution from the encoded one obtained by the encoder on the latent layer; (ii) joint training of the encoder and decoder as an autoencoder; (iii) training the encoder to compress the data in such a way that they look like a normal distribution.

To validate the developed autoencoder, the library of the MACCS fingerprints (http://www.dalkescientific.com/writings/NBN/fingerprints.html) for 21 325 567 compounds from the Drug Like subset of the ZINC15 database was generated using the RDKit software package and molecular fingerprints of 5 ligands were constructed by the autoencoder at a threshold value of the binding free energy equal to –5.0 kcal/mol. As a result of virtual screening of this library, the ligands with fingerprints similar to those generated by the neural network were found. To do this, the Hamming distance and the Tanimoto coefficient were used as a measure of fingerprint similarity.

3 Results

Analysis of the obtained data showed that the combined use of the developed neural network with virtual screening of the library of molecular fingerprints allows one to identify the ligands with the lower values of binding free energy compared to a given threshold value. Furthermore, the compound with code ZINC000026430653 in the ZINC15 database exhibits the value of binding free energy comparable with that of -9.5 ± 0.1 kcal/mol measured for the CD4-gp120 complex by isothermal titration calorimetry [13]. This value is also very close to that predicted by QuickVina 2 for NBD-11021 (Table 1), the lead HIV-1 entry antagonist [9] which was used in the calculations as a positive control. The data on the high-affinity binding of ZINC000026430653 to gp120 are also supported by the values of dissociation constants (K_d) calculated for the docking models of gp120 with this molecule and NBD-11021 using the neural network-based scoring function NNScore 2.0 [14] (Table 1). Furthermore, the above conclusions are in agreement with the values of binding free energy of the analyzed molecules to gp120 calculated from the predicted values of K_d with the formula $\Delta G_{Kd} = R \times T \times \ln(K_d)$ (where ΔG_{Kd} is the binding free energy, R is the universal gas constant, T is the absolute temperature equal to 310 K) [15]. Taking together, the data of Table 1 suggest that the identified compound may expose strong attachment to the hydrophobic Phe-43 cavity of the HIV-1 CD4-binding site of gp120, in line with the low values of binding free energy and K_d comparable with those predicted for NBD-11021.

Table 1. Values of binding free energy ΔG and K_d calculated for the docking complexes of ZINC000026430653 and NBD-11021 with gp120.

Compound	ZINC000026430653	NBD-11021
ΔG, kcal/mol	−8.8	−8.4
K_d (μM)	0.788	0.948
ΔG_{Kd}, kcal/mol	−8.65	−8.53

Examination of the ligang-gp120 complex reveals multiple van der Waals contacts of ZINC000026430653 with the gp120 Phe-43 cavity that plays a key role in the HIV-1 binding to cellular receptor CD4 (Fig. 2). By analogy with CD4 [8], one of the aromatic systems of ZINC00002643065 is buried into this hydrophobic pocket and mimics the pivotal interactions of the benzene ring of Phe-43$_{CD4}$ with the functionally important residues of gp120. These gp120 residues are Val-255, Ser-256, Thr-257, Glu-370, Ile-371, Asp-368, Ser-375, Phe-376, Phe-382, Met-426, Trp-427, and Met-475 (Fig. 2). From the crystal gp120-CD4 structure, all these amino acids are involved in the direct interatomic contacts with CD4, and the highly conserved Asp-368, Glu-370 and Trp-427 are the dominant contributors to the gp120-CD4 interaction [8].

Along with van der Waals contacts, analysis of the docking ligand-gp120 model indicates a high probability of a special type of weak hydrogen bond C–H···O [16] associated with the CH group of the aromatic ring of ZINC000026430653 and the carboxyl group of Asp-368 that is located within the vestibule of the Phe-43 cavity of

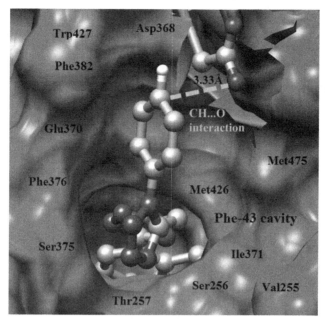

Fig. 2. The docking model of ZINC000026430653 with gp120. The Phe-43 cavity of gp120 and the residues located within the vestibule of this hydrophobic pocket are shown. The residues of gp120 forming van der Waals contacts with the CD4-mimetic candidate are indicated. C–H···O interaction of the CH group of the aromatic ring of ZINC000026430653 and the carboxyl group of Asp-368$_{gp120}$ is marked by dotted line.

gp120 (Fig. 2). These are evidenced by the values of the distance d between the O and C atoms ($d = 3.33$ Å) and the angle α between the C, H and O atoms (α is close to 90°) (Fig. 2) which support the assumption that these atoms may be involved in a dipole-dipole C–H···O interaction [17]. These findings are of considerable interest because hydrogen bonding with Asp-368 of gp120 is highly important for potential CD4 mimetics to show the properties characteristic for the small-molecule CD4 antagonists [9, 10].

Thus, the validation of the developed generative adversarial autoencoder using a wide range of chemical compounds from the ZINC15 database shows that this neural network combined with virtual screening of chemical databases by molecular finger-prints and traditional methods of molecular modeling forms a productive platform for discovery of promising anti-HIV-1 agents able to target CD4-binding site of the viral envelope gp120 protein. According to the data of molecular modeling, the small molecule ZINC000026430653 identified when the autoencoder validation exhibits a high-affinity binding to gp120 (Table 1) which is provided by interaction modes dominating the gp120-CD4 interface (Fig. 2). In addition, this molecule fully satisfies the criteria imposed on potential drug by Lipinski's "rule of five" [11]. This drug-like molecule may be used therefore as a good starting point for design of novel potent antiviral drugs inhibit-ing the early stages of HIV-1 infection. In this context, the further advancement of the current study proposes the use of this CD4-mimetic candidate as the basic structure for computer-assisted generation of its chemical modifications with the higher anti-HIV activity and improved ADMET parameters followed by synthesis and detailed biochemical assays.

4 Conclusion

The data obtained indicate the great opportunities of applying deep learning methods in pharmaceutical research which allow one not only to predict novel promising drug candidates but also significantly decrease resources, time and financial investments. This is supported by the recent reports of successful using deep learning approaches to drug discovery and development (reviewed in [18]) and, in particular, by the findings of a newly study [19] in which a deep learning network was developed and used for predicting compounds that are structurally distant from known antibiotics and exhibit antibactericidal activity against a wide phylogenetic spectrum of pathogens.

Acknowledgments. This study was supported by a grant of the State Program of Scientific Research "Convergence 2020" (subprogram "Consolidation", project 3.08).

References

1. Cherkasov, A., et al.: QSAR modeling: where have you been? Where are you going to? J. Med. Chem. **57**, 4977–5010 (2014). https://doi.org/10.1021/jm4004285
2. Ali, S.M., Hoemann, M.Z., Aubé, J., Georg, G.I., Mitscher, L.A., Jayasinghe, L.R.: Butitaxel analogues: synthesis and structure-activity relationships. J. Med. Chem. **40**, 236–241 (1997). https://doi.org/10.1021/jm960505t
3. Vamathevan, J., et al.: Applications of machine learning in drug discovery and development. Nat. Rev. Drug Discov. **18**(6), 463–477 (2019). https://doi.org/10.1038/s41573-019-0024-5
4. Kadurin, A., et al.: The cornucopia of meaningful leads: Applying deep adversarial autoencoders for new molecule development in oncology. Oncotarget **8**, 10883–10890 (2017). https://doi.org/10.18632/oncotarget.14073
5. Durrant, J.D., McCammon, J.A.: AutoClickChem: click chemistry in silico. PLoS Comput. Biol. **8**, e1002397 (2012). https://doi.org/10.1371/journal.pcbi.1002397
6. Sterling, T., Irwin, J.J.: ZINC 15–ligand discovery for everyone. J. Chem. Inf. Model. **55**, 2324–2337 (2015). https://doi.org/10.1021/acs.jcim.5b00559
7. Sander, T., Freyss, J., von Korff, M., Rufener, C.: DataWarrior: an open-source program for chemistry aware data visualization and analysis. J. Chem. Inf. Model. **55**, 460–473 (2015). https://doi.org/10.1021/ci500588j
8. Kwong, P.D., Wyatt, R., Robinson, J., Sweet, R.W., Sodroski, J., Hendrickson, W.A.: Structure of an HIV gp120 envelope glycoprotein in complex with the CD4 receptor and a neutralizing human antibody. Nature **393**, 648–659 (1998). https://doi.org/10.1038/31405
9. Curreli F., et al.: Structure-based design of a small molecule CD4-antagonist with broad spectrum anti-HIV-1 activity. J. Med. Chem. **58**, 6909–6927 (2015). https://doi.org/10.1021/acs.jmedchem.5b00709
10. Curreli, F., et al.: Synthesis, antiviral potency, in vitro ADMET, and X-ray structure of potent CD4 mimics as entry inhibitors that target the Phe43 cavity of HIV-1 gp120. J. Med. Chem. **60**, 3124–3153 (2017). https://doi.org/10.1021/acs.jmedchem.7b00179

11. Lipinski, C.A., Lombardo, F., Dominy, B.W., Feeney, P.J.: Experimental and computational approaches to estimate solubility and permeability in drug discovery and development settings. Adv. Drug Deliv. Rev. **46**, 3–26 (2001). PMID: 11259830

12. Handoko, S.D., Ouyang, X., Su, C.T.T., Kwoh, C.K., Ong, Y.S.: QuickVina: accelerating autodock vina using gradient-based heuristics for global optimization. IEEE/ACM Trans. Comput. Biol. Bioinform. **9**, 1266–1272 (2012). https://doi.org/10.1109/tcbb.2012.82

13. Myszka, D.G., et al.: Energetics of the HIV gp120-CD4 binding reaction. Proc. Natl. Acad. Sci. USA **97**, 9026–9031 (2000). https://doi.org/10.1073/pnas.97.16.9026

14. Durrant, J.D., McCammon. J.A.: NNScore 2.0: a neural-network receptor–ligand scoring function. J. Chem. Inf. Model. **51**, 2897–2903 (2011). https://doi.org/10.1021/ci2003889

15. Sharma, G., First, E.A.: Thermodynamic analysis reveals a temperature-dependent change in the catalytic mechanism of Bacillus stearothermophilus tyrosyl-tRNA synthetase. J. Biol. Chem. **284**, 4179–4190 (2009). https://doi.org/10.1074/jbc.m808500200

16. Weiss, M.S., Brandl, M., Sühnel, J., Pal, D., Hilgenfeld, R.: More hydrogen bonds for the (structural) biologist. Trends Biochem. Sci. **26**, 521–523 (2001). https://doi.org/10.1016/s0968-0004(01)01935-1

17. Steiner, T.: C–H···O hydrogen bonding in crystals. Cryst. Rev. **9**, 177–228 (2003). https://doi.org/10.1080/08893110310001621772

18. Lavecchia, A.: Deep learning in drug discovery: opportunities, challenges and future prospects. Drug Discov. Today **24**, 2017–2032 (2019). https://doi.org/10.1016/j.drudis.2019.07.006

19. Stokes, J.M., et al.: A deep learning approach to antibiotic discovery. Cell **180**, 688–702 (2020). https://doi.org/10.1016/j.cell.2020.01.021

In *Silico* Design and Evaluation of Novel Triazole-Based Compounds as Promising Drug Candidates Against Breast Cancer

Alexander M. Andrianov[1]([✉]), Grigory I. Nikolaev[2], Yuri V. Kornoushenko[1], and Sergei A. Usanov[1]

[1] Institute of Bioorganic Chemistry, National Academy of Sciences of Belarus, Minsk, Republic of Belarus
alexande.andriano@yandex.ru

[2] United Institute of Informatics Problems, National Academy of Sciences of Belarus, Minsk, Republic of Belarus

Abstract. Computational development of novel triazole-based aromatase inhibitors (AIs) was carried out followed by investigation of the possible interaction modes of these compounds with the enzyme and prediction of the binding affinity by tools of molecular modeling. In doing so, *in silico* design of potential AIs candidates fully satisfying the Lipinski's "rule of five" was performed using the concept of click chemistry. Complexes of these drug-like molecules with the enzyme were then simulated by molecular docking and optimized by semiempirical quantum chemical method PM7. To identify the most promising compounds, stability of the PM7-based ligand/aromatase structures was estimated in terms of the values of binding free energies and dissociation constants. At the final stage, structures of the top ranking compounds bound to aromatase were analyzed by molecular dynamic simulations and binding free energy calculations. As a result, eight hits that specifically interact with the aromatase catalytic site and exhibit the high-affinity ligand binding were selected for the final analysis. The selected AIs candidates show strong attachment to the enzyme active site, suggesting that these small drug-like molecules may present good scaffolds for the development of novel potent drugs against breast cancer.

Keywords: Aromatase · Molecular docking · Quantum chemistry · Molecular dynamics · Aromatase inhibitors · Breast cancer

1 Introduction

In women organism during the fertile phase, estrogen synthesis occurs mainly in the ovaries. However, the intensity of estrogen synthesis in the ovaries decreases in post-menopause associated with about a third of cases of breast cancer [1–3]. At this phase, estrogens synthesized in the peripheral tissues using the cytochrome P450 complex, called aromatase. This complex consists of the heme-containing cytochrome P450

© Springer Nature Switzerland AG 2020
Z. Cai et al. (Eds.): ISBRA 2020, LNBI 12304, pp. 312–318, 2020.
https://doi.org/10.1007/978-3-030-57821-3_29

(CYP19A1) protein and flavoprotein NADPH-cytochrome P450 reductase [1–3]. Aromatase encoded by a single large gene, CYP19A1, catalyzes conversion of androgens to estrogens and exhibits biological activity in both peripheral target tissues and the mammary tumor tissues, providing a high level of estrogen concentration [1–3]. In estrogen-dependent malignant neoplasms, estrogens act as growth factors for tumor development. Therefore, inhibition of aromatase results in a decrease of estrogens level in the organism and prevents to growth and spread of cancer cells [1–3].

There are three generations of AIs among the drugs for treating hormone-dependent breast cancer. The disadvantage of the drugs of the first two generations (aminoglutethimide, fadrozole, formestane) is the lack of selectivity of action: besides aromatase, these drugs inhibit a number of other enzymes. The third-generation AIs vorozole, letrozole, anastrozole, and exemestane approved for clinical use by the USA Food and Drug Administration show greater specificity and efficacy. These inhibitors include drugs of two categories, namely i) irreversible steroidal inhibitor exemestane that is an androstenedione derivative and ii) reversible non-steroidal inhibitors vorozole, anastrozole and letrozole. Steroidal AIs and, in particular, exemestane are transformed by aromatase into compounds that irreversibly bind to the enzyme active site, completely disrupting its activity as a biocatalyst. After the termination of the action of these inhibitors, aromatase needs considerable time to be synthesized in the tissues again. Reversible nonsteroidal AIs vorozole, letrozole and anastrozole are triazole compounds that bind to the catalytic site of the enzyme by coordinating the heme iron of the CYP19A1 through a heterocyclic nitrogen lone pair. The third generation AIs are now the front-line drugs for treating the early and advanced stages of breast cancer in postmenopausal women.

Despite significant progress in the treatment of hormone-dependent breast cancer, this problem has not been completely solved. Unfortunately, the third-generation of aromatase inhibitors (AIs) cause a number of serious side effects, such as inhibition of muscle growth, arthralgia, decreased bone strength, impaired blood lipid profile, drop *in libido*, as well as deterioration of the general condition. In addition, resistance acquired after long-term therapy with these drugs also occurs. In this context, development of novel, more effective and less toxic AIs is of great value.

2 Methods

In this study, computational development of novel triazole-based AIs was carried out followed by evaluation of their antitumor activity by tools of molecular modeling. In doing so, the following studies were performed: i) *in silico* design of potential aromatase inhibitor candidates by the AutoClickChem techniques [4]; ii) identification of compounds satisfying the Lipinski's "rule of five" [5]; iii) molecular docking of these drug-like compounds with the enzyme active site using the QuickVina 2 program [6]; iv) refinement of the ligand-binding poses by the PM7 semiempirical quantum chemical method [7]; v) prediction of the interaction modes dominating the binding; vi) calculation of the values of binding free energy and dissociation constant (K_d) for the ligand/CYP19A1 complexes; vii) prediction of the binding affinity between the identified compounds and aromatase by molecular dynamics simulations and binding free energy calculations [8]; and viii) selection of molecules most promising for synthesis and biochemical trials.

3 Results

Based on the analysis of the data obtained, eight top-ranked compounds that exhibited the low values of binding free energy (< -7 kcal/mol) in the PM7-based ligand/aromatase complexes were selected for the final analysis. Depending on the mechanism of binding to the active site of CYP19A1, these compounds were divided into two structural groups designated as groups 1 and 2. The data of molecular modeling indicate that each of the identified compounds of group 1 (Fig. 1) shows peculiar interactions with the enzyme binding pocket, the interaction being realized between the triazole ring and the heme iron, van der Waals interactions with the hydrophobic pocket lined by Arg-115, Ile-133, Phe-134, Trp-224, Thr-310, Val-370, Met-374, Leu-477, Ser-478, and, except compound II, the hydrogen bond with Met-374, which is also involved in hydrogen bonding with the natural substrate androstenedione. In addition, some identified compounds form van der Waals contacts with the heme of CYP19A1, and π-conjugated systems of individual molecules participate in specific π-π interactions with the pyrrole rings of the heme group. Finally, the selected AIs candidates expose strong attachment to the enzyme active site, in line with the low values of binding free energy and K_d (Table 1). In summary, the conclusions that can be made by the new identified AIs from group 1 are that, in addition to the interaction between the triazole rings and the heme iron, hydrophobic contacts play a pivotal role in the ligand binding, and hydrogen bond involving Met-374 is essential for the ligand recognition (Fig. 1).

Unlike the molecules of group 1, the ligands of group 2 demonstrate a mechanism of binding to aromatase uncharacteristic for triazole-based compounds generally coordinating the iron atom of the heme via a heterocyclic nitrogen lone pair. According to the calculated data (Fig. 2), the ligands of interest coordinate the iron atom of the CYP19A1 heme group through the lone pairs of their oxygen atoms. Similarly to the molecules of group 1, all these compounds target the well-conserved hotspots of the aromatase catalytic site using multiple van der Waals interactions with the critically important residues of this hydrophobic pocket.

Molecular dynamics insights into the ligand/aromatase complexes validate the main findings derived from the analysis of their static structures. These complexes are relatively stable during the MD simulations, which is supported by the averages of binding free energies, their enthalpic components, and corresponding standard deviations (Table 3). The averages of binding free energy predicted for the designed compounds in the complexes with aromatase are comparable with the value calculated for the letrozole/aromatase complex by the identical computational protocol (Table 3). Furthermore, these averages are also comparable with the values estimated for the static models of the ligand/aromatase complexes (Tables 1, 2) as well as to that of -12.06 kcal/mol obtained with the formula $\Delta G = R \times T \times \ln(K_d)$ [9] at temperature T = 310 K using the experimental value of K_d for letrozole bound to aromatase [10]

Besides, compound II participates in specific π-π interactions with the pyrrole rings of the CYP19A1 heme group and form hydrogen bonds with residues Met-374 and Thr-310, and compound I forms a salt bridge with the enzyme heme.

Finally, analysis of the data of Table 2 indicates that, like the molecules from group 1, the compounds of group 2 exhibit a high binding affinity in the complexes with

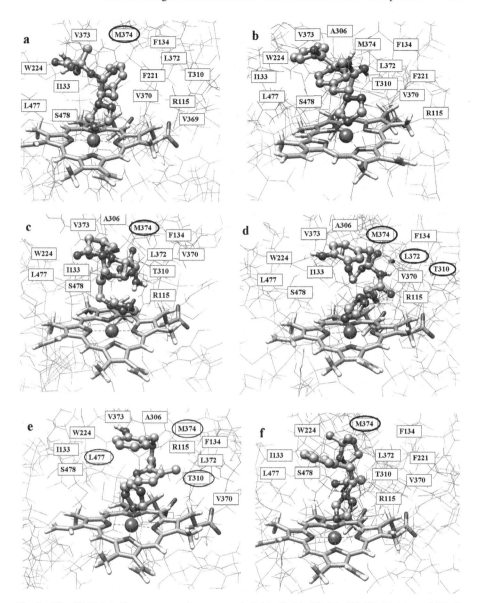

Fig. 1. The PM7-based structures of compounds I (a), II (b), III (c), IV (d), V (e), and VI (f) from group 1 bound to aromatase. The residues of CYP19A1 forming van der Waals contacts with ligands are located in rectangles. The residues involved in the hydrogen bonding are marked by ellipses.

Table 1. Values of binding free energy (ΔG) and K_d for the compounds of group 1 bound to aromatase.

Compound	I	II	III	IV	V	VI
ΔG, kcal/mol	−7.8	−9.2	−8.8	−8.1	−8.7	−8.0
K_d (nM)	12.21	38.35	43.24	51.07	64.39	73.72

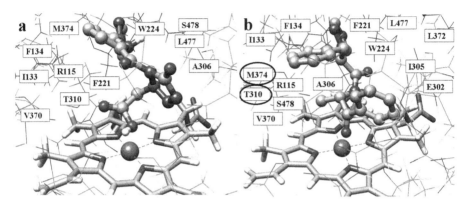

Fig. 2. The PM7-based structures of compounds I (a) and II (b) from group 2 bound to aromatase. The residues of CYP19A1 forming van der Waals contacts with ligands are located in rectangles. The residues involved in hydrogen bonding are marked by ellipses.

aromatase, as evidence with the low values of dissociation constant and binding free energy.

Table 2. Values of binding free energy (ΔG) and K_d for the compounds of group 2 bound to aromatase.

Compound	I	II
ΔG, kcal/mol	−8.7	−8.6
K_d (nM)	22.38	30.30

Unlike the molecules of group 1, the ligands of group 2 demonstrate a mechanism of binding to aromatase uncharacteristic for triazole-based compounds generally coordinating the iron atom of the heme via a heterocyclic nitrogen lone pair. According to the calculated data, the ligands of interest coordinate the iron atom of the CYP19A1 heme group through the lone pairs of their oxygen atoms. Similarly to the molecules of group 1, all these compounds target the well-conserved hotspots of the aromatase catalytic site using multiple van der Waals interactions with the critically important residues of this hydrophobic pocket. Besides, compound II participates in specific π-π interactions with

Table 3. Averages of binding free energy ($<\Delta G>$) for the complexes of the AIs candidates and letrozole with aromatase and their standard deviations (ΔG_{STD})[a]

Compound	$<\Delta H>$ kcal/mol	$(\Delta H)_{STD}$ kcal/mol	$<T\Delta S>$ kcal/mol	$(T\Delta S)_{STD}$ kcal/mol	$<\Delta G>$ kcal/mol	ΔG_{STD} kcal/mol
Compounds of group 1						
I	−32.2	4.0	−22.6	2.9	−9.6	3.4
II	−28.3	5.7	−16.4	4.0	−11.9	4.8
III	−35.2	5.6	−23.9	6.2	−11.3	5.9
IV	−26.8	4.5	−17.9	3.4	−8.9	3.9
V	−25.7	5.5	−16.1	3.6	−9.6	4.4
VI	−28.4	4.6	−19.7	3.6	−8.7	4.0
Compounds of group 2						
I	−27.9	3.5	−18.5	3.7	−9.4	3.6
II	−24.8	3.6	−16.5	5.5	−8.3	4.5
Letrozole						
	−37.3	4.3	−27.0	9.6	−10.3	6.4

[a] $<\Delta H>$ and $<T\Delta S>$ are the mean values of enthalpic and entropic components of free energy, respectively; $(\Delta H)_{STD}$ and $(T\Delta S)_{STD}$ are standard deviations corresponding to these values.

the pyrrole rings of the CYP19A1 heme group and form hydrogen bonds with residues Met-374 and Thr-310, and compound I forms a salt bridge with the enzyme heme.

Finally, analysis of the ligand/aromatase complexes indicates a high binding affinity between the identified compounds and the enzyme, in agreement with the low values of binding free energy calculated both for their static and dynamic models.

4 Conclusions

Thus, the data of molecular modeling indicate that each of the identified compounds of group 1 shows peculiar interactions with the enzyme binding pocket, the interaction being realized between the triazole ring and the heme iron, van der Waals interactions with the hydrophobic pocket lined by Arg-115, Ile-133, Phe-134, Trp-224, Thr-310, Val-370, Met-374, Leu-477, Ser-478, and, except compound II, the hydrogen bond with Met-374, which is also involved in hydrogen bonding with the natural substrate androstenedione. In addition, some identified compounds form van der Waals contacts with the heme of CYP19A1, and π-conjugated systems of individual molecules participate in specific π-π interactions with the pyrrole rings of the heme group. Finally, the selected AIs candidates expose strong attachment to the enzyme active site, in line with the low values of binding free energy and K_d. In summary, the conclusions that can be made by the new identified AIs are that, in addition to the interaction between the triazole rings and the heme iron, hydrophobic contacts play a pivotal role in the ligand binding, and hydrogen bond involving Met-374 is also essential for the ligand recognition.

Acknowledgments. This study was supported by a grant of the State Program of Scientific Research "Chemical Technologies and Materials" (subprogram 2.2 "Biologically active compounds", project 2.15).

References

1. Macedo, L.F., Sabnis, G., Brodie, A.: Aromatase inhibitors and breast cancer. Ann. N.Y. Acad. Sci. **1155**, 162–173 (2009)
2. Ghosh, D., Griswold, J., Erman, M., Pangborn, W.: Structural basis for androgen specificity and oestrogen synthesis in human aromatase. Nature **457**, 219–223 (2009)
3. Hong, Y., Chen, S.: Aromatase inhibitors: structural features and biochemical characterization. Ann. N.Y. Acad. Sci. **1089**, 237–251 (2006)
4. Durrant, J.D., McCammon, J.A.: AutoClickChem: click chemistry *in silico*. PLoS Comput. Biol. **8**, e1002397 (2012)
5. Lipinski, C.A., Lombardo, F., Dominy, B.W., Feeney, P.J.: Experimental and computational approaches to estimate solubility and permeability in drug discovery and development settings. Adv. Drug Deliv. Rev. **46**, 3–26 (2001)
6. Handoko, S.D., Ouyang, X., Su, C.T.T., Kwoh, C.K., Ong, Y.S.: QuickVina: accelerating AutoDock vina using gradient-based heuristics for global optimization. IEEE/ACM Trans. Comput. Biol. Bioinform. **9**, 1266–1272 (2012)
7. Stewart, J.J.P.: Optimization of parameters for semiempirical methods VI: more modifications to the NDDO approximations and re-optimization of parameters. J. Mol. Model. **19**, 1–32 (2013)
8. Case, D.A., et al.: AMBER 2016. University of California, San Francisco (2016)
9. Sharma, G., First, E.A.: Thermodynamic analysis reveals a temperature-dependent change in the catalytic mechanism of bacillus stearothermophilus tyrosyl-tRNA synthetase. J. Biol. Chem. **284**, 4179–4190 (2009)
10. Adamchik, S., et al.: Synthesis and properties of new triazole aromatase inhibitors. In: Proceedings of Belarusian State University, vol. 11, pp. 280–290 (2016). (in Russian)

Identification of Essential Genes with NemoProfile and Various Machine Learning Models

Yangxiao Wang[✉] and Wooyoung Kim[✉]

Division of Computing and Software Systems School of Science, Technology, Engineering, and Mathematics University of Washington Bothell, 18115 Campus Way NE, Bothell, WA 98011-8246, USA
{wyxiao,kimw6}@uw.edu

Abstract. Genes are sequences of nucleotide in DNA that encode proteins. Essential genes are a type of genes that are critical and indispensable for an organism's survival. Many network-based algorithms have been developed to identify essential genes. We introduce a novel approach to predict essential genes that are based on network motif profiles (NemoProfile) and various machine learning models. Experimental results show that NemoProfile is an effective data feature generated from biological networks, and balanced data is a critical factor to improve the overall performance.

Keywords: NemoProfile · Machine learning · Essential genes · PPI network

1 Introduction

Essential genes are the genes that are required to sustain the life of an organism in nutritious conditions [10,18]. They play significant roles in synthetic biology, and identification of them can help design new therapies for infectious diseases. While there are numerous computational methods to identify essential genes with graph theory, we suggest an alternative approach using network motif profile (NemoProfile) and various machine learning models.

Network motifs are subgraphs patterns that are statistically significant in a graph. Network motif has been applied in various biological and medical problems [2], including predicting protein interactions, determining protein functions, detecting breast-cancer susceptibility genes, and discovering essential proteins. Kim and Haukap introduced "NemoProfile" and "NemoCollect" as additional output options to the traditional output of "NemoCount" [9]. While NemoCount fails to provide network motif instances, NemoCollect provides whole set of all network motif instances, and NemoProfile provides the number of times a node involved in each network motif as shown in Fig. 1. For example, the node 6141 belonged 2538 times to the instances of *motif1*, 306 times to *motif2*, 9 times to *motif3*, 8 times to *motif4*, and once to *motif5*.

© Springer Nature Switzerland AG 2020
Z. Cai et al. (Eds.): ISBRA 2020, LNBI 12304, pp. 319–326, 2020.
https://doi.org/10.1007/978-3-030-57821-3_30

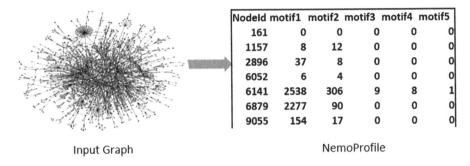

NodeId	motif1	motif2	motif3	motif4	motif5
161	0	0	0	0	0
1157	8	12	0	0	0
2896	37	8	0	0	0
6052	6	4	0	0	0
6141	2538	306	9	8	1
6879	2277	90	0	0	0
9055	154	17	0	0	0

Input Graph   NemoProfile

Fig. 1. Example input graph and its corresponding NemoProfile data feature. Nemo-Profile is a 2D matrix format of n by m, where n is the number of nodes, and m is the number of network motifs. The first column lists all the node ids from the input graph. The (i, j) cell indicates i^{th} node's involvement on the jth motif.

In this paper, we used the NemoProfile as data feature to find the essential genes. Through the experiments with various machine learning models, we show that NemoProfile is an innovative and effective data feature for detecting essential genes.

2   Data Feature Extraction

For the experiment, We preprocessed two data sets from the organism *Bacillus subtilis 168* and extracted features using network motif analysis before the actual training process, as Fig. 2 summarizes the experimental processes as a flowchart.

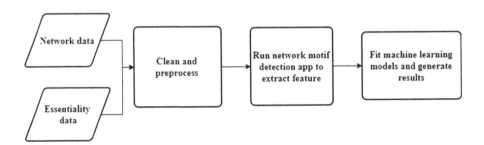

Fig. 2. Flowchart of experiment process.

The first data set is from STRING database consisting of a protein-protein interaction (PPI) network [17]. The second data set is the information of essential genes obtained in the work by Kobayashi et al. [11]. The PPI network includes 1,021,787 edges and 4,169 vertices with confidence scores. The essentiality data contains 4,176 tested genes with their essentiality, out of which 228 are essential genes. We processed a number of steps to extract data features.

First, we excluded isolated genes (0.1% of genes) in the PPI, and removed edges with a low confidence score. The confidence score measures the reliability of each interaction based on the nature and quality of the supporting evidence. The threshold of confidence score was determined to 750 because the number of essential genes dropped out abruptly after 750. Consequently, there are 41,290 edges, 3,700 vertices, and 227 essential genes.

Next, we detected four network motifs which are labeled as Cr, CN, C˜, and C^(see Fig. 3). Therefore, we obtained a NemoProfile where each gene is a vector of a dimension four. Then we normalized the feature vectors using standardization (Eq. (1)) and min-max normalization (Eq. (2)).

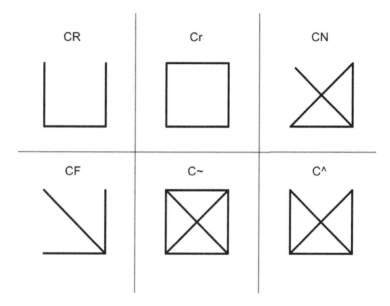

Fig. 3. There are six non-isomorphic subgraph patterns for size 4 undirected graphs. The canonical label of each pattern shown at top of each graph. The detected network motifs from our datasets are Cr, CN, C˜, and C^.

$$x' = \frac{x - \bar{x}}{\sigma} \tag{1}$$

$$x' = \frac{x - min(x)}{max(x) - min(x)} \times (max - min) + min \tag{2}$$

where $\bar{x}$ is the mean and σ is the standard deviation, and max and min are set as 1 and 0, respectively.

3   Methods

We used five machine learning models for the prediction of essential genes: *Decision Tree, K-nearest neighbors, Support-vector Machines, Logistic Regression,* and *Convolutional Neural Network*. We chose those five models because we were unclear which model would be the best fit with this new data feature (Nemo-Profile) since there is no previous work that we could learn from. We used scikit-learn [13] to implement the machine learning models. To test, we used 10 fold cross-validation to find the best target score.

Through parameter tuning, we selected the best parameters for each model with our data sets. These parameters are provided in Table 1.

Table 1. Best parameter settings for each model

Model	Parameter setting
Decision tree	min_samples_leaf = 8, min_samples_split = 2, criterion = entropy
KNN	number of neighbors = 9, leaf size = 30, metric = Manhattan distance
SVM	kernel = linear, decision_function_shape = one vs rest, degree = 3, C = 5
Logistic regression	solver = liblinear, max_iter = 100
CNN	activation = relu, solver = sgd, max_iter = 200, learning_rate = constant, hidden_layer_sizes = 90

The Decision Tree model [15] builds a tree based on training data, and the best split is decided by impurity, entropy, and information gain. We tried different random seeds to reach the suboptimal solution. K-Nearest Neighbor (KNN) [5] model, which is a nonparametric lazy learning algorithm, considers close points belong to the same class and the class of a test data point is decided by a majority vote of its K nearest neighbors. Support Vector Machine (SVM) [4] is another nonparametric learning algorithm designed for binary classification. Logistic Regression [3,6] provides the probability of the essentiality of each node. Convolutional Neural Network (CNN) [12,16] is one of the deep neural network methods which has a long developing history and became popular over the past seven years. CNNs are composed of regularized multi-layer perceptrons, which are connected networks where some regularization is added.

4   Results

We compared the results of our experiments with various measurements. This section illustrates the performance metrics and provides the results.

4.1   Performance Metrics

The prediction of essential genes is a binary classification problem. Since unbalanced data can produce poor prediction accuracy [1,14], we used five different performance metrics: Accuracy, precision, recall, F1 score, and AUC_ROC score.

Accuracy is the measure of the correctly predicted true positive and true negative overall input dataset.

$$Accuracy = \frac{TP + TN}{TP + TN + FP + FN} \tag{3}$$

Precision is the measure of proportion of correctly predicted true positive over the positively predicted data.

$$Precision = \frac{TP}{TP + FP} \tag{4}$$

Recall is the measure of correctly predicted true positive and among all actual positive data.

$$Recall = \frac{TP}{TP + FN} \tag{5}$$

F1 Score is the harmonic mean of precision and recall.

$$F1\_Score = \frac{2 \times Precision \times Recall}{Precision + Recall} \tag{6}$$

ROC curve plots TPR vs. FPR where TPR on y-axis and FPR on x-axis. AUC_ROC measures the area underneath the ROC curve. It measures the overal quality of the model's prediction.

$$TPR = \frac{TP}{TP + FN} \tag{7}$$

$$Specificity = \frac{TN}{TN + FP} \tag{8}$$

$$FPR = 1 - Specificity \tag{9}$$

4.2   Experimental Results

We experimented with the balanced and unbalanced data sets to see the performance differences. Since the original data is unbalanced, we generated balanced data sets, where the same number of essential and non-essential genes are randomly selected. All tests are performed both on unbalanced and balanced datasets. As Table 2 and 3 show, tests with balanced datasets provide better results in general. The values are the average of 10-fold cross-validation.

We can see that the decision tree model shows the best performance with the balanced dataset from Table 2, while logistic regression and CNN models have better results with the unbalanced dataset as shown in Table 3. SVM model has the lowest AUC_ROC values for both datasets, indicating that its overall low performance with our datasets. It seems that the fact of PPI's incompleteness and noisiness affected the performance of SVM. The logistic regression model, although the precision is comparable with others, performs poorly with the lowest recall value. Considering that the limited parameters may have caused bad performance, we could improve the performance with different parameters and random seeds. With balanced datasets, CNN model does not perform well. The low performance might be because of the limited time, limited hyper-parameters, and resources.

Table 2. Testing with balanced dataset

Model	Accuracy	Precision	Recall	F1 Score	AUC_ROC
Decision tree	0.811	0.824	0.794	0.796	0.874
KNN	0.786	0.761	0.797	0.768	0.849
SVM	0.774	0.784	0.646	0.695	0.841
Logistic regression	0.759	0.782	0.584	0.641	0.865
CNN	0.726	0.755	0.514	0.585	0.844

Table 3. Testing with unbalanced dataset

Model	Accuracy	Precision	Recall	F1 Score	AUC_ROC
Decision tree	0.931	0.651	0.34	0.391	0.748
KNN	0.945	0.541	0.282	0.339	0.773
SVM	0.948	0.546	0.274	0.336	0.658
Logistic regression	0.950	0.588	0.295	0.360	0.882
CNN	0.951	0.582	0.304	0.375	0.873

5   Conclusion

We proposed a novel approach to predicting essential genes based on network motif analysis. We used NemoProfile as data features and applied five machine learning models to compare the performance. We built models with the balanced data set and tested them with both the balanced and unbalanced data sets. We used 10-cross fold evaluation methods and the results are average of 10 experiments. The performance of the balanced model is significantly better than the unbalanced one. Our experiment with NemoProfile data feature can be considered successful compared with other previous works shown in Table 4 [7,8,10]. From our experiments, the AUC_ROC values range from 0.841 to 0.874, while previous works have values ranging from 0.711 to 0.818.

Although we had decent results, there are many improvements that can be made. For example, using other machine learning models such as Naive Bayes,

Random Forest, and Gradient Boosting algorithms; or more parameter tuning for each model. Additionally, different size of network motifs can be tried for the improvement.

Table 4. The performances of each method with accuracy, area under ROC, area under PR [8]

Method	Accuracy	AUC_ROC	AUC_PR
CENT-GO(ALL)	0.727	0.784	0.781
DCGO	0.716	0.760	0.733
BCGO	0.682	0.711	0.682
CCGO	0.701	0.742	0.711
SCGO	0.682	0.732	0.719
ECGO	0.675	0.718	0.720
SoECCGO	0.702	0.743	0.741
LACGO	0.689	0.743	0.738
MCGO	0.710	0.749	0.747
CENT-GO	0.727	0.784	0.781
ING-GO	0.723	0.793	0.784
CENT-ING-GO(CENT-GO + ING-GO)	0.753	0.818	0.804

References

1. Almas, A., Farquad, M.A.H., Avala, N.S.R., Sultana, J.: Enhancing the performance of decision tree: a research study of dealing with unbalanced data. In: Seventh International Conference on Digital Information Management (ICDIM 2012), pp. 7–10, August 2012
2. Andersen, A., Kim, W., Fukuda, M.: MASS-based nemoprofile construction for an efficient network motif search. In: 2016 IEEE International Conferences on Big Data and Cloud Computing, pp. 601–606, October 2016
3. Berkson, J.: Application of the logistic function to bio-assay. J. Am. Stat. Assoc. **39**(227), 357–365 (1944)
4. Chapelle, O., Haffner, P., Vapnik, V.: Support vector machines for histogram-based image classification. IEEE Trans. Neural Netw. **10**(5), 1055–1064 (1999)
5. Cover, T., Hart, P.: Nearest neighbor pattern classification. IEEE Trans. Inf. Theory **13**(1), 21–27 (1967)
6. Cramer, J.S.: The Origins of Logistic Regression. Technical Report, ID 360300, Social Science Research Network, Rochester, NY, December 2002
7. Deng, Y.Y., Guo, F.B.: Applications of four machine learning algorithms in identifying bacterial essential genes based on composition features. In: ChinaSIP, pp. 821–825, July 2015

8. Kim, W.: Prediction of essential proteins using topological properties in GO-pruned PPI network based on machine learning methods. Tsinghua Sci. Technol. **17**(6), 645–658 (2012)

9. Kim, W., Haukap, L.: NemoProfile as an efficient approach to network motif analysis with instance collection. BMC Bioinform. **18**(12), 423 (2017)

10. Kim, W., Li, M., Wang, J., Pan, Y.: Essential protein discovery based on network motif and gene ontology. In: 2011 IEEE International Conference on Bioinformatics and Biomedicine, pp. 470–475, November 2011

11. Kobayashi, K., et al.: Essential Bacillus subtilis genes. Proc. Nat. Acad. Sci. U.S.A. **100**(8), 4678–4683 (2003)

12. Krizhevsky, A., Sutskever, I., Hinton, G.E.: ImageNet classification with deep convolutional neural networks. Commun. ACM **60**(6), 84–90 (2017)

13. Pedregosa, F., et al.: Scikit-learn: machine learning in Python. J. Mach. Learn. Res. **12**, 6 (2011)

14. Provost, F.J., Fawcett, T., Kohavi, R.: The case against accuracy estimation for comparing induction algorithms. In: International Conference on Machine Learning, pp. 445–453. San Francisco (1998)

15. Quinlan, J.R.: Induction of decision trees. Mach. Learn. **1**(1), 81–106 (1986)

16. Schmidhuber, J.: Deep learning in neural networks: an overview. Neural Netw. **61**, 85–117 (2015)

17. Szklarczyk, D., et al.: STRING v10: protein-protein interaction networks, integrated over the tree of life. Nucleic Acids Res. **43**(D1), D447–D452 (2015)

18. Zhao, B., Wang, J., Li, M., Wu, F.X., Pan, Y.: Prediction of essential proteins based on overlapping essential modules. IEEE Trans. NanoBiosci **13**(4), 415–424 (2014)

NemoLib: Network Motif Libraries for Network Motif Detection and Analysis

Wooyoung Kim$^{(\boxtimes)}$ and Zachary Arthur Brader$^{(\boxtimes)}$

Division of Computing and Software Systems,
School of Science, Technology, Engineering, and Mathematics,
University of Washington Bothell,
18115 Campus Way NE, Bothell, WA 98011-8246, USA
{kimw6,brader}@uw.edu

Abstract. Network motifs are frequent and unique subgraph patterns located inside networks, and have been applied to solve various biological problems. Due to the high computational costs of performing network motif analysis, various tools have been created to make the process more efficient. However, existing tools lack extensible functionality and provide limited output formats. This restricts the ability to use network motif analysis for extensive and exhaustive experiments in real problems. We provide NemoLib (Network Motif Libraries) as a general purpose tool for detection and analysis of network motifs. It is an easily adoptable and highly accessible tool with a focus on efficiency and extensibility.

Keywords: NemoLib · ESU · Biological network · Network motif

1 Introduction

The interconnections between biological components (such as species, molecules, or neurons) are often represented as biological networks to explain complex biological systems. There are many types of networks at various levels including ecological, neurological, metabolic, and protein-protein interaction networks.

Network motif analysis is a graph-based algorithm used for finding biological functions using unique patterns (network motifs) that exist within biological networks [6]. Network motifs are defined as statistically frequent and unique subgraph patterns, which have been applied to various problems such as predicting protein interactions [1], determining protein functions [3], detecting breast-cancer susceptibility genes [18], and discovering essential proteins [10]. Being able to perform network motif analysis quickly and efficiently will allow for the process to be more available and applicable to more problems in the future.

Detection of a network motif requires expensive computational power. Therefore, multiple heuristic methods and parallel algorithms have been proposed to alleviate the computational burden [9]. Currently, there are various tools available for network motif detection, including [9]: MFinder [8], FANMOD [16],

© Springer Nature Switzerland AG 2020
Z. Cai et al. (Eds.): ISBRA 2020, LNBI 12304, pp. 327–334, 2020.
https://doi.org/10.1007/978-3-030-57821-3_31

Kavosh [7], Mavisto [15], NeMoFinder [2], LaMoFinder [3], Grochow's [4], and MODA [14].

However, current tools suffer from multiple drawbacks that restrict their range of application, and consequently cause a certain amount of skepticism regarding their structure-based and pattern-based analyses. The drawbacks include the following: some tools are unavailable; most tools fail to collect instances of network motifs due to heavy memory overhead; many tools only accept the input graphs whose nodes are integers; some lack of functionality; and some are non-extensible as they work only for smaller sized motifs [12].

To deal with these drawbacks and provide a more accessible way of performing network motif analysis, NemoLib (Network Motif Libraries) was created and published. This library alleviates the problems aforementioned for the following reasons: it is well-optimized to run faster than other tools; it can provide different output options including frequency, motifs' profiles, and instances of motifs; it runs on graphs with various type of nodes and edges (directed and undirected); and it is designed to be flexible and easy for developers to incorporate or extend. NemoLib is a powerful tool for developers to efficiently perform network motif analysis.

Source codes of NemoLib are available at https://github.com/Kimw6/NemoLib-Java-V2 and at https://github.com/Kimw6/NemoLib-C-V2-StaticLib.

2   Network Motif

A network motif is a connected subgraph pattern M of size k that is frequent and unique in a target network G. The uniqueness is determined by P-value (Eq. (1)) or Z-score (Eq. (2)) of the frequency of M in a randomly generated pool.

$$\text{P-value}(M) = \frac{1}{N} \sum_{n=1}^{N} c(n), \text{ where } c(n) = \begin{cases} 1, \text{if } f_R(M) \geq f_G(M) \\ 0, \text{otherwise} \end{cases} \tag{1}$$

$$\text{Z-score}(M) = \frac{f_G(M) - \mu(f_R(M))}{\sigma(f_R(M))} \tag{2}$$

Here, N is the number of random graphs, and $f_G(M)$ is the frequency of M in the target network, while $f_R(M)$ is the frequency of M in the random network. $\mu(f_R(M))$ is the mean of frequencies of M in the random networks and $\sigma(f_R(M))$ is the standard deviation of frequencies of M in the random networks. Generally, subgraphs with P-value < 0.05 or Z-score > 2.0 are considered network motifs.

Figure 1 illustrates the process of detection of network motifs. A directed or undirected network (a) is provided as an input, then all subgraph of size k is enumerated and categorized into different non-isomorphic patterns to find their frequencies (b). Each pattern's frequency is compared in a random pool to determine whether the frequency is statistically high (c), which will provide

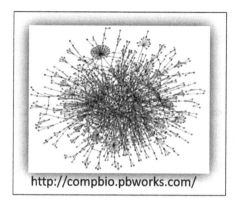

(a) PPI network

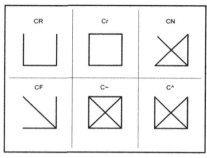

(b) Enumeration: search all possible size-4 subgraphs from input (a)

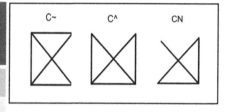

C~	Freq(O)	Freq(R)	P-value
C~	50	2	0.005
Cr	1000	2000	0.9

(c) Random graph pool

(d) Detected network motifs

Fig. 1. The process of detecting network motifs: (a) is the example of a target network; (b) shows all the possible non-isomorphic patterns of size four; (c) shows the example output that compares each pattern's frequency in a target, average frequency in random graphs, and its P-value. (d) shows the determined network motifs

the statically verified frequent patterns, that is, network motifs (d). From the example, the subgraphs labeled as C, C^A, CN are network motifs from the input graph (a).

There are two approaches to detecting network motifs: network-centric and motif-centric. The former approach will search all possible patterns from the input graph, while the latter finds all instances of given patterns in the input graph. Currently, NemoLib includes a network-centric method that implements the ESU (Enumerate Subgraph) algorithm [17], and plans to add a motif-centric method in near future.

3   NemoLib: Network Motif Library

NemoLib is a library that can be used as a better alternative to contemporary tools. It is developed with the idea to abstract the complexity of the detection

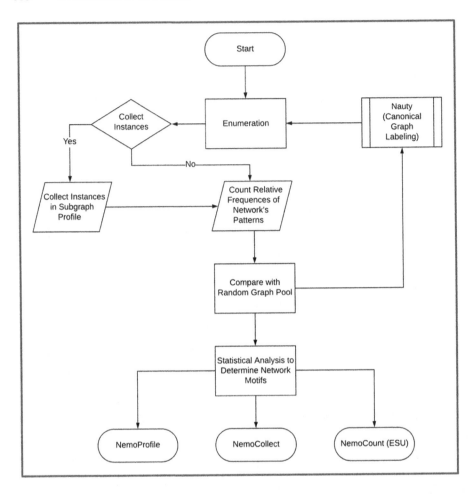

Fig. 2. This shows the essential processes that make up NemoLib. Starting from a file in graph format, the relative frequencies are gathered using the ESU algorithm. Depending on our output type, the instances might be collected alongside the ESU algorithm. The program will then generate and enumerate the selected number of random graphs and develop a set of relative frequencies based on the results. The final results are used to determine the network's network motifs in the statistical analysis process. The results and instances will be shown based on the selected output type.

of a network motif while allowing for easy customization. The design follows the principle of "open for extension, closed for modification" by exposing an API of common tools used for network motif detection. A user can use NemoLib to develop a web-based network motif detection program, as an example. Currently NemoLib is available through a public Github repository in Java and C++. Both versions have been tested to work in Linux and Windows.

The network-centric method used by NemoLib for network analysis includes three steps as described in Fig. 2: enumeration, random graph generation, and statistical analysis. First, enumeration searches the network to find each pattern to create a map of the relative frequencies for each pattern from the target graph. Then, random graph generation produces a pool of random graphs with the same degree distribution as the input graph, and then provide the average frequency of each pattern. Finally, statistical analysis takes the generated maps from the previous steps and computes the P-value and Z-score for each pattern and then returns their associated values. Network motifs are then determined by the P-value and Z-score whose threshold can be also customized.

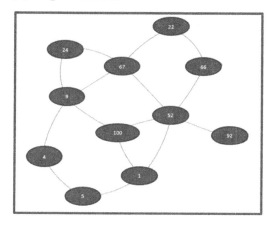

(a) Target network

(b) Input format

```
1  5
1  100
1  52
5  4
4  9
9  100
......
```

(c) Output Options

NemoCount

Label	Freq	MeanFreq	Z-Score	P-Value
&CA@o	1.754%	0.988%	0.384	0.244
...				
&C?gW	**7.018%**	**1.428%**	**2.284**	**0.044**
...				
&C?gO	21.053%	16.526%	0.539	0.265

NemoProfile

Node	&C?gW
1	3
100	3
22	1
24	1
25	......

NemoCollect

&C?gW
[67,52,1,100]
&C?gW
[1,52,100,66]
...

Fig. 3. NemoLib: (a) is the example of a target network; (b) is the input file format, where each line is an edge which consists of two end nodes; (c) shows the three output options which NemoLib provides.

NemoLib can read undirected or directed graph with any type of nodes (integer or string), however does not read in weights. Figure 3 describes some features of NemoLib. The target network (a) is given as a graph file shown in (b), which a list of edges, and each edge is a pair of two end nodes. When implementing a program with NemoLib, there are three possible output formats as shown in

Fig. 3 (c). The *NemoCount* option, the most basic option, displays each node's frequency, P-value, and Z-value. Choosing *NemoProfile* shows each node's participation in every pattern as well as the *NemoCount* results [11]. *NemoProfile* results are stored inside as an object matrix in memory during run time, and users will have the choice of outputting the results as a file. Finally, *NemoCollect* provides the *NemoCount* results and the list of instances for each motif. *Nemo-Collect* results are always saved in a file to avoid heavy memory overhead due to the volume of results that accompany this operation.

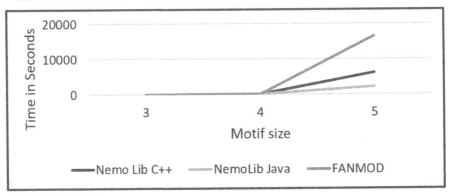

Fig. 4. It shows the performance of NemoLib is better than FANMOD program as the size of motif grows.

NemoLib was designed to provide efficient and effective results. Figure 4 compares the time efficiency of NemoLib with that of the FANMOD program, which is implementing ESU algorithm. After testing on various target graphs of different sizes, we could conclude that both the Java and C++ versions appear to be significantly faster than FANMOD. This result indicates that, along with providing additional functionality, NemoLib will provide a faster alternative to other tools.

4   Conclusion

NemoLib provides a combination of extensibility, efficiency, and functionality that the modern network motif detection software programs lack. It is designed to be an ideal tool for testing novel network motif detection algorithms and for learning about network motif detection. Currently, Nemolib is available online in both C++ and Java as a library, able to run faster than contemporary tools, and able to provide multiple ways to view network motifs.

Going forward, there are multiple ideas to improve the NemoLib family. The first proposed work is the development of a java version of the Nauty program

[13], since the current version, NemoLib-Java, spends additional time to externally call the Nauty program (which is a C-library) to get the canonical label for each non-isomorphic subgraph pattern. The next step is incorporating parallel and multi-threaded client programs into NemoLib for performance enhancement. Further along, another undertaking would be completing the Python version, which is not yet fully optimized. The final planned improvement is adding a motif-centric option by extending the NemoMap algorithm [5]. NemoLib fills a void in network motif detection, and should be an ideal tool for testing and learning about network motif detection.

Acknowledgements. We would like to thank the division of Computing and Software Systems at the University of Washington Bothell for the support. We also would like to thank members of the bioinformatics research group who are involved in developing various versions of NemoLib.

References

1. Albert, I., Albert, R.: Conserved network motifs allow protein-protein interaction prediction. Bioinformatics **20**(18), 3346–3352 (2004)
2. Chen, J., Hsu, W., Lee, M., Ng, S.: Nemofinder: dissecting genome-wide protein-protein interactions with meso-scale network motifs. In: Proceedings of the 12th ACM SIGKDD International Conference on Knowledge Discovery and Data Mining, New York, NY, pp. 106–115 (2006)
3. Chen, J., Hsu, W., Lee, M.L., Ng, S.K.: Labeling network motifs in protein interactomes for protein function prediction. In: International Conference on Data Engineering, pp. 546–555 (2007)
4. Grochow, J.A., Kellis, M.: Network motif discovery using subgraph enumeration and symmetry-breaking. In: Speed, T., Huang, H. (eds.) RECOMB 2007. LNCS, vol. 4453, pp. 92–106. Springer, Heidelberg (2007). https://doi.org/10.1007/978-3-540-71681-5_7
5. Huynh, T., Mbadiwe, S., Kim, W.: Nemomap: improved motif-centric network motif discovery algorithm. Adv. Sci. Technol. Eng. Syst. J. **3**(5), 186–199 (2018)
6. Junker, B.H., Schreiber, F.: Analysis of Biological Networks. Wiley, Hoboken (2008)
7. Kashani, Z., et al.: Kavosh: a new algorithm for finding network motifs. BMC Bioinform. **10**(1), 318 (2009)
8. Kashtan, N., Itzkovitz, S., Milo, R., Alon, U.: Efficient sampling algorithm for estimating sub-graph concentrations and detecting network motifs. Bioinformatics **20**, 1746–1758 (2004)
9. Kim, W., Diko, M., Rawson, K.: Network motif detection: algorithms, parallel and cloud computing, and related tools. Tsinghua Sci. Technol. **18**(5), 469–489 (2013)
10. Kim, W.: Prediction of essential proteins using topological properties in Go-pruned PPI network based on machine learning methods. Tsinghua Sci. Technol. **17**(6), 645–658 (2012)
11. Kim, W., Haukap, L.: Nemoprofile: effective representation for network motif and their instances. In: 12th International Symposium on Bioinformatics Research and Applications (ISBRA), 5–8 June 2016 (2016)
12. Konagurthu, A.S., Lesk, A.M.: On the origin of distribution patterns of motifs in biological networks. BMC Syst. Biol. **2**(1), 1–8 (2008)

13. McKay, B.: Practical graph isomorphism. Congr. Numer. **30**, 45–87 (1981)
14. Omidi, S., Schreiber, F., Masoudi-Nejad, A.: Moda: an efficient algorithm for network motif discovery in biological networks. Genes Genet. Syst. **84**(5), 385–395 (2009)
15. Schreiber, F., Schwobbermeyer, H.: Mavisto: a tool for the exploration of network motifs. Bioinformatics **21**, 3572–3574 (2005)
16. Wernicke, S., Rasche, F.: Fanmod: a tool for fast network motif detection. Bioinformatics **22**, 1152–1153 (2006)
17. Wernicke, S.: Efficient detection of network motifs. IEEE/ACM Trans. Comput. Biol. Bioinform. **3**(4), 347–359 (2006)
18. Zhang, Y., Xuan, J., de los Reyes, B.G., Clarke, R., Ressom, H.W.: Network motif-based identification of breast cancer susceptibility genes. In: 2008 30th Annual International Conference of the IEEE Engineering in Medicine and Biology Society, pp. 5696–5699. IEEE, August 2008 (2008)

Estimating Enzyme Participation in Metabolic Pathways for Microbial Communities from RNA-seq Data

F. Rondel[1(✉)], R. Hosseini[1], B. Sahoo[1], S. Knyazev[1], I. Mandric[2],
Frank Stewart[3], I. I. Măndoiu[4], B. Pasaniuc[5], and A. Zelikovsky[1]

[1] Department of Computer Science, Georgia State University,
Atlanta, USA
{frondel1,ahosseini3,bsahoo1,skniazev1}@student.gsu.edu, alexz@gsu.edu
[2] Department of Computer Science,
University of California Los Angeles, Los Angeles, CA, USA
imandric@ucla.edu
[3] Department of Microbiology and Immunology, Montana State University,
Bozeman, MT, USA
frank.stewart@montana.edu
[4] Computer Science and Engineering Department,
University of Connecticut, Storrs, CT, USA
ion@engr.uconn.edu
[5] Bioinformatics Interdepartmental Program,
University of California, Los Angeles, CA, USA

Abstract. Metatranscriptome sequence data analysis is necessary for understanding biochemical changes in the microbial community and their effects. In this paper, we propose a methodology to estimate activities of individual metabolic pathways to better understand the activity of the entire metabolic network. Our novel pipeline includes an expectation-maximization based estimation of enzyme expression and simultaneous estimation of pathway activity level and enzyme participation level in each pathway. We applied our novel pipeline to metatranscriptome data generated from surface water planktonic communities sampled over a day-night cycle in the Northern Gulf of Mexico (Louisiana Shelf). Our results show the estimated enzyme expression, pathway activity levels as well as enzyme participation levels in each pathway are robust and stable across all data points. In contrast to expression of enzymes, the estimated activity levels of significant number of metabolic pathways strongly correlate with the environmental parameters.

Keywords: NGS · Enzyme expression · Pathway activity level · Microbial community · Metatranscriptome · Enzyme participation in pathways

ⓒ Springer Nature Switzerland AG 2020
Z. Cai et al. (Eds.): ISBRA 2020, LNBI 12304, pp. 335–343, 2020.
https://doi.org/10.1007/978-3-030-57821-3_32

1   Introduction

Measuring the functional activity, enrichment, and interaction of metabolic pathways in microbial communities is essential for understanding the biochemical and ecological contributions of microorganisms. Despite many advances in using microbial biomolecules (DNA, RNA, proteins) to assess the biochemical contributions of microbes, it remains challenging to quantify how the expression of individual enzymes contributes to the activity of multi-enzyme metabolic pathways. In this study, we analyze time-series metatranscriptomic (community RNA) data to generate an efficient model for understanding metabolic pathway activity in depth [4, 12, 14, 15]. Even though advances in high-throughput sequencing have aided the exploration of RNA sequencing data, particularly for single organisms, it is often challenging to disentangle community-level data [3, 12, 16], notably as existing pathway analysis tools (e.g., MEGAN4, MetaPathways, MinPath) often yield variable conclusions about the activity of pathways based on RNA data [6, 9, 13, 17]. To overcome the current challenges, we developed a workflow that uses a Maximum Likelihood-based model and annotations based on the KEGG [7] database to estimate transcript frequency, enzyme expression, enzyme participation in pathways, and metabolic pathway activity. In this paper, we test this model using metatranscriptomic data from a marine microbial community. The data span multiple time points with different environmental parameters to elucidate the complex metabolic pathway activity in the microbial community, generally challenging to mimic in the laboratory.

The proposed methodology is the first to use a likelihood model to infer the pathway activity considering an enzyme's expression and participation coefficient. First, we filtered the microbial community-specific metabolic pathways from the KEGG database and merged the expression of enzymes sharing the same contigs and having sequence homologs. We implemented a novel Expectation-Maximization algorithm to estimate the enzyme participation level in each pathway and then used these estimations for more accurate predictions of pathway activity. Increased correlation between estimated metabolic pathway activity and environmental parameters validated our approach. Our contributions include the following:

- An EM-based algorithm for estimating enzyme expression.
- A novel EM-based algorithm for estimating metabolic pathway activity levels using estimation of enzyme participation level in each pathway.
- Validation of enzymes expression and pathways activity using their correlation with the environmental parameters.

The rest of the paper is organized as follows. In the next section we describe the pipeline of our software framework and several EM-based algorithms for estimating enzyme expression and metabolic pathway activity in microbial communities. Then we describe our datasets including sequencing data, and extraction of metabolic enzymes and pathways. Finally our results statistically validate the proposed pipeline.

2   Methods

We first describe the pipeline containing the previous version of our software
and an alternative flow with three new EM algorithms. Then each of these three
new EMs are described separately and the global loop for pathway activity level
estimation concludes description of our software.

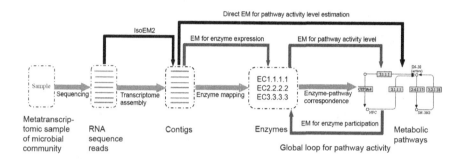

Fig. 1. Pipeline of metabolic pathway analysis for a microbial community sample. The
metatranscriptomic data obtained from microbial community samples are sequenced,
and raw reads are assembled into contigs. The genes containing obtained contigs are
further mapped into the enzyme-pathway database. Contig frequencies are obtained
using IsoEM2 [11]. Instead of estimating pathway activity levels using direct EM [10],
we propose to first estimate enzyme expression. Then the pathway activity level and
enzyme participation coefficients are estimated in a single EM feedback loop.

2.1   Pipeline for Estimating Pathway Activity Levels

This paper proposes to enhance the pipeline proposed in [10] (see Fig. 1) with
the inference of enzyme expressions and enzyme participation levels in metabolic
pathway repeatedly applying the maximum likelihood model. These models are
resolved using the Expectation-Maximization (EM) algorithm. The proposed
inferences are highlighted in red (see Fig. 1). The first step is to estimate the
abundances of the assembled contigs. The abundances can be inferred by any
RNA-seq quantification tool, but we suggest using IsoEM [11] since it is sufficiently fast to handle Illumina Hiseq data and more accurate than Kallisto [1].
We propose to estimate the enzyme expressions based on contig abundances
and mapping of contigs onto enzymes (EM for enzyme expression in Fig. 1).
The EM for pathway activity levels is based on inferred enzyme expressions and
metabolic pathway annotation. Each enzyme is initially assigned a participa-
tion level of $1/|w|$, where $|w|$ is the total amount of enzymes in the pathway
w. The *Global loop for pathway activity* updates the enzyme participation level
by fitting expected enzyme expressions to the expressions estimated by *EM for
enzyme expression*.

2.2 EM for Enzyme Expression and Pathway Activity Level Estimation

Let T be a random variable with values from the set of observed contigs, and let E be a random variable whose values belong to the set of relevant metabolic enzymes from the KEGG database. The probability of observing a contig t is given by the following formula: $P(T = t) = \sum_{w \in W} f_w \, P(T = t \mid E = e)$, where f_e stands for the expression of the relevant metabolic enzyme e. Thus, in our model we adopt the following likelihood function:

$$L(f_e) = \prod_{t \in T} \left(\sum_{e \in E} f_e \, P(T = t \mid E = e) \right)^{a_t}$$

where a_t denotes the abundance of t estimated by IsoEM2. Following [10] we estimate the probability of contig t coming from enzyme e as follows:

$$P(T = t | E = e) = p_{te} = \frac{b_{te}}{\sum_{t' \in e} b_{t'e}} \qquad (1)$$

where b_{te} is the best bit-score obtained from the alignment of t to the protein that have a function of the enzyme e.

The details of the EM for enzyme expression are as follows. We initialize estimates for each enzyme with a random number $f_e \in [0, 1]$, $e \in E$. Then, we iterate the following two steps until a convergence criteria is satisfied:

The E-step. We first compute the expected number of reads n_e emitted by each enzyme e through the following formula:

$$n_e = \sum_{t \in T} a_t \cdot \frac{p_{te} f_e}{\sum_{e' \in E} p_{te'} f_{e'}}$$

The M-step. The new estimates are provided based on a standard normalization step:

$$f_e^{new} = \frac{n_e}{\sum_{e' \in E} n_{e'}}$$

The algorithm halts when the change in estimates between iterations is small enough: $||f_E^{new} - f_E|| \leq \epsilon$, where $\epsilon \ll 1$

The EM algorithm for estimating pathway activity levels $f_W = \{f_w | w \in W\}$ based on frequencies of enzymes $f_E = \{f_e | e \in E\}$ is similar to the EM algorithm above. The only difference is that insted of Eq. (1) we use the uniform probability distribution over the set of enzymes/enzyme groups participating in each pathway:

$$P(E = e | W = w) = p_{ew} = \begin{cases} \frac{1}{|w|}, & \text{if } e \in w \\ 0, & \text{otherwise} \end{cases} \qquad (2)$$

2.3   Global Loop for Pathway Activity Level Estimation

The initial estimate (2) of the participation level of enzyme e in the pathway w can be very far from reality. More accurate estimates of the enzyme participation levels can lead to more accurate estimates for the pathway activity levels. The algorithm below estimates pathway activity levels Steps (1–3) and then checks how well the computed activities f_w's fit the enzyme expressions (step (4)). If the fit is not good enough, then EM-based algorithm is applied to update the enzyme participation levels p_{ew}'s (Steps (5–6)) and then f_w's are recomputed according to updated p_{ew}'s in Step (3).

1. Find expression $f(e)$ of each enzyme e running EM from Sect. 2.2.
2. According to (2), initialize $p_{ew} = \frac{1}{|w|}$ for $e \in w$ and $p_{ew} = 0$, otherwise.
3. Find activity levels f_w for each pathway $w \in W$ running EM from Sect. 2.3.
4. Find expected frequency of each enzyme e according to formula $f_e^{exp} = \sum_{w \in W} p_{ew} f_w$ If expected and observed enzymes frequencies are close to each other: $\|f_{e \in E}^{exp} - f_{e \in E}\| = \sum_{e \in E} (f_e^{exp} - f_e)^2 < \epsilon \ll 1$, then exit, i.e. go to step 7.
5. Find better fitted p_{ew}''s by using the following EM algorithm:
   The E-step. Compute expected p_{ew}^{exp}'s that will make $f_e = f_e^{exp}$ for each $e \in E, w \in W$,

$$p_{ew}^{exp} = p_{ew} \times \frac{f_e}{f_e^{exp}}$$

   The M-step. Provide the new estimates by normalization for each $e \in E, w \in W$,

$$p_{ew}^{new} = \frac{p_{ew}^{exp}}{\sum_{e \in E} p_{ew}^{exp}}$$

   The algorithm halts when the change in estimates between iterations is small enough:
   $\|p^{new} - p\| = \sum_{e \in E, w \in W} (p_{ew}^{new} - p_{ew})^2 \leq \epsilon \ll 1$
6. For each $e \in E, w \in W$, update $p_{ew} \leftarrow p_{ew}'$ and go to step 3
7. Output $\{f_w | w \in W\}$ and $\{p_{ew} | e \in E, w \in W\}$

3   Datasets

Samples. The study uses metatranscriptome data from 26 samples collected from surface water (depths of 2 and 18 m) on the Louisiana Shelf in the Gulf of Mexico. These samples were collected via Niskin water at the same site (28.867N -90.476W) over a 3-day period in July 2015. Furthermore, six environmental parameters - including PAR (photosynthetic active radiation) and seawater dissolved oxygen concentration, density, salinity, temperature, and chlorophyll concentration - were measured for each sample. One liter of seawater was pumped onto a 0.22 um Sterivex filter. Filtered biomass was then preserved using 1.8 ml of RNA-later and flash frozen. Filters were stored at -80 C until RNA extraction.

RNA was extracted via the mirVana™ Total RNA Isolation kit, with residual DNA removed via DNase treatment. RNA samples were then sequenced via the Illumina HiSeq 2500 1TB sequencing protocol following cDNA preparation at the Department of Energy Joint Genome Institute (DOE-JGI). All datasets are publicly available through the JGI Genomes Online (GOLD) database via GOLD ID Gs0110190. Out of 26 samples three samples (Day1, 12:00, 18m; Day 2, 20:00, 2m; Day 3, 08:00, 2m) were discarded as they did not contain enough reads to assemble transcripts for our pipeline.

Microbial-Specific Metabolic Pathway Identification. The KEGG pathway database has information on all metabolic pathways that occur in the living organisms. However, the scope of the current tool is to analyse metabolic pathways in microbial communities. We extracted metabolic pathways that play a significant role in microbial communities which is confirmed by literature referenced in PUBMED. Furthermore, we remove from consideration the high-level metabolic pathways including ec01100, ec01110, ec01120, and ec01130. As a result, we extracted 69 microorganism-relevant pathways out of 152 metabolic pathways. The reduced number of pathways increased the efficiency and performance of the algorithm.

Metabolic Enzyme Dataset Identification and Modification. We restrict ourselves to enzymes that belong to microbial metabolic pathways and remove the unlikely enzyme matches. Since the same set of contigs assembled from reads can match multiple metabolic enzymes, the EM for enzyme expression cannot differentiate between them. Therefore, we identified the enzymes sharing the same set of contigs and grouped them. For detecting such groups of enzymes, we use an essential property that the individual enzyme expression can vary across randomly initialized EM runs, while the sum of the expression of all enzymes in the group does not change. We collapsed the enzymes belonging to a single group and rerun EM to get an accurate and stable enzyme expression. After applying the above method, we obtain expressions of 1446 enzymes and enzyme groups for the metabolic pathway activity analysis.

4   Results

Our results consist of empirical and statistical validation of estimated enzyme expression, enzyme participation levels, and pathway activity level estimations. We first analyze the stability of enzyme participation levels and then wcheck how many enzyme expressions and pathway activities correlate with environmental parameters.

We estimate the participation level of each enzyme in each pathway separately for each data point. Table 1 presents the participation level of all expressed enzymes in the pathway ec00020. We can see that the participation level does not significantly change from one data point to another, i.e. the standard deviation is significantly smaller than the mean for all enzymes. Note that if an enzyme is not expressed in a sample, then the participation is not defined and the participation level is reported as 0. This means that we need to take in account

Table 1. Enzyme participation levels for all enzymes across all data points for 2 m depth in the metabolic pathway ec00020. Two rightmost columns are means and standard deviations of enzyme participation levels.

ec00020	D1:12	D1:16	D1:20	D2:00	D2:04	D2:08	D2:12	D2:16	D3:00	D3:04	D3:12	AVE	STD
EC:1.2.4.1	12.82	21.68	20.64	33.71	35.76	30.38	21.78	23.71	32.40	28.07	21.98	25.72	6.60
EC:1.2.7.1	0.51	6.18	15.43	6.69	4.97	9.32	13.14	9.61	7.87	12.95	2.54	8.11	4.37
EC:1.2.7.3	13.99	21.46	20.32	26.74	28.96	24.87	21.26	22.22	27.08	24.44	26.70	23.46	4.02
EC:1.8.1.4	7.61	12.92	11.24	16.94	16.65	14.39	12.93	16.92	19.16	14.03	22.16	15.00	3.78
EC:2.3.1.12	12.82	21.68	20.64	33.71	35.76	30.38	21.78	23.71	32.40	28.07	21.98	25.72	6.60
EC:4.1.1.32	12.82	21.68	20.64	33.71	35.76	30.38	21.78	23.71	32.40	28.07	21.98	25.72	6.60
EC:4.1.1.49	14.78	23.66	23.38	32.19	36.13	37.34	26.62	28.41	35.90	33.66	25.61	28.88	6.60
EC:1.1.1.37	18.14	19.76	26.62	17.90	18.93	30.78	20.27	20.43	22.97	22.13	44.21	23.83	7.43
EC:1.1.1.41	72.88	72.85	70.78	71.20	68.42	38.66	45.68	60.11	62.77	61.29	27.09	59.25	14.74
EC:1.1.1.42	19.96	24.06	22.58	21.52	23.68	19.95	22.48	22.32	22.95	21.92	42.38	23.98	5.95
EC:1.1.5.4	0.00	0.00	0.00	29.35	0.00	0.00	0.00	20.53	0.00	0.00	0.00	24.94	4.41
EC:1.2.4.2	10.10	13.02	10.76	11.91	10.91	11.72	12.75	14.08	14.74	10.13	25.75	13.26	4.21
EC:1.3.5.1	21.35	27.74	28.74	34.65	39.51	30.74	29.40	29.56	36.38	33.32	46.73	32.56	6.43
EC:2.3.1.61	10.10	13.02	10.76	11.91	10.91	11.72	12.75	14.08	14.74	10.13	25.75	13.26	4.21
EC:2.3.3.1	86.31	41.26	66.16	28.14	39.20	260.4	209.0	93.27	70.39	107.9	96.40	99.85	68.92
EC:2.3.3.8	19.96	24.06	22.58	21.52	23.68	19.95	22.48	22.32	22.95	21.92	42.38	23.98	5.95
EC:4.2.1.2	14.54	18.81	19.68	23.77	28.00	20.30	19.67	20.16	24.74	22.70	32.79	22.29	4.72
EC:4.2.1.3	33.31	29.83	34.13	23.43	28.96	41.10	44.43	37.46	35.39	8.11	69.02	37.74	11.35
EC:6.2.1.4	19.96	24.06	22.58	21.52	23.68	19.95	22.48	22.32	22.95	21.92	42.38	23.98	5.95
EC:6.4.1.1	14.54	18.81	19.68	23.77	28.00	20.30	19.67	20.16	24.74	22.70	32.79	22.29	4.72

only data points with non-zero participation levels when computing mean and standard deviation over all data points.

Table 2. 1. The number of enzymes significantly correlated with each of 6 environmental parameters and their linear combination (via multiple linear regression (MLP)). 2. The number of enzymes strongly correlated with randomly permuted parameter values (95% CI). 3. The EC number of the metabolic enzyme which is the most strongly correlated with the corresponding parameter.

	Salinity	Temp	Oxygen	Chl	PAR	Density	MLP
1. # enzymes	146	110	117	93	97	138	156
2. 95% CI	80–190	79–114	62–94	58–92	36–63	82–123	70–107
3. EC number	1.2.1.59	2.6.1.1	3.1.3.11	2.2.1.7	3.5.1.16	2.4.1.16	1.1.1.136

The goal of regression-based validation is to check our hypothesis that there exist enzymes and pathways whose expression and activity level variation across data points can be explained (i.e. correlate with) certain environmental parameters. For each environmental parameter, we check whether it significantly correlates ($P < 5\%$) with each enzyme across 11 data points for the 2-meter depth (see Table 2). In the row 2 we give 95% CI for the number of significantly correlated enzymes withe randomly permuted parameter. Since the upper bound of 95% CI for salinity is 190 (row 2), we conclude that there is no evidence of enzymes significantly correlated with salinity. We also report the enzyme that correlates the most with salinity, i.e. EC 1.2.1.59. From Table 2 we see that most parameters do not correlate well with enzymes, except perhaps PAR.

Table 3 is the same as Table 2 but reports correlation significance of pathway activities instead of enzyme expressions. In contrast to enzymes it is clear that the many metabolic pathways correlate with each environmental parameter and this correlation is not by chance. Indeed, pathway activity is supposed to be more stable than enzyme expression since generally metabolism is much less affected by the current. For each environmental parameter, we also cross-check the PubMed database whether the most correlated pathway is known to depend on this parameter. For instance, fatty acid degradation is well correlated with salinity, and several studies reported that fatty acid degradation is often altered by salinity at sea surface environments [2,5,8].

Table 3. 1. The number of pathways significantly correlated with each of 6 environmental parameters and correlated via multiple linear regression. 2. The number of pathways strongly correlated with randomly permuted parameter values (95% CI). 3. The EC number of the metabolic pathway which is the most strongly correlated with the corresponding parameter.

	Salinity	Temp	Oxygen	Chl	PAR	Density	Multiple
1. # pathways	31	22	19	18	14	30	22
2. 95% CI	1–8	0–8	0–6	0–6	0–6	1–8	(0–7)
3. Pathway	ec00071	ec00195	ec00622	ec00460	ec00360	ec00071	ec00626

5   Discussion

This paper proposes a maximum likelihood model for the estimation of metabolic pathway activity in the microbial community using the KEGG pathway database. Specifically, the proposed approach uses an EM-based pipeline to estimate enzyme expression, enzyme participation levels in pathways, and metabolic pathway activity from metatranscriptomic data. The proposed metabolic pathway analysis was applied to the metatranscriptomic data of 26 samples collected with different environmental parameters. The key findings of the study are as follows:

- The participation levels of enzymes in pathways do not significantly vary across the data samples.
- The enzyme expression and metabolic pathway activities were validated using regression with each environmental parameter: salinity, temperature, oxygen, chlorophyll, and PAR.
- In contrast to enzyme expressions, pathway activity levels significantly correlate with environmental parameters, e.g. 31 out of 61 metabolic pathways significantly correlate with salinity.

References

1. Bray, N.L., Pimentel, H., Melsted, P., Pachter, L.: Near-optimal probabilistic RNA-seq quantification. Nat. Biotechnol. **34**(5), 525–527 (2016)
2. de Carvalho, C.C.C.R., Caramujo, M.: The various roles of fatty acids. Molecules **23**(10), 2583 (2018)
3. Donato, M., et al.: Analysis and correction of crosstalk effects in pathway analysis. Genome Res. **23**(11), 1885–1893 (2013)
4. Efron, B., Tibshirani, R.: On testing the significance of sets of genes. Ann. Appl. Stat. **1**(1), 107–129 (2007)
5. Heinzelmann, S.M., et al.: Comparison of the effect of salinity on the D/H ratio of fatty acids of heterotrophic and photoautotrophic microorganisms. FEMS Microbiol. Lett. **362**(10), fnv065 (2015)
6. Huson, D.H., Mitra, S., Ruscheweyh, H.-J., Weber, N., Schuster, S.C.: Integrative analysis of environmental sequences using MEGAN4. Genome Res. **21**(9), 1552–1560 (2011)
7. Kanehisa, M.: KEGG: kyoto encyclopedia of genes and genomes. Nucleic Acids Res. **28**(1), 27–30 (2000)
8. Kaye, J.Z.: Halomonas neptunia sp. nov., halomonas sulfidaeris sp. nov., halomonas axialensis sp. nov. and halomonas hydrothermalis sp. nov.: halophilic bacteria isolated from deep-sea hydrothermal-vent environments. Int. J. Syst. Evol. Microbiol. **54**, 499–511 (2004)
9. Konwar, K.M., Hanson, N.W., Pagé, A.P., Hallam, S.J.: MetaPathways: a modular pipeline for constructing pathway/genome databases from environmental sequence information. BMC Bioinform. **14**, 202 (2013)
10. Mandric, I., Knyazev, S., Padilla, C., Stewart, F., Măndoiu, I.I., Zelikovsky, A.: Metabolic analysis of metatranscriptomic data from planktonic communities. In: Cai, Z., Daescu, O., Li, M. (eds.) ISBRA 2017. LNCS, vol. 10330, pp. 396–402. Springer, Cham (2017). https://doi.org/10.1007/978-3-319-59575-7_41
11. Mandric, I., Temate-Tiagueu, Y., Shcheglova, T., Al Seesi, S., Zelikovsky, A., Mandoiu, I.I.: Fast bootstrapping-based estimation of confidence intervals of expression levels and differential expression from RNA-Seq data. Bioinformatics **33**(20), 3302–3304 (2017)
12. Mitrea, C., et al.: Methods and approaches in the topology-based analysis of biological pathways. Front Physiol. **4**, 278 (2013)
13. Sharon, I., Bercovici, S., Pinter, R.Y., Shlomi, T.: Pathway-based functional analysis of metagenomes. J. Comput. Biol. **18**(3), 495–505 (2011)
14. Shen, M., Li, Q., Ren, M., Lin, Y., Wang, J., Chen, L., Li, T., Zhao, J.: Trophic status is associated with community structure and metabolic potential of planktonic microbiota in plateau lakes. Front. Microbiol. **10**, 2560 (2019)
15. Subramanian, A., et al.: Gene set enrichment analysis: a knowledge-based approach for interpreting genome-wide expression profiles. Proc. Natl. Acad. Sci. U. S. A. **102**(43), 15545–15550 (2005)
16. Tarca, A.L., Draghici, S., Bhatti, G., Romero, R.: Down-weighting overlapping genes improves gene set analysis. BMC Bioinform. **13**, 136 (2012)
17. Ye, Y., Doak, T.G.: A parsimony approach to biological pathway reconstruction/inference for genomes and metagenomes. PLoS Comput. Biol. 5(8), e1000465 (2009)

Identification of Virus-Receptor Interactions Based on Network Enhancement and Similarity

Lingzhi Zhu[1,2], Cheng Yan[1,3(✉)], and Guihua Duan[1]

[1] Hunan Provincial Key Lab on Bioinformatics, School of Computer Science and Engineering, Central South University, Changsha 410083, China
`yancheng01@mail.csu.edu.cn`
[2] School of Computer and Information Science, Hunan Institute of Technology, Hengyang 421008, China
[3] School of Computer and Information, Qiannan Normal University for Nationalities, Duyun 558000, China

Abstract. As a main composition of the human-associated microbiome, viruses are directly associated with our health and disease. The receptor-binding is critical for the virus infection. So identifying potential virus-receptor interactions will help systematically understand the mechanisms of virus-receptor interactions and effectively treat infectious diseases caused by viruses. Several computational models have been developed to identify virus-receptor interactions based on assumption that similar viruses show similar interaction patterns with receptors and vice versa, but the performance need to be improved. Furthermore, the virus network and the receptor network are also noisy. Therefore, we present a new prediction model (NERLS) to identify potential virus-receptor interactions based on Network Enhancement, virus sequence information and receptor sequence information by Regularized Least Squares. Firstly, the virus network is constructed based on the virus sequence similarity and Gaussian interaction profile (GIP) kernel similarity of viruses by a mean method. They are calculated based on the viral RefSeq genomes downloaded from NCBI and known virus-receptor interactions, respectively. Similarly, we also use the same mean method to construct the receptor network based on the amino acid sequence similarity and known virus-receptor interactions. Then Network Enhancement is applied to denoise the virus network and the receptor network. Finally, we employ the regularized least squares algorithm to identify potential virus-receptor interactions. The 10-fold cross validation (10CV) experimental results indicate that an average Area Under Curve (AUC) values of NERLS is 0.8930, which is superior to other computing models of 0.8675 (IILLS), 0.7959 (BRWH), 0.7577 (LapRLS), and 0.7128 (CMF). Furthermore, the Leave One Out Cross Validation (LOOCV) experimental results also show that NERLS can achieve the AUC values of 0.9210, which is better than other models (IILLS: 0.9061, BRWH: 0.8105, LapRLS: 0.7713, CMF: 0.7491). In addition, a case study also confirms the effectiveness of NERLS in predicting potential virus-receptor interactions.

© Springer Nature Switzerland AG 2020
Z. Cai et al. (Eds.): ISBRA 2020, LNBI 12304, pp. 344–351, 2020.
https://doi.org/10.1007/978-3-030-57821-3_33

Keywords: Virus-receptor interactions · Network enhancement · Similarity · Regularized least squares · Gaussian interaction profile (GIP) kernel

1   Introduction

Viruses can infect humans and cause hundreds of diseases [1]. Above all, some emerging and epidemic-prone viral diseases, such as Coronavirus Disease 2019 (COVID-19) [2], Severe Acute Respiratory Syndrome (SARS) [3], and Middle East Respiratory Syndrome (MERS) [4], directly threaten human health and become the public health concern [5]. For example, the current outbreak of COVID-19 in Wuhan, China [2] has leaded to more and more Chinese people to start staying at home to prevent contagion. A recent study shows that SARS-CoV-2 is the pathogen of COVID-19 and can use its matching receptor, angiotensin converting enzyme 2, to enter human cells [2]. The binding of viruses to their matching receptors is thought to have started on the virus infection. To explore the interaction mechanism of viruses and receptors, Yan et al. selected 104 viruses, 74 receptors and 211 virus-receptor interactions as a benchmark dataset from viralReceptor [6,7]. Based on this dataset, NERLS is presented for identifying virus-receptor interactions.

The advantages of NERLS are as below: (i) The virus network is constructed with the virus sequences information and known virus-receptor interactions. (ii) Network Enhancement is used to denoise the virus network and the receptor network, and the regularized least squares (RLS) algorithm is applied to predict virus-receptor interactions.

2   Methods

2.1   Construct Virus Network

Let $\{v_1, v_2, v_3, ..., v_{nv}\}$ denote the set of viruses V, and $\{r_1, r_2, r_3, ..., r_{nr}\}$ denote the set of receptors R. Then, Y is an adjacency matrix of the nv rows and nr columns and $Y \in \mathbb{R}^{nv \times nr}$ denotes known virus-receptor interactions. The virus network can be constructed with the virus sequence similarity and virus GIP kernel similarity. First of all, the virus sequence similarity is computed by d_2^* oligonucleotide frequency measures based on genomic sequences. The correlation of genomic sequences is inferred by the dissimilarity measure approach based on genomic k-mer frequencies. According to the assumption that similar k-mer patterns are shared between similar viruses [8], we can use the k-mer similarity to measure the correlation of genomic sequences. So the distance of k-mer frequency vectors between each virus-virus pair is computed by d_2^* oligonucleotide frequency measures [8]. According to the existing research [8], we can obtain the correlation $V_{seq}(v_i, v_j)$ between each pair of virus v_i and virus v_j when k is set to be 6.

Afterwards, the virus GIP kernel similarity $V_{GIP}(v_i, v_j)$ between virus v_i and virus v_j is computed as below:

$$V_{GIP}(v_i, v_j) = \exp(-\gamma_v ||y_{v_i} - y_{v_j}||^2) \tag{1}$$

$$\gamma_v = \gamma_v' / (\frac{1}{nv} \sum_{i=1}^{nv} ||y_{v_i}||^2) \tag{2}$$

Here y_{v_i} and y_{v_j} denote the interaction profiles of virus v_i and virus v_j, respectively. γ_v regulates the normalized kernel bandwidth by the original bandwidth γ_v'. Based on other similar research [9], we can set γ_v' to be 1.

Finally, we can construct the virus network SV based on two virus similarity matrices by a linear weighted method. In the virus network, the weight of edge $SV(v_i, v_j)$ between virus v_i and virus v_j can be calculated as follows:

$$SV(v_i, v_j) = \frac{V_{GIP}(v_i, v_j) + V_{seq}(v_i, v_j)}{2} \tag{3}$$

2.2 Construct Receptor Network

To construct the receptor network, the receptor sequence similarity and receptor GIP kernel similarity are computed. Firstly, We can use their normalized Smith-Waterman score $SW(r_i, r_j)$ to compute the sequences similarity of receptors. Then the receptor sequence similarity $R_{seq}(r_i, r_j)$ between receptor r_i and receptor r_j is computed as below:

$$R_{seq}(r_i, r_j) = SW(r_i, r_j) / \sqrt{SW(r_i, r_i)} \sqrt{SW(r_j, r_j)} \tag{4}$$

According to the above method of the virus GIP kernel similarity, we can calculate the receptor GIP kernel similarity $R_{GIP}(r_i, r_j)$ between receptor r_i and receptor r_j based on the known virus-receptor interactions.

Finally, two receptor similarity matrices are also used to construct the receptor network by the linear weighted method. The weight of edge $SR(r_i, r_j)$ in the receptor network is computed as follows:

$$SR(r_i, r_j) = \frac{R_{GIP}(r_i, r_j) + R_{seq}(r_i, r_j)}{2} \tag{5}$$

2.3 Network Enhancement

Inspired by the Network Enhancement (NE) method [10], we adopt NE to denoise the virus network and the receptor network, respectively. For example, the virus network is chosen as an input network. Let W be a weighted virus matrix of the nv rows and nv columns, N_m be a set with K-nearest neighbors of virus m and τ be a novel localized network based on the weighted virus matrix. Then $\tau_{m,n}$ of virus m and its neighbors n can be calculated as follows:

$$P_{m,n} \leftarrow \frac{W_{m,n}}{\sum_{k \in N_m} W_{n,k}} \mathbb{I}_{\{n \in N_m\}}, \quad \tau_{m,} \leftarrow \sum_{k=1}^{nv} \frac{P_{m,k} P_{n,k}}{\sum_{u=1}^{nv} P_{u,k}} \tag{6}$$

where P is a row-normalized transition probability matrix, and $\mathbb{I}_{\{.\}}$ denotes a indicator function. If n is one of K-nearest neighbors of virus m, the value of this indicator function is 1, otherwise is 0. Based on zhou et al. a diffusion process can be expressed by random walk with restart and the regularization information flow in NE.

$$W_{t+1} = \alpha\tau \times W_t \times \tau + (1-\alpha)\tau \tag{7}$$

Here α denotes a regularized parameter, t denotes iterative step and the initial value of W_t is the virus network SV. For each input, the update rule is expressed as follows:

$$(W_{t+1})_{m,n} = \alpha \sum_{k \in N_m} \sum_{l \in N_n} \tau_{m,k}(W_t)_{k,l}\tau_{l,n} + (1-\alpha)\tau_{m,n} \tag{8}$$

The final virus network SV_{NE} is obtained based on this diffusion process. Similarly, we can use the same method to obtain the final receptor network SR_{NE}.

2.4   Initialized Interaction Profiles for New Viruses and Receptors

Considering all neighbors of viruses and receptors directly affects the ability of NERLS, if a new virus (receptor) has no known association with all receptors (viruses), we introduce the known interactions profiles of all neighbors to initialize an interaction score for new virus (receptor). Then the initial score of this new virus v_i and a certain receptor r_j is computed as below:

$$y(v_i, r_j) = \frac{\sum_{l=1}^{nv} SV_{NE}^{(il)} y_{lj}}{\sum_{l=1}^{nv} SV_{NE}^{(il)}} \tag{9}$$

As mentioned above, the initial score of a new receptor can be computed by the same mothod. We can compute the initial interaction profile between this new receptor r_j and a certain virus v_i as follows:

$$y(v_i, r_j) = \frac{\sum_{l=1}^{nr} SR_{NE}^{(jl)} y_{il}}{\sum_{l=1}^{nr} SR_{NE}^{(jl)}} \tag{10}$$

2.5   Regularized Least Squares

In our study, RLS is applied to identify virus-receptor interactions based on the virus sequence information and the receptor sequence information. To make full use of the feature of viruses and receptors, SV_{NE} and SR_{NE} are normalized for two laplacian similarity matrixes LV and LR by the laplacian operation, respectively. Let $||.||$ be the Frobenius norm, β_v and β_r be the trade-off parameters, and $tr(.)$ be the matrix trace. According to the Regularized Least Squares, we can compute two matrixes FV and FR with minimizing the cost functions.

$$FV^* = arg_{FV} min[||Y - FV^T||_F^2 + \beta_v tr(FV^T \cdot LV \cdot FV)] \tag{11}$$

$$FR^* = arg_{FR}min[||Y - FR||_F^2 + \beta_r tr(FR^T \cdot LR \cdot FR)] \tag{12}$$

Based on the existing studies [6, 11], FV^* and FR^* can be expressed as:

$$FV^* = SV_{NE}(SV_{NE} + \beta_v \cdot LV \cdot SV_{NE})^{-1}Y \tag{13}$$

$$FR^* = SR_{NE}(SR_{NE} + \beta_r \cdot LR \cdot SR_{NE})^{-1}Y^T \tag{14}$$

Finally, two prediction matrixes FV^* and FR^* are transformed into an integrated prediction matrix with a linear mean method as follows:

$$F^* = \frac{FV^* + (FR^*)^T}{2} \tag{15}$$

3 Experimental Results and Discussion

3.1 Performance Evaluation

To assess the prediction performance of NERLS, we use 10CV and LOOCV to verify potential virus-receptor interactions. In 10CV, we randomly divide all known virus-receptor interactions ($\mathbb{S}^+$) into ten roughly equal subsets. Each subset in turn can be selected as the testing data, while other subsets as the training data. Furthermore, all the rest are the candidate interactions. At the very least, 10CV is run 100 times to get the average as the final result.

$$\mathbb{S}^+ = \mathbb{S}_1^+ \cup \mathbb{S}_2^+ \cup \cdots \cup \mathbb{S}_{10}^+ \tag{16}$$

with

$$\emptyset = \mathbb{S}_1^+ \cap \mathbb{S}_2^+ \cap \cdots \cap \mathbb{S}_{10}^+ \tag{17}$$

$$|\mathbb{S}_1^+| \approx |\mathbb{S}_2^+| \approx \cdots \approx |\mathbb{S}_{10}^+| \tag{18}$$

where $\cup$ is a union of the set, $\cap$ is a intersection of the set, and $\emptyset$ is the empty set.

In LOOCV, each known virus-receptor interaction is in turn selected as a testing data, and other known virus-receptor interactions can be treated as the training data. In addition, all the remaining interactions are the candidate interactions.

3.2 Comparison with Other Methods

NERLS is compared with four recent models: IILLS [6], BRWH [12], LapRLS [11], and CMF [13]. Figure 1 shows that NERLS outperforms other models in 10CV. Specifically, NERLS obtains AUC values of 0.8930, while IILLS, BRWH, LapRLS, and CMF have 0.8675, 0.7959, 0.7577, and 0.0.7128 in 10CV, respectively. In LOOCV, the AUC value of NERLS is also higher than the AUC values of other models. As be shown in Fig. 2, NERLS can achieve 0.9210 when IILLS, BRWH, LapRLS, and CMF is 0.9061, 0.8105, 0.7713, and 0.7491 in LOOCV, respectively. And it is obvious that NERLS is better than other four models.

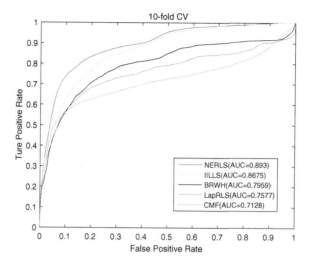

Fig. 1. The AUC curves of NERLS among different similarities in 10-fold CV

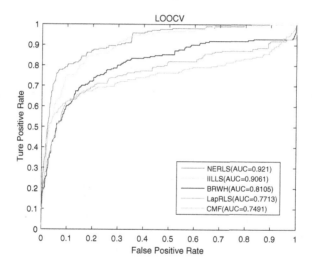

Fig. 2. The AUC curves of NERLS among different similarities in LOOCV

3.3   Case Study

In this section, a case study is selected to validate the prediction ability of
NERLS. As shown in the Table 1, 4 of top 10 hidden interactions are validated
by existing literatures. For example, C-type lectin domain family 4 member M is
also named as L-SIGN or CD209L. Its carbohydrate recognition domain mediates the recognition of fucose and high-mannose glycan, and these carbohydrate
structures are discovered in Lassa virus [14,15]. Human coronaviruses and 229E

can use CD209 for viral infection [16]. Rift valley fever virus and uukuniemi virus can use L-SIGN to infects cells of the expression lectin abnormally [14,17].

Table 1. Top-10 Predicted Results of NERLS

Rank	Virus name	Receptor name	References
1	Lassa mammarenavirus	C-type lectin domain family 4 member M	Garcia-Vallejo et al. (2015), Sakuntabhai et al. (2005)
2	Lymphocytic choriomeningitis mammarenavirus	C-type lectin domain family 4 member M	unknown
3	Human coronavirus 229E	CD209 molecule	Lo et al. (2006)
4	Marburg marburgvirus	C-type lectin domain family 4 member G	unknown
5	Uukuniemi virus	C-type lectin domain family 4 member M	Lger et al., (2016), Sakuntabhai et al., (2005)
6	Rift Valley fever virus	C-type lectin domain family 4 member M	Lger et al., (2016), Sakuntabhai et al., (2005)
7	Ebola virus	dystroglycan 1	unknown
8	Marburg marburgvirus	NPC intracellular cholesterol transporter 1	unknown
9	Lassa mammarenavirus	MER proto-oncogene, tyrosine kinase	unknown
10	Human immunodeficiency virus 2	C-type lectin domain family 4 member M	unknown

4 Conclusion

Accumulating evidences show that systematically understanding the interaction mechanisms of viruses and receptors will be helpful to the prevention, diagnosis, treatment of human infectious diseases caused by viruses. To explore this mechanisms, a prediction model (NERLS) is proposed to identify interactions of viruses and receptors based on Network Enhancement, virus and receptor sequence information. Furthermore we already have confirmed the effectiveness of NERLS by cross validation and a case study. In addition, the efficient integrated approaches should be explored in the future, and more computational models should be considered to improve the prediction performance.

Acknowledgement. This work is supported in part by the National Natural Science Foundation of China (No. 61772552, No. 61420106009, No. 61832019 and No. 61962050), 111 Project (No. B18059), Hunan Provincial Science and Technology Program (No. 2018WK4001), the Science and Technology Foundation of Guizhou Province of China under Grant NO. [2020]1Y264, the Scientific Research Foundation of Hunan Provincial Education Department (No. 18B469), and the Aid Program Science and Technology Innovative Research Team of Hunan Institute of Technology.

References

1. Geoghegan, J.L., Senior, A.M., Di Giallonardo, F., Holmes, E.C.: Virological factors that increase the transmissibility of emerging human viruses. Proc. Natl. Acad. Sci. **113**(15), 4170–4175 (2016)
2. Zhou, P., et al.: A pneumonia outbreak associated with a new coronavirus of probable bat origin. Nature **579**(7798), 270–273 (2020)
3. Ge, X.Y., et al.: Isolation and characterization of a bat SARS-like coronavirus that uses the ACE2 receptor. Nature **503**(7477), 535–538 (2013)
4. Breban, R., Riou, J., Fontanet, A.: Interhuman transmissibility of middle east respiratory syndrome coronavirus: estimation of pandemic risk. Lancet **382**(9893), 694–699 (2013)
5. Wu, F., et al.: A new coronavirus associated with human respiratory disease in China. Nature **579**(7798), 265–269 (2020)
6. Yan, C., Duan, G., Wu, F.X., Wang, J.: IILLS: predicting virus-receptor interactions based on similarity and semi-supervised learning. BMC Bioinform. **20**(23), 651 (2019)
7. Zhang, Z., et al.: Cell membrane proteins with high n-glycosylation, high expression and multiple interaction partners are preferred by mammalian viruses as receptors. Bioinformatics **35**(5), 723–728 (2019)
8. Ahlgren, N.A., Ren, J., Lu, Y.Y., Fuhrman, J.A., Sun, F.: Alignment-free oligonucleotide frequency dissimilarity measure improves prediction of hosts from metagenomically-derived viral sequences. Nucleic Acids Res. **45**(1), 39–53 (2017)
9. Zhu, L., Duan, G., Yan, C., Wang, J.: Prediction of microbe-drug associations based on katz measure. In: 2019 IEEE International Conference on Bioinformatics and Biomedicine (BIBM), pp. 183–187. IEEE (2019)
10. Wang, B., et al.: Network enhancement as a general method to denoise weighted biological networks. Nat. Commun. **9**(1), 1–8 (2018)
11. Xia, Z., Wu, L.Y., Zhou, X., Wong, S.T.: Semi-supervised drug-protein interaction prediction from heterogeneous biological spaces. BMC Syst. Biol. **4**, S6 (2010)
12. Luo, H., Wang, J., Li, M., et al.: Drug repositioning based on comprehensive similarity measures and bi-random walk algorithm. Bioinformatics **32**(17), 2664–2671 (2016)
13. Zheng, X., Ding, H., Mamitsuka, H., Zhu, S.: Collaborative matrix factorization with multiple similarities for predicting drug-target interactions. In: Proceedings of the 19th ACM SIGKDD International Conference on Knowledge Discovery and Data Mining, pp. 1025–1033. ACM (2013)
14. Sakuntabhai, A., et al.: A variant in the CD09 promoter is associated with severity of dengue disease. Nat. Genet. **37**(5), 507–513 (2005)
15. Garcia-Vallejo, J.J., van Kooyk, Y.: DC-SIGN: the strange case of Dr. Jekyll and Mr. Hyde. Immunity **42**(6), 983–985 (2015)
16. Lo, A.W., Tang, N.L., To, K.F.: How the sars coronavirus causes disease: host or organism? J. Pathol. J. Pathol. Soc. Great Br Irel. **208**(2), 142–151 (2006)
17. Lger, P., et al.: Differential use of the C-type lectins L-SIGN and DC-SIGN for phlebovirus endocytosis. Traffic **17**(6), 639–656 (2016)

Enhanced Functional Pathway Annotations for Differentially Expressed Gene Clusters

Chun-Cheng Liu[1], Tao-Chuan Shih[2], Tun-Wen Pai[1,2(✉)], Chin-Hwa Hu[3], and Lee-Jyi Wang[4]

[1] Department of Computer Science and Engineering, National Taiwan Ocean University, Keelung, Taiwan
twp@csie.ntut.edu.tw
[2] Department of Computer Science and Information Engineering, National Taipei University of Technology, Taipei, Taiwan
[3] Department of Bioscience and Biotechnology, National Taiwan Ocean University, Keelung, Taiwan
[4] Deptartment of Information Technology, Tungnan University, New Taipei City, Taiwan

Abstract. Biological pathway enrichment analysis is mainly applied to interpret correlated behaviors of activated gene clusters. In traditional approaches, significant pathways were highlighted based on hypergeometric distribution statistics and calculated P-values. However, two important factors are ignored for enrichment analysis, including fold-change levels of gene expression and gene locations on biological pathways. In addition, several reports have shown that noncoding RNAs could inhibit/activate target genes and affect the results of over-representation analysis. Hence, in this study, we provided an alternative approach to enhance functional gene annotations by considering different fold-change levels, gene locations in a pathway, and non-coding RNA associated genes simultaneously. By considering these additional factors, the ranking of significant P-values would be rearranged and several important and associated biological pathways could be successfully retrieved. To demonstrate superior performance, we used two experimental RNA-seq datasets as samples, including Birc5a and HIF2α knocked down in zebrafish during embryogenesis. Regarding Birc5a knock-down experiments, two biological pathways of sphingolipid metabolism and Herpes simplex infection were additionally identified; for HIF2α knock-down experiments, four missed biological pathways could be re-identified including ribosome biogenesis in eukaryotes, proteasome, purine metabolism, and complement and coagulation cascades. Thus, a comprehensive enrichment analysis for discovering significant biological pathways could be overwhelmingly retrieved and it would provide integrated and suitable annotations for further biological experiments.

Keywords: RNA-seq · Differential expression · Biological pathway · Long-noncoding RNA

1 Introduction

Next generation sequencing (NGS) is a high-throughput technology for RNA-seq datasets and provides authoritative operation to discover gene functions of species

© Springer Nature Switzerland AG 2020
Z. Cai et al. (Eds.): ISBRA 2020, LNBI 12304, pp. 352–363, 2020.
https://doi.org/10.1007/978-3-030-57821-3_34

through genome and transcriptome analysis in recent years [1]. Contrary to microarray, NGS provides efficient and effective performance through powerful sequencing techniques so that it is easy to discover biological reactions between two different biological experiments under various environmental settings. Sequenced NGS raw data requires standard pipelines to assemble short sequenced reads into contigs through assembling tools and referred the assembled contigs to known genes annotated in authoritative databases. Once the assembled sequences and corresponding expression levels were obtained, the following enrichment analysis for differentially expressed gene clusters is performed for genes function annotations.

A biological pathway describes the interactive relationships between chemical substances and genes, and identified over-representative pathways are considered as similar to real biological responses. Regarding to biological pathway functional annotations analysis, the developed approaches in current years would be roughly generalized into two ways. One is Over-Representation Analysis (ORA) which focuses on only differential expressed genes clusters. This method mainly applied thresholdingon statistics theory such as P-values. This operation is easy to distinguish the significant biological pathways, but some considerable biological pathways might be discarded by using the ambiguous limitation [2]. The other method is Functional Class Scoring (FCS) which considers all genes clusters. It uses the specific scoring mechanisms and sorts the rankings of importance for each biological pathway by the scores. In contrast to ORA, all the possible situations of biological pathways would be concerned and the rough limitation is not required either. However, how to pick the final results of biological pathways based on the scoring mechanisms becomes the crucial problems [3]. In this study, we adopt different ways to improve functional annotations. The method of calculation of P-values and the thresholds from ORA were continuously applied, but we tied to take the scoring mechanisms from FCS to avoid the problems with the situation of ignored important biological pathways. The purposed method focuses on the gene expression levels and locations of associated genes within a pathway map.

In this study, a mechanism of enhanced functional enrichment analysis was designed for utilizing KEGG (Kyoto Encyclopedia of Genes and Genomes) pathways database. For RNA-seq datasets, all differentially expressed genes would be grouped according to various fold-change levels of gene expression and gene location within a pathway map, and each stratified gene class would imply their importance index respectively through additionally assigned weightings. According to the principles, it becomes easy to realize and annotate behaviors of various genes within a pathway and gains more flexibility and supporting to decide the suitable key genes for following biological experiment design. It is expected that we could appropriately re-adjust relative P-values for all pathways and try to recall significant biological pathways based on differentially expressed transcripts. Besides, to involve differentially expressed long non-coding RNAs in this study, the new modified parameters for the hypergeometric distribution analysis was proposed to retrieve associated genes that did not possess significant fold-change representation. Hence, hidden regulation responses by lncRNAs would be revealed in each associated pathway. Briefly, the proposed system provides a comprehensive functional enrichment analysis on biological pathways and considers additional characteristics of gene responses. To demonstrate the proposed mechanism, we adopted two scenarios of RNS-seq experiments, including Birc5a and HIF2α knocked down experiments in zebrafish embryo stages respectively. For the first experiment, it was already verified that Birc5a knock down would strongly impact cell survival, involve the reaction of apoptosis, and influence nerve growth and development during embryogenesis [4]. For the second experiment of knocking down HIF2α, the hypoxia-inducible factor regulated genes related to angiogenesis and anaerobic metabolism of cells exposed to hypoxic stress [5]. The additionally obtained significant pathways and corresponding importance indices will be compared to the identified significant pathways by traditional approaches.

2   Materials and Methods

2.1   KEGG Database and Data Processing

KEGG (Kyoto Encyclopedia of Genes and Genomes) is a biological database system [7], which provides the information of biological pathways with corresponding genes. A pathway is a series of interactions among molecules in cell that leads to a certain production of a change in cell. Besides, a pathway can turn genes on and off or spur a cell to move. KEGG is easy to observe chemical interactions among various molecular components. In the study, some global and overview KEGG pathways were discarded and a total of 388 biological pathways were considered in this study. Here we added two additional factors to enhance enrichment analysis, including gene expression fold-change and gene location. Furthermore, modifying P-values to discover missed biological pathways which were not considered as significant pathways. The DFS (depth-first search) algorithm was applied to count how many downstream substances connected to each gene element in a pathway map, and therefore, the locations and relative impact factor of each specific gene in a biological pathway could be annotated. If a gene contains abundant leaf nodes as downstream elements within a pathway, then the gene must locate near the source regions and vice versa. With all associated number of leaf nodes in a pathway, the location and corresponding impact factor for each gene could be roughly classified into three categories including upstream, midstream and downstream regions in each biological pathway. For the second factor of gene expression fold-change, the acquired differentially expressed gene dataset might possess problems of different contigs mapped with an identical gene in KEGG. Therefore, the system calculates the average coverage from all different contigs mapped to the identical gene, and the average expressed levels are applied to calculate corresponding fold-change. Fold-change values are divided into two types, increasing and declining tendencies with high, medium, and low levels. Thus, there were a total of six different types regarding fold-change levels. To enhancing functional enrichment analysis in terms of KEGG pathways, three classes of distinctive gene locations and six subcategories of fold-change levels of differentially expressed genes were applied as additional impact factors for calculating corresponding P-values of all pathways. Therefore, a total 18-class important indices by integrating these two additional features would be obtained for distinguishing impact factor of each differential expressed gene. The formula to calculate corresponding feature scores for each biological pathway is formulated as following Eq. (1).

$$S_j = \frac{\sum_{i=1}^{i=n} \left(f_{i,j}(x) g_i(x) \right)}{n} \tag{1}$$

If n differentially expressed genes were mapped to the jth biological pathway. Then, $f_{i,j}(x)$ represents a categorized score of the ith gene located in j^{th} pathway and $g_i(x)$ represents a categorized score of ith gene regarding its fold-change level of gene expression. In both categorized features, three various values of a, b, and c are used to represent the ith gene under various conditions. For all identified significant genes within the jth biological pathway, the average impact factor of S_j could be obtained for the jth pathway. In this research, the default impact factors were default settled as $a = 9$, $b = 6$, $c = 3$, and the obtained average impact scores for each biological pathway could be calculated. According to these guidelines, both features of fold-change level and gene location were simultaneously involved within the functional enrichment analysis for discriminating and ranking true significance of identified pathways.

2.2 Functional Enrichment Analysis

For biological pathway enrichment analysis, we applied hypergeometric distribution statistics theory to calculate P-value for representing corresponding significance of each pathway. To reflect the conditions of possible hidden interactions within a pathway, several genes without significant fold-change expression could be recalled according to differentially expressed long-noncoding RNAs. An additional dataset was built for annotating associated genes with neighbor and/or overlapped long-noncoding RNAs. The additionally recalled gene set due to differentially expressed long-noncoding RNAs are applied to design a new mechanism for functional enrichment analysis. The modified formula is represented in Eq. (2), which is mainly based on the total numbers of mapped genes, differentially expressed genes, and the additionally recalled genes due to differentially expressed long noncoding RNAs.

$$P_j(at\ least\ i\ genes|N,n,K,r) = \sum_{k=i}^{\min(n+r,K)} \frac{\binom{K}{k+r}\binom{N-K}{(n+r)-(k+r)}}{\binom{N}{n+r}} \quad (2)$$

According to additionally retrieved gene dataset, the parameters are dynamically modified when proximity genes are additionally identified, where the N is the number of mapped unique genes, n is the number of the differentially expressed genes, K is the number of all genes which were annotated in the jth pathway, k is the number of differentially expressed genes which were successfully mapped in the jth pathway, and r is the number of extra genes which were recalled on j^{th} pathway due to differentially expressed long noncoding RNAs. Therefore, we could calculate a new P-value for each pathway and select significant biological pathways according to a threshold of equal to or smaller than 0.05. To avoid controversial results between traditional and the novel proposed approaches due to false positive recalled genes and/or two impact factors, the calculated P-values from traditional enrichment analysis were divided into two conditions, smaller and larger than 0.05. No matter considering the additional lncRNA associated genes or not, the pathways with traditional P-values less than 0.05 are primarily considered as significant pathways, and these pathways are only ranked in order according to the proposed two impact factors. For the pathways with traditional P-values larger than 0.05, a new P-value could be recalculated by subtracting a proportional factor according to the ranking positions of pathways.

For the first group identified as significant pathways with P-values less than 0.05, we rebuilt their significant ranking positions according to their corresponding impact scores instead of P-values. Hence, an identified biological pathway with a larger P-value (less significant pathway) might be revised and recommended due to better gene impact scores. In contrast, an identified biological pathway with a smaller P-value (very significant pathway) might be switched to lower ranking position due to the pathway possessing low average impact score obtained from less influential gene locations or lower fold-change of gene expression. By this constraint, we could maintain all original identified significant pathways with P-values less than 0.05 and provide gene expression and gene location associated attributes for ranking significance order for primarily detected pathways. For the second group identified as insignificant pathways with P-values larger than 0.05, the impact scores of fold-change gene expression and gene location might play a key to retrieve the pathways as significant pathways by adjusting the P-values. These additionally identified pathways were similar to consider additional genes regulated by lncRNAs, even these genes are not with significant changes in gene expression for themselves. The way to diminish the P-values for changing the significance status of a specific pathway, the original number of identified pathways with small P-values (less than 0.05) and the statistical analysis of P-values of each pathway were considered. The statistical attributes of P-values of each pathway was analyzed in advance for constructing a look-up table as a reference for proportional adjustment. According to traditional P-value calculation, for a specific organism, both parameters of N and K are fixed for a specific species. For various ratios of differentially expressed genes (n) and various ratios of mapped differentially expressed genes in the corresponding pathway (k), all combinations of (n, k) with different ratios could be calculated for all 388 KEGG pathways. Therefore, an average P-values and a standard deviation from 388 pathways under different ratios of differential expressed genes could be constructed as a reference table for P-value adjustment. In general, twice of standard deviation was considered as the largest subtracting values for adjusting the P-values according to a normal distribution. To guarantee the modified P-values are in accordance with the ranking order by considering proposed impact feature factors, decreasing subtraction values is applied. The decreasing interval is obtained by taking the standard deviance divided by the number of ranking from impact scores.

3   Results

To validate the effectiveness of the enhanced functional enrichment analysis, we compared gene expression profiles of the transcriptome of zebrafish with Birc5a gene knocked down against the wild type (WT) transcriptome. Both RNA-seq datasets were sequenced by NextSeq 500 (Illumina) sequencer with paired-end short read fragments of length 100 bps, and raw reads were assembled through a standard pipeline. The sequenced raw reads were filtered and trimmed to remove low quality reads for sequence mapping. The public assembling tools include TopHat2 and cufflinks. TopHat2 aligned filtered reads to zebrafish genome sequences (Ensembl release 79 version Zv9) and the cufflinks analyzed and normalized mapped reads for corresponding expression values. The expression values were normalized by FPKM (Fragments Per Kilobase of transcript per Million). For long noncoding RNA analysis, the GTF file from zflncRNApedia [6] was applied for annotating lncRNAs in zebrafish. To compare differentially expressed changes of genes and lncRNAs between Birc5a knock-down and WT datasets, the Cufcompare [7] was applied by using average FPKM of repeated samples, and the differential FPKM values of each gene were checked according to the following rules: FPKM > 1 and fold change >1.5 or <−1.5. Genes nearby filtered differentially expressed lncRNAs within 5 kb upstream and downstream regions were identified and considered as lncRNA associated genes (lncGenes). All annotated transcripts and differentially expressed genes were mapped to KEGG through the function Retrieve and ID mapping of Uniprot [8], and a total of 2162 mapped genes involving 88 differentially expressed genes, and 67 additional genes (lncGenes) were identified. We applied the proposed P-value formula and compared the differences between traditional approaches and employing additional lncGenes. Focused on the identified significant biological pathways with P-values smaller than 0.05, the former datasets without considering lncGenes identified 7 over-representation biological pathways, while the later employing additional lncGenes increased to 28 significant biological pathways. For example, the P-value of ko00790 (Folate biosynthesis) was improved (from 391.E-02 to 3.25E-03) because the gene of gch1 was concerned from the lncGene dataset. Besides, some additional biological pathways could be identified as significant pathways when additional lncGenes were applied. In contrast to only signaling transductions pathway, ko04330 (Notch signaling pathway) detected, several signaling transductions pathways which actually related to cell Apoptosis and neurogenesis were appeared when DE genes and DE lncRNAs associated genes were both concerned. These pathways include ko04210 (Apoptosis), ko04010 (MAPK signaling pathway), ko04068 (FoxO signaling pathway), ko04150 (mTOR signaling pathway), ko04020 (Calcium signaling pathway), ko04371(Apelin signaling pathway), and ko04070 (Phosphatidylinositol signaling system). Among them, FoxO and MAPK signaling pathway were main pathways which connect to apoptosis. There were three subnetworks of MAPK pathway, including MAP/ERK, JNK/p38, and ERK5. It was reported that MAP/ERK and JNK/p38 pathways related to cell proliferation and apoptosis. In addition, MAP/ERK was associated with CNS development and JNK/p38 played an important role in neural stem cells [9, 10]. The rest of pathways: ko4150, ko04020 and ko04070 are proved to have indirect connection with apoptosis by specific mechanisms [11–13]. The results showed that differentially expressed long-noncoding RNAs indeed interacted with proximity or target genes in certain mechanisms. According to the results, the biological pathways were divided into two group with P-values smaller or larger than a threshold of 0.05. We compared the ranking orders of traditional approach and the adjusted orders after adding impact scores based on gene location and fold-change of differential expressed genes. The 28 identified significant pathways with P-value smaller than 0.05 were shown in Table 1. In the table, ranking order of P-values and corresponding impact score were shown in the right columns. It can be observed that ko00531 (Glycosaminoglycan degradation) is ranked as the last position by sorting original P-values and rearranged as the most important pathway according to the proposed impact feature score. It is mainly due to the gene HPSE in this biological pathway possessing high fold-change level of gene expression and located at the source (upstream) region of the pathway, which provides the highest average impact score compared to other identified significant pathways. It is noticed that HPSE located at the beginning position of the biological pathway with large fold change and decreasing trend. For the identified pathway of ko04261 (Adrenergic signaling in cardiomyocytes) was ranked as the first according to the original P-value calculation. However, the significant genes of ATP1B (INak), CACNB4 (DHPR), AKT, and ATF4 (CREBP) located at downstream region (ending region) within the pathway, so the newly adjusted orders could not be changed to a higher-ranking position among all identified significant pathways.

For the rest biological pathways with P-values larger than 0.05 were also re-checked for their significance. A standard deviation of 1.48E-02 was selected from the look-up table which was obtained by taking various ratios of differentially expressed genes compare to all collected genes in KEGG pathways. According to the original P-values, these biological pathways (with P-values larger than 0.05) were also ranked according to corresponding impact feature scores. The adjusted P-values were obtained by subtracting a proportional standard deviation according to the ranked orders. We found that two additional biological pathways could be marked as significant pathways due to standard deviation subtraction and possessing new P-value less than 0.05. The results were shown in Table 2. For the pathway of ko00600 (Sphingolipid metabolism), a reference standard deviation of 1.48E-02 based on 4% of n against N and 14% of k against K. A new P-value could be obtained as 0.026 by taking a subtraction of two-fold of standard deviation. The P-value was adjusted from 0.0510 to 0.0268, and it could be considered as a significant pathway due to high gene fold-change level and gene locations. The associated gene of GAL3ST1 [EC:2.8.2.11]] located at upstream region and the gene of PPAP2 [EC:3.1.3.4] possessing high fold-change levels are the main reason in this case. The sphingolipid is particularly abundant in the central nervous system and it is considered as important components in membrane neurons [14]. These additionally identified biological pathways are actually corresponding to this experiment regarding neurons development function. The other biological pathway of ko05168 (Herpes simplex infection) was also additionally selected because of P-value changed from 0.0542 to 0.0421. It is due to the gene of IRF9 possessed higher gene fold-change expression. This pathway mainly showed cell apoptosis and the influence of viral Herpes. However, the viral Herpes were verified with relation to neurons infection. It would establish a quiescent or latent infection in peripheral neurons and actually caused severe infection with neurological impairment and high mortality [15, 16].

Taking the same analytical steps, our proposed mechanisms were applied to another HIF2α knock-down experiments. However, there were different ways for selecting model species in KEGG. The biological pathways were only appeared for choosing specific specie. In this case, the pathways would be only shown for zebrafish, but some pathways would might be missed. Hence, if we would like to expect that other interesting results were highlighted or not, the reference specie needs to be changed. Here, we choose Human to be the only reference specie. Following the processes for gene annotation, there were 2686 mapped genes, 32 differentially expressed genes, and 14 additional significant genes perhaps regulated by differentially expressed long non-coding RNAs.

Table 1. Compare two ranked pathways based on P-values and impact scores. ko00531 is considered as the last significant pathway according to the adjusted ranking, but which is was considered as the first significant pathway based on impact feature score analysis. In contrast to the pathway of ko04261, it was ranked as the 13rd significant pathway without considering the gene location and gene expression profiles.

Path_id.	Pathway name	P-value	Score	Rank P-value	Rank Score
ko00531	Glycosaminoglycan degradation	5.00E–02	81	28	1
ko00270	Cysteine and methionine metabolism	1.28E–04	81	2	1
ko00561	Glycerolipid metabolism	3.02E–02	67.5	21	2
ko00564	Glycerophospholipid metabolism	1.48E–03	52.5	7	3
ko04330	Notch signaling pathway	2.48E–02	51	19	4
ko04210	Apoptosis	4.58E–04	42	3	5
ko03008	Ribosome biogenesis in eukaryotes	9.60E–03	41.4	13	6
ko00760	Nicotinate and nicotinamide metabolism	3.91E–02	36	25	7
ko04020	Calcium signaling pathway	3.21E–02	36	23	7
ko00240	Pyrimidine metabolism	2.58E–02	36	20	7
ko00565	Ether lipid metabolism	2.48E–03	33.75	8	8
ko00790	Folate biosynthesis	3.25E–03	31.5	9	9
ko00562	Inositol phosphate metabolism	2.00E–02	30	17	10
ko00510	N-Glycan biosynthesis	4.76E–02	27	27	11
ko04150	mTOR signaling pathway	3.14E–02	27	22	11
ko04371	Apelin signaling pathway	2.27E–02	27	18	11
ko04621	NOD-like receptor signaling pathway	7.90E–04	27	5	11
ko04260	Cardiac muscle contraction	6.44E–03	24	10	12
ko04261	Adrenergic signaling in cardiomyocytes	1.08E–04	23.4	1	13
ko04010	MAPK signaling pathway	1.09E–02	23.4	14	13
ko04141	Protein processing in endoplasmic reticulum	5.34E–04	20.25	4	14
ko04068	FoxO signaling pathway	1.16E–02	20.25	15	14
ko04620	Toll-like receptor signaling pathway	3.48E–02	18	24	15
ko04912	GnRH signaling pathway	1.26E–03	18	6	15

(*continued*)

Table 1. (*continued*)

Path_id.	Pathway name	P-value	Score	Rank P-value	Rank Score
ko04070	Phosphatidylinositol signaling system	8.06E–03	18	12	15
ko00534	Glycosaminoglycan biosynthesis - heparan sulfate/heparin	1.17E–02	18	16	15
ko04914	Progesterone-mediated oocyte maturation	4.73E–02	9	26	16
ko04270	Vascular smooth muscle contraction	7.02E–03	9	11	16

Table 2. The two ignored biological pathways by original enrichment analysis. Both ko00600 (Sphingolipid metabolism) and ko05168 (Herpes simplex infection) were retrieved as significant pathway by adjusting P-values according to the ranking order impact features.

Path_id.	Pathway name	n/N (%)	k/K (%)	SD	P-value (1)	P-value (2)
ko00600	Sphingolipid metabolism	4	14	1.48E–02	5.10E–02	2.68E–02
ko05168	Herpes simplex infection	4	9	1.33E–02	5.42E–02	4.21E–02

The examining pathways with P-values smaller than 0.05, we compared the results with additional lncGenes to the original approaches. The traditional approach identified only 3 significant biological pathways, while the number of significant pathways was 21 by considering additional lncGene sets.

Next, we focused on the identified biological pathways (with P-values less than 0.05) and tried to rank the enrichment pathways according to corresponding impact feature scores. The pathway of ko05322 (Systemic lupus erythematosus) possessed a P-value of 0.029 (consider lncGenes) and considered as the 10th significant pathway, but the pathway became the first ranked significant pathway due to the gene of C5 (MAC) possessing high gene fold-change expression and located at midstream regions. In contrast to ko04976 (Bile secretion), the biological pathway was annotated as the most significant pathway due to the lowest P-value. However, the genes of SLC2A1 located at downstream region and possessing low fold-change expression and the gene of ATP1A also located in the downstream region as well. Thus, the new orders obtained by impact score were readjusted the pathway to the significant position.

Next, original P-values larger than 0.05 were tuned by standard deviation based on impact factors. In Table 3, there were four additional biological pathways could be retrieved as significant pathways. The adjusted P-values became smaller than 0.05 based on subtraction of various proportion of standard deviation. Among them, ko03008 (Ribosome biogenesis in eukaryotes) showed both genes of WDR75 (NAN1) and smfn (Rex1/2) had high gene fold-change expression and located at midstream regions. Then, the gene of C5 in ko04610 (Complement and coagulation cascades) also possesses high gene fold-change expression and located at midstream regions. For the other two biological pathways, including ko03050 (Proteasome) and ko00230 (Purine metabolism), even though the impact scores in both pathways ranked in middle range, the previous P-values from traditional approach were already near the threshold of 0.05, so that they also had chances to be considered as significant pathways. In addition, both additionally identified Proteasome and Purine metabolism pathway had been experimentally verified as important and associated biological pathways within hypoxia environments [17, 18], and which possibly triggered the situation of cancer development. In summary, we proposed a novel approach to obtain the better performance in biological pathway enrichment analysis. Not only concern additional genes regulated by differentially expressed lncR-NAs, but also using a heuristic approach to rank the significant index by gene expression levels and location, this proposed mechanism provides more integral and comprehensive results compared to traditional approaches. It is very helpful for biologists to presume the real situations before designing biological experiments.

Table 3. The P-values of four ignored biological pathways were retrieved as significant pathways, and the P-values were readjusted from P-value (1) to P-value (2) through a standard deviation mechanism

Path_id.	Pathway name	n/N (%)	k/K (%)	SD	P-values (1)	P-value (2)
ko03008	Ribosome biogenesis in eukaryotes	1	6	1.33E–02	6.09E–02	3.76E–02
ko04610	Complement and coagulation cascades	1	6	1.33E–02	6.10E–02	4.77E–02
ko03050	Proteasome	1	6	1.33E–02	5.16E–02	4.49E–02
ko00230	Purine metabolism	1	4	1.41E–02	5.48E–02	4.42E–02

4   Discussion

In this study, we focused on enhancing the biological pathways enrichment analysis for RNA-seq dataset comparisons. In tradition approach for over-representation analysis, Only the number of mapped genes, differentially expressed genes, mapped genes in a specific pathway and mapped differentially expressed genes in specific pathway were required for calculating P-value. However, there were some gene characteristics including the quantity of gene response levels and influence index of gene location in a pathway were ignored. Besides, most of the hidden and frequent interactions between genes and long non-coding RNAs were yet not shown in most pathways. Therefore, we proposed a concept of gene impact scores affecting the ranking order of identified significant pathways. We constructed a look-up tables for different species according the ratios of n/N and k/K regarding hypergeometric distribution among KEGG databases. The set of pathways with p-values over 0.05 were not identified as significant pathways through traditional approaches. The original P-value would be subtracted by a value of various proportion of standard deviations from a look-up table according to their impact score ranking order. Only original P-values closed to 0.05 were possible too be retrieved and consider as significant pathways. In the results, we took two RNA-seq experiments for illustration and comparisons. Both Bric5a knocked-down and HIF2αa knocked-down experiments during zebrafish embryo stages were used for demonstration. In Bric5a datasets, some additional significant biological pathways related to cell apoptosis would be actually identified when concerned the regulated genes (lncGenes), including Apoptosis, MAPK signaling pathway and FoxO signaling pathways. In addition, two biological pathways of Sphingolipid metabolism and Herpes simplex infection were re-highlighted through our mechanism. Regarding the HIF2α knock-down experiment, some biological pathways related to cancer development could be annotated as significant pathways through adopting our proposed mechanisms. Furthermore, we would find additional significant biological pathways including Ribosome biogenesis in eukaryotes, Complement and coagulation cascades, Proteasome, and Purine metabolism. It illustrates that additionally identified significant pathways were affected under hypoxia environments. From the illustrated examples, we believed that our proposed approaches provide flexible and comprehensive analysis for gene enrichment analysis and provide suitable mechanisms in terms of gene characteristics. Therefore, the systematic analysis is helpful to discover significant biological pathways and provide biologists an effective and efficient experimental design.

Acknowledgements. Genetic and statistical analysis was supported by the Academia Silica Biomedical Research Centre, Sinica, Taiwan. The publication charges of this article were funded by the Ministry of Science and Technology, Taiwan (MOST 103-2627-B-019 -003 and MOST 103-2221-E-019 -037 to Tun-Wen Pai).

References

1. Ansorge, W.J.: Next-generation DNA sequencing techniques. New Biotechnol. **25**, 195–203 (2009)
2. Git, A., et al.: Systematic comparison of microarray profiling, real-time PCR, and next-generation sequencing technologies for measuring differential microRNA expression. RNA **16**, 991–1006 (2010)
3. Khatri, P., Sirota, M., Butte, A.J.: Ten years of pathway analysis: current approaches and outstanding challenges. PLoS Comput. Biol. **8**(2), e1002375 (2012)
4. Delvaeye, M., et al.: Role of the 2 zebrafish survivin genes in vasculo-angiogenesis, neurogenesis, cardiogenesis and hematopoiesis. BMC Dev. Biol. **9**, 25 (2009)
5. Ko, C., et al.: Integration of CNS survival and differentiation by HIF2 α. Cell Death Differ. **18**, 1757–1770 (2011)
6. Dhiman, H., Kapoor, S., Sivadas, A., Sivasubbu, S., Scaria, V.: zflncRNApedia: a comprehensive online resource for zebrafish long non-coding RNAs. PLoS ONE **10**(6), e0129997 (2015)
7. Trapnell, C., et al.: Differential gene and transcript expression analysis of RNA-seq experiments with TopHat and Cufflinks. Nat. Protoc. **7**, 562–578 (2012)

8. UniProt Consortium: UniProt: a hub for protein information. Nucleic Acids Res. **43**, D204–D212 (2015)

9. Samuels, I.S., Saitta, S.C., Landreth, G.E.: MAP'ing CNS development and cognition: an ERKsome process. Neuron **61**, 160–167 (2009)

10. Yang, S.-R., et al.: The role of p38 MAP kinase and c-Jun N-terminal protein kinase signaling in the differentiation and apoptosis of immortalized neural stem cells. Mutat. Res., Fundam. Mol. Mech. Mutagen. **579**, 47–57 (2005)

11. Vasco, V.R.L.: Phosphoinositide pathway and the signal transduction network in neural development. Neurosc. Bull. **28**, 789–800 (2012)

12. Tee, A.R., Sampson, J.R., Pal, D.K., Bateman, J.M.: The role of mTOR signalling in neurogenesis, insights from tuberous sclerosis complex. In: Seminars in Cell & Developmental Biology, pp. 12–20. Elsevier (2016)

13. Duchen, M.R.: Mitochondria, calcium-dependent neuronal death and neurodegenerative disease. Pflügers Arch. Eur. J. Physiol. **464**, 111–121 (2012)

14. Sabourdy, F., et al.: Monogenic neurological disorders of sphingolipid metabolism. Biochim. Biophys. Acta (BBA) Mol. Cell Biol. Lipids **1851**(8), 1040–1051 (2015)

15. Harkness, J.M., Kader, M., DeLuca, N.A.: Transcription of the herpes simplex virus 1 genome during productive and quiescent infection of neuronal and nonneuronal cells. J. Virol. **88**, 6847–6861 (2014)

16. Zheng, K., et al.: Epidermal growth factor receptor-PI3K signaling controls cofilin activity to facilitate herpes simplex virus 1 entry into neuronal cells. MBio **5**, e00958–e00913 (2014)

17. Zhuang, Z., et al.: Somatic HIF2A gain-of-function mutations in paraganglioma with polycythemia. New Engl. J. Med. **367**, 922–930 (2012)

18. Wang, X., et al.: Purine synthesis promotes maintenance of brain tumor initiating cells in glioma. Nat. Neurosci. **20**, 661 (2017)

Automated Detection of Sleep Apnea from Abdominal Respiratory Signal Using Hilbert-Huang Transform

Xingfeng Lv[1], Jinbao Li[2(✉)], and Qin Yan[3]

[1] School of Computer Science and Technology, Heilongjiang University, Harbin, China
lvxingfeng@hlju.edu.cn
[2] Qilu University of Technology (Shandong Academy of Sciences), Shandong Artificial Intelligence Institute, Jinan, China
lijinb@sdas.org
[3] Qingdao Maternal and Child Health and Family Planning Service Center, No. 6 Tongfu Road, Shibei District, Qingdao, China
qdjsyj@126.com

Abstract. Sleep Apnea (SA) seriously affects human life and health. In recent years, many studies use polysomnography (PSG) to detect sleep apnea, but it is expensive and inconvenient. In order to solve this problem, this paper proposes a method to detect sleep apnea automatically by using a single Abdominal Respiratory Signal. In this method, Hilbert-Huang Transform (HHT) is used to extract frequency domain features, and combined with time domain features. Then sleep apnea is detected by machine learning methods such as Support Vector Machine, AdaBoosting and Random Forest (RF). The experimental results show that HHT can extract significant frequency domain features, and the accuracy of sleep apnea detection can reach 95% using Random Forest method. This method is better than the existing methods in the convenience and accuracy of detection. It is more suitable for family environment, and has a wide range of application prospects.

Keywords: Sleep Apnea · Abdominal Respiratory Signal · Hilbert-Huang Transform · Random Forest

1 Background

Sleep Apnea is a common sleep disorder, affecting 2–4% of the adult population. It is characterized by the occurrence of breathing pauses of at least 10 s

Supported by the Harbin science and technology bureau innovation under Grants No. 2017RAQXJ131, the Basic research project of scientific research operating expenses of Heilongjiang provincial colleges and universities under Grants No. KJCX201815, Heilongjiang Province Natural Science Foundation key project of China under Grant No. ZD2019F003.

Z. Cai et al. (Eds.): ISBRA 2020, LNBI 12304, pp. 364–371, 2020.
https://doi.org/10.1007/978-3-030-57821-3_35

during the night. If left untreated, sleep apnea could lead to serious problems such as heart failure and stroke [1]. The gold standard to detect sleep apnea is polysomnography (PSG) in a specialized sleep laboratory. But PSG is discomfortable and expensive for the patient.

Recently, many automatic sleep apnea detection approches have been developed using various biological signals [2–4]. Respiratory signals can directly reflect the breathing condition during sleep. They can be obtained directly from the airflow, chest and abdomen. Many studies detect sleep apnea using these respiratory signals [5–15]. But some challenges still need to be addressed. One of the major challenges is that process of respiratory signal collection affects human sleep. Respiratory signals are susceptible to interference from other physiological signals. For example, the respiratory signals of nasal airflow tend to cause discomfort to human beings, the respiratory signals of chest are easily interfered by Electrocardiogram (ECG) and Electromyograph (EMG). Another challenge is that respiratory signals are non-stationary. It is very difficult to extract frequency domain features. Although wavelet transform can extract frequency domain features, it is very sensitive to the selection of wavelet basis.

To tackle the above challenges, we propose a sleep apnea detection method based on single abdominal respiratory signal using Hilbert-Huang Transform. This method only uses a single abdominal respiratory signal to detect sleep apnea, avoiding the problems of poor comfort and easy interference. For the nonstationary abdominal respiratory signal, HHT overcomes the shortcomings of Fourier transform which can not reflect the local features of time-frequency, and is not limited by the selection of wavelet basis. It extracts the frequency domain features of significant changes in normal sleep and sleep apnea, and then combines the time domain features to detect sleep apnea automatically by machine learning method. It does not require sleep experts to analyze each sleep epoch manually, reducing the cost of sleep detection.

2   Methods

In this paper, a method of automatic sleep apnea detection from abdominal respiratory signal using HHT is proposed, as shown in Fig. 1. Firstly, the method needs to preprocess the abdominal respiratory signal due to electrical inference and measurement noise. Secondly, the time domain and frequency domain features of respiratory signal are extracted. Considering that abdominal respiratory signal belongs to non-stationary signal, HHT is used to extract frequency domain correlation features.Thirdly, because of the redundancy between features, feature selection is needed. Finally, sleep apnea is detected by using different machine learning algorithms, and the number of sleep apnea is counted according to the detection results of each epoch, so as to detect whether the subject is a sleep apnea patient.

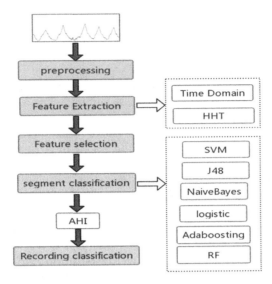

Fig. 1. Schematic illustration of automatic Sleep Apnea detection.

2.1 Dataset and Proprocessing

In order to evaluate the performance of the proposed approach, we have applied this algorithm to Apnea-ECG datasets. Apnea-ECG datasets consists of eight records which contain nasal airflow, chest respiratory and abdominal respiratory signals. In this paper, only a single abdominal respiratory signal was selected to perform automatic sleep apnea detection. These records from the participants who are aged 27–63 years old, weighed 35–135 kg and slept for 7–10 h. The respiratory signal is sampled at 100 Hz and the resolution is 16 bits. The abdominal respiratory signal was divided into epochs of 60s. Each epoch contained 6000 points of data. Each epoch was labelled as Normal(N) and Apnea(A), which were scored by sleep experts according to the standard of American Academy of Sleep Medicine.

In order to improve the accuracy of classification, it is necessary to filter the noise. Since the frequency recognition interval of the spectrum is $100/6000 = 0.017$ Hz, the respiratory frequency of adults ranges from 12 to 25 times per minute, resulting in a respiratory frequency range of 0.2-0.4167 Hz, while the respiratory frequency of infants can reach up 0.748 Hz. For this purpose, this paper uses a eighth-order low-pass Butterworth filter. The cut-off frequency is set as 0.8 Hz.

2.2 Feature extraction and selection

After proprocessing, the features of abdominal respiratory signals of each epoch should be extracted. This paper mainly extracts two kinds of features: time domain features and frequency domain features.

Time Domain Features: It is the statistical features in time domain, including maximum value, minimum value, mean value and variance of peak and trough.

Frequency domain features: Since the signals of abdominal respiration are nonstationary, HHT was used to analyze the frequency information of respiratory signals. HHT decomposes the signals by empirical mode decompositon (EMD), obtains a finite number of intrinsic mode functions (IMF). The EMD process is shown in formula (1). Each IMF performs the Hilbert transform to get the instantaneous frequency and instantaneous amplitude of respiratory signals.

$$data(t) = \sum_{i=1}^{n} IMF_i + r(t) \tag{1}$$

Hilbert transformation is carried out for each IMF to obtain the corresponding Hilbert spectrum, and the marginal spectrum of the respiratory signal is obtained by integrating the Hilbert spectrum in time. The marginal spectrum of normal sleep and sleep apnea are different, as shown in Fig. 2. (a) the curve in the red rectangle is the marginal spectrum of respiratory signals during normal sleep, and (b) the curve in the red rectangle is the marginal spectrum of respiratory signals during sleep apnea, and the two curves are greatly different between normal sleep and apnea sleep.

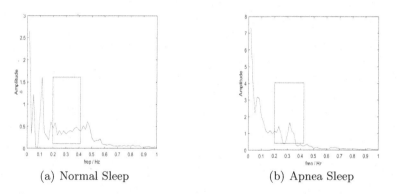

(a) Normal Sleep   (b) Apnea Sleep

Fig. 2. Marginal spectrum of abdominal respiratory signal.

For abdominal respiratory signals, 27 time domain features and 14 frequency domain features were extracted in this paper. There is redundancy among these features, which affects the accuracy of classification, so features need to be selected before classification. By evaluating the predictive ability of each feature and the redundancy between features, some features were selected which high correlation with classification and low correlation between features.

2.3   Classification Methods

After selecting 16 features, sleep apnea was detected using Naive Bayes, Support Vector Machine, Logistic regression, decision tree J48 algorithm, Adaboosting algorithm and Random Forest algorithm. According to PhysioNet Apnea-ECG data set, the performance of each classification algorithm was analyzed from the accuracy, recall rate, f-score, ROC curve, etc. Choose a best performance of machine learning algorithm to detect sleep apnea.

Table 1. Evaluation the performance of different classifiers.

Method	Accuracy	Precision	Recall	F-score	MCC	ROC
LibSVM	0.832	0.841	0.832	0.833	0.665	0.838
J48	0.931	0.932	0.931	0.931	0.858	0.926
NaiveBayes	0.932	0.935	0.932	0.932	0.864	0.962
Logistic	0.947	0.947	0.947	0.947	0.890	0.985
Adaboosting	0.932	0.934	0.932	0.932	0.862	0.974
RF	0.950	0.951	0.950	0.951	0.898	0.990

3   Experiments

In order to verify the effectiveness of the method, training and testing were carried out on the Apnea-ECG data set. The datas used in the experiment included 8 abdominal respiratory records, or 3955 epochs of data. The feature subset was evaluated to reduce the redundancy between features. 16 features were selected from 41 features, including 9 time domain features and 7 features extracted by Hilbert-Huang transformation. These features were significantly different during normal sleep and sleep apnea. As shown in Fig. 3, the percentage of high frequency IMF energy is higher in normal sleep than sleep apnea, and the mean and standard deviation of IMF3 and IMF4 are higher in sleep apnea than in normal sleep. The quartile distribution of these features is significantly different between normal sleep(N) and sleep apnea(A).

A variety of machine learning algorithms are used to detect sleep apnea and different performance indicators are used for evaluation. The detection results are shown in Table 1. The experimental results show that the random forest method can achieve 95% accuracy in detecting sleep apnea, which is better than other methods. For this purpose, Random Forest (RF) is selected to detect sleep apnea.

According to the detection results of sleep Apnea at each epoch, the number of sleep Apnea per patient could be counted to calculate the Apnea Hypopnea Index (AHI). According to the AHI value, the severity of sleep apnea in patients can be classified.The number of sleep apnea per patient divided by the total

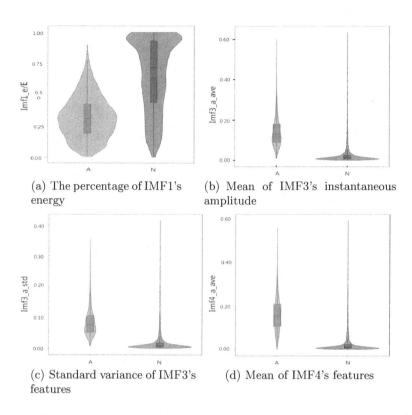

(a) The percentage of IMF1's energy

(b) Mean of IMF3's instantaneous amplitude

(c) Standard variance of IMF3's features

(d) Mean of IMF4's features

Fig. 3. Quartile map of features in sleep apnea and normal sleep

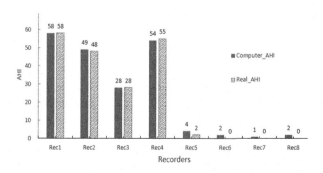

Fig. 4. Comparision so real AHI and computed AHI

number of hours of sleep to obtain the value of AHI, compared with the actual AHI of each patient. The experimental results are shown in Fig. 4, and it can be found that the two AHI are approximately equal. If AHI=5 is taken as the basis to determine whether a record is a patient with sleep apnea, this method can correctly determine the record whether is sleep apnea.

Table 2. A comparision of algorithmic performance on respiration signals.

Authors	Signals	Classifier	Accuracy%	Sensitivity%	Specificity%
Maali [11]	AF AE TE	SVM	89	87	90
Cafer [12]	AF AE TE	RF	98.6	—	—
Haidar [13]	AF AE TE	CNN	83.4	—	—
Kagawa [14]	AE TE	Threshold	94.0	96.0	100
Koley [6]	AF	SVM	89.6	92.8	88.9
Gutierrez [7]	AF	AB-CART	86.5	89.0	80.0
Ours	AE	RF	95.0	94.4	96.0

The comparison of the accuracy, sensitivity and specificity of the detection of sleep apnea using respiratory signals in the existing literature is shown in Table 2. AF denotes nasal airflow respiratory signal, TE denotes Thoracic Effort respiratory signal, and AE denotes Abdominal Effort respiratory signal. It can be seen from the table that three kinds of respiratory signals can get 98.6% accuracy. Both kinds of respiratory signals yields 96.0% sensitivity and 100% specificity. However, the using of multiple respiratory signals at the same time can increase the cost, and affect the patient's normal sleep. In this study, singal abdominal respiratory signals are used to detect sleep apnea, with an accuracy of 95.0%. Although the accuracy of using respiratory signals in three parts at the same time is not achieved, the signal collection is convenient, the normal sleep of patients is not affected, and the monitoring cost is reduced. This sleep monitoring process can be applied to family environment.

4   Conclusion

In order to reduce the cost and the impact of human sleep, and facilitate patients to monitor sleep in the home environment, this paper only used the single abdominal respiratory signal to detect sleep apnea. Firstly, features were extracted from time domain and frequency domain, and then features were selected. Sleep apnea was detected by Random Forest method. Experimental results showed that the accuracy of this method can reach 95.0% for each epoch detection of sleep apnea, and the accuracy of each recording could correctly determine whether the patient was a sleep apnea patient. In the future work, larger datasets will be tested to verify the robustness of the algorithm.

References

1. Siobhan, B., Dinges, D.F.: Behavioral and physiological consequences of sleep restriction. J. Clin. Sleep Med. Jcsm Off. Publ. Am. Acad. Sleep Med. **3**(5), 519–28 (2007)
2. Chen, L., Zhang, X., Song, C.: An automatic screening approach for obstructive sleep apnea diagnosis based on single-lead electrocardiogram. IEEE Trans. Automat. Sci. Eng. **12**(1), 106–115 (2015)
3. Song, C., Liu, K., Zhang, X., Chen, L., Xian, X.: An obstructive sleep apnea detection approach using a discriminative hidden markov model from ecg signals. IEEE Trans. Biomed. Eng. **63**(11), 1532–1542 (2016). https://doi.org/10.1109/TBME.2015.2498199
4. Morillo, D.S., Gross, N.: Probabilistic neural network approach for the detection of sahs from overnight pulse oximetry. Med. Biol. Eng. Compu. **51**(3), 305–315 (2013)
5. Koley, B., Dey, D.: Automated detection of apnea and hypopnea events. In: 2012 Third International Conference on Emerging Applications of Information Technology (EAIT) (2012)
6. Koley, B., Dey, D.: Real-time adaptive apnea and hypopnea event detection methodology for portable sleep apnea monitoring devices. IEEE Trans. Biomed. Eng. **60**(12), 3354–3363 (2013)
7. Gutierrez-Tobal, G.C., Alvarez, D., Del Campo, F., Hornero, R.: Utility of adaboost to detect sleep apnea-hypopnea syndrome from single-channel airflow. IEEE Trans. Biomed. Eng. **63**(3), 636–646 (2016)
8. Gutiérrez-Tobal, G.C., Alvarez, D., Marcos, J.V., Campo, F.D., Hornero, R.: Pattern recognition in airflow recordings to assist in the sleep apnoea-hypopnoea syndrome diagnosis. Med. Biol. Eng. Compu. **51**(12), 1367–1380 (2013)
9. Selvaraj, N., Narasimhan, R.: Detection of sleep apnea on a per-second basis using respiratory signals. In: Conference Proceedings - IEEE Engineering in Medicine and Biology Society Membership 2013, pp. 2124–2127 (2013)
10. Ciolek, M., Niedzwiecki, M., Sieklicki, S., Drozdowski, J., Siebert, J.: Automated detection of sleep apnea and hypopnea events based on robust airflow envelope tracking in the presence of breathing artifacts. IEEE J. Biomed. Health Informat. **19**(2), 418–429 (2015)
11. Maali, Y., Al-Jumaily, A.: Automated detecting sleep apnea syndrome: A novel system based on genetic SVM. In: International Conference on Hybrid Intelligent Systems (2011)
12. Cafer, A., Ahmet, A., Feng, L., Dong-Hoon, L., Ricardo, L., Sandeep, K.: Sleep apnea classification based on respiration signals by using ensemble methods. Bio Med. Mater. Eng. **26**(s1), S1703–S1710 (2015)
13. Haidar, R., McCloskey, S., Koprinska, I., Jeffries, B.: Convolutional neural networks on multiple respiratory channels to detect hypopnea and obstructive apnea events, pp. 1–7 (07 2018). https://doi.org/10.1109/IJCNN.2018.8489248
14. Kagawa, M., Tojima, H., Matsui, T.: Non-contact diagnostic system for sleep apnea-hypopnea syndrome based on amplitude and phase analysis of thoracic and abdominal doppler radars. Med. Biol. Eng. Compu. **54**(5), 789–798 (2016)
15. Hammer, J., Newth, C.J.L.: Assessment of thoraco-abdominal asynchrony. Paediatr. Respir. Rev. **10**(2), 75–80 (2009)

Na/K-ATPase Glutathionylation: *in silico* Modeling of Reaction Mechanisms

Yaroslav V. Solovev[1](✉) (iD), Daria S. Ostroverkhova[2,3] (iD), Gaik Tamazian[4] (iD),
Anton V. Domnin[5] (iD), Anastasya A. Anashkina[6] (iD), Irina Yu. Petrushanko[6] (iD),
Eugene O. Stepanov[7,8,9] (iD), and Yu. B. Porozov[2,10,11](✉) (iD)

[1] Laboratory of Bioinformatics Approaches in Combinatorial Chemistry and Biology,
Institute of Bioorganic Chemistry, Russian Academy of Sciences, 119991 Moscow, Russia
yaroslavsolovev78@gmail.com
[2] Laboratory of Bioinformatics, I.M. Sechenov First Moscow State Medical University
(Sechenov University), 119991 Moscow, Russia
yuri.porozov@gmail.com
[3] Department of Bioengineering, M.V. Lomonosov Moscow State University, 119991 Moscow,
Russia
[4] Theodosius Dobzhansky Center for Genome Bioinformatics, St. Petersburg State University,
199034 Saint-Petersburg, Russia
[5] TheoMAT Group, Saint Petersburg National Research University of Information
Technologies, Mechanics and Optics, 191002 Saint-Petersburg, Russia
[6] Engelhardt Institute of Molecular Biology, Russian Academy of Sciences, 119991 Moscow,
Russia
[7] St. Petersburg Branch of the Steklov Mathematical Institute of the Russian
Academy of Sciences, 191023 Saint-Petersburg, Russia
[8] Faculty of Mathematics, Higher School of Economics, 119048 Moscow, Russia
[9] Department of Mathematical Physics, Faculty of Mathematics and Mechanics,
St. Petersburg State University, Saint-Petersburg, Russia
[10] Department of Food Biotechnology and Engineering, Saint Petersburg National Research
University of Information Technologies, Mechanics and Optics, 197101 Saint-Petersburg, Russia
[11] Department of Computational Biology, Sirius University of Science and Technology,
354340 Sochi, Russia

Abstract. Na,K-ATPase is a redox-sensitive transmembrane protein. Understanding the mechanisms of Na,K-ATPase redox regulation can help to prevent impairment of Na,K-ATPase functioning under pathological conditions and reduce damage and death of cells. One of the basic mechanisms to protect Na,K-ATPase against stress oxidation is the glutathionylation reaction that is aimed to reduce several principal oxidized cysteines (244, 458, and 459) that are involved in Na,K-ATPase action regulation. In this study, we carried out in silico modeling to evaluate glutathione affinity on various stages of Na,K-ATPase action cycle, as well as to discover a reaction mechanism of disulfide bond formation between reduced glutathione and oxidized cysteine. To achieve this goal both glutathione and Na,K-ATPase conformer sampling was applied, the reliability of the protein-ligand complexes was examined by MD assay, the reaction mechanism was studied using semi-empirical PM6-D3H4 approach that could have a deal with large organic systems optimization.

Keywords: Na/K-ATPase · Glutathionylation · In silico modeling ·
PM6-D3H4 · Hybrid MD · Conformer search

1 Introduction

Na,K-ATPase creates a transmembrane gradient of Na+ and K+ ions and acts as a receptor for cardiotonic steroids (CTS). Its functioning is necessary for the viability of all animal cells. Na,K-ATPase is redox-sensitive and interruption of its activity in some pathologies is associated with a change in the redox status of cells. The optimum activity of Na,K-ATPase is observed in a specific range of redox conditions. The enzyme activity has a maximum at the physiological concentration of oxygen, decreasing both under hypoxia and hyperoxia. Understanding the mechanisms of Na,K-ATPase redox regulation can help to prevent the interruption of its functioning under pathological conditions and reduce cell damage.

The functional monomer of Na,K-ATPase consists of α- and β-subunits. The α-subunit is catalytic subunit, which contains binding sites for ATP, K^+, Na^+ and CTS. The β-subunit is a regulatory subunit. The α-subunit contains 23 cysteine residues, 15 of which are cytosolic and available for redox modification. It was found that one of the main reasons for the redox sensitivity of Na,K-ATPase is the S-glutathionylation of the α-subunit of the enzyme [9]. S-glutathionylation is a binding of the tripeptide glutathione to the thiol group of the cysteine residue. Induction of the α-subunit glutathionylation leads to inhibition of the enzyme, up to its complete inactivation due to the disruption of adenine nucleotides binding [9]. The ratio of reduced (GSH) and oxidized (GSSG) form of glutathione determines the thiol redox status of the cell and changes under pathological conditions. It was shown that the reason for the inhibition of Na,K-ATPase during hypoxia is the induction glutathionylation of α-subunit due to an increase in the level of GSSG. It was found that residues of Cys 244 in the actuator domain and residues of Cys 454, 458, 459 in the nucleotide-binding domain of the protein undergo regulatory glutathionylation [12]. Glutathionylation of the Cys 244 residue plays a major role in the Na,K-ATPase inhibition [8]. The enzyme with Cys244Ala replacement is not inhibited by GSSG and becomes insensitive to hypoxia [11]. In addition, the viability of cells expressing the α1 subunit Na,K-ATPase with Cys244Ala or Cys244-454-458-459Ala substitutions decreases under hypoxia [8]. Thus glutathionylation of the α-subunit of Na,K-ATPase plays an important role in the adaptation of cells to hypoxia and helps prevent ATP depletion in the cell. Under hypoxia, not only the transport but also the receptor function of Na, K-ATPase is disturbed [8]. It was found that substitutions Cys454-458-459Ala disrupt the receptor function of Na,K-ATPase. In cells with these substitutions, Src kinase is not activated in response to the CTS ouabain and the oxygen sensitivity of the receptor function of Na,K-ATPase is impaired [8]. Using simulation, it was shown that the glutathionylation of Cys 458 and 459 prevents the inhibitory binding of Src kinase to the nucleotide-binding domain of Na,K-ATPase. Thus the regulation of receptor function is closely related to the glutathionylation of residues of Cys 458 and 459.

One of the key questions is the availability of the described regulatory residues for glutathionylation in different conformations of the enzyme. Ligands and partner proteins, increasing the residence time of the enzyme in a certain conformation, can affect the effectiveness of glutathionylation. It was experimentally shown that the greatest efficiency of α-subunit glutathionylation is observed for the enzyme in the E1 conformation, it decreases in the E2 conformation and becomes minimal in the E2P conformation [10]. It correlates with the availability of cysteine residues for the solvent [10]. However, the availability of regulatory residues for glutathionylation in different conformations was not evaluated. The solution to this problem is important because redox regulation under different conditions can be carried out in different ways and glutathionylation can be involved in the regulation of the binding of Na, K-ATPase to ligands and partner proteins.

2   Materials and Methods

2.1   Coarse-Grained Modeling of Conformational Movement

Full-atomic 29 conformations of Na,K-ATPase were modeled by using the coarse-grained method PROMPT [3, 15]. These conformations represent a time-lapse of a transition from E2P (PDB ID: 3B8E) to E1P (PDB ID: 3WGU) state.

2.2   ATPase Conformations Clustering

Previously obtained full-atomic 29 conformations of Na,K-ATPase were additionally optimized using Protein Preparation Wizard pipeline, MacroModel package [13]. Due to their significant geometries similarity, we came to the decision to reduce the number of studied geometries employing clustering assay. 29 conformations were aligned to the first structure and clustered using their heavy atom coordinates. The fact that each structure contains an equal number of heavy atoms allowed us to use atomic coordinates as the features for clustering. To strictly reduce the number of considered features, the PCA algorithm (Python 3, sklearn library) was applied, 95% of input dispersion was hold down. The obtained matrix was clusterized using the Ward algorithm with the centroid linkage method, 4 principle clusters were separated which were applied in the further studies.

2.3   Glutathione Conformer Search

To achieve glutathione conformer distribution close to the native Maxwell-Boltzmann one in water media, the self-developed experimental conformer sampling algorithm was applied. The glutathione reduced form (monomer) was placed in the orthorhombic box $60 \times 60 \times 60$ Å filled with TIP3P water molecules. To gain an extended set of geometries the ligand and solvent were thermostated individually, both using the Nose-Hoover chain algorithm with coupling strength 20 ps [1]. Glutathione was heated to 700 K, at the same time solvent temperature was 300 K only - this hybrid MD approach ("cold solvent-hot solute") allowed to eliminate all steric barriers in ligand molecule and at the same time to conserve water molecules parametrization (e.g. hydrogen bonds, viscosity, self-diffusion, etc.). The MD simulation was run during 50 ns, the NVT ensemble was emulated. After that, the MD trajectory was separated to 10000 frames, from which the ligand and its first solvate shell (water molecules around 4 Å) were extracted. In the first step, obtained water-ligand complexes were minimized using the OPLS3e force field [12] to achieve better bond length and dihedral angles in glutathione and to eliminate abnormal location of water molecules near the ligand (so-called 'boiling effect'). In the next step, water-ligand complexes were optimized using the semi-empirical PM6-D3H4 method [11] in MOPAC [14]. We applied the two-stage optimization algorithm, on the first stage the ligand was optimized in the frozen water shell, on the second - water molecules were preset to the frozen ligand geometry. This complicated approach allowed us to maintain ligand geometry polymorphism obtained in MD simulation. To make the calculations less time-consuming, input complexes were clusterized by ligand torsional angles and the optimization time for each water-ligand complex was also strictly limited to 2000 minimization steps; the geometries with non-converged SCF were dropped from further studies. Obtained water-ligand complexes were grouped by the number of water molecules and in every group, the structure with the lowest energy was selected. In a few cases when the energy difference between the most profitable complexes was less than 5 kJ/mol, the best-in-group structure was chosen after additional B97-3c/PCM(water) optimization [2] in the ORCA program [7]. Finally, the best complexes were clusterized by ligand torsional angles once again, when geometry differences were too low the structure with smaller solvent shell was selected due to the higher statistical existence probability. Final ligand-water complexes were applied in the further docking calculations.

2.4 Molecular Docking

The Top 20 conformers were obtained from the advanced glutathione conformer search. The five Na,K-ATPase clusters (corresponding conformations) were taken for performing docking procedures. Receptor grids for docking were generated using the default algorithm. For each structure, $30 \times 30 \times 30$ Å outer grid box (with the $20 \times 20 \times 20$ Å inner box) was centered on 244 and on 458, 459 cysteines. The flexible ligand docking was performed using the Glide v8.1 in the extra precision (XP) mode [5]. The Top 20 conformers of glutathione were docked into every five clusters of Na,K-ATPase. We used the extra precision Glide score (XPG Score) tool to evaluate ligand docked poses, which enables calculating Gscore and Emodel scoring functions. Protein-glutathione complexes with the best scoring function were chosen for further analysis. To characterize the three-dimensional protein-binding mode, we applied the Interaction fingerprint assay [4]. This method included grouping the compounds with similar binding modes into the same clusters. It includes the following interaction types: any contact, backbone and sidechain contact, polar contacts, hydrogen bond acceptors, hydrogen bond donors, and aromatic. We performed the clustering of all Na,K-ATPase-glutathione docking poses using their associated interaction fingerprints. Finally, the 6 protein-ligand structures were selected for the next molecular dynamics simulations.

2.5 Molecular Dynamics

For MD simulations Desmond v5.4 program package was applied [1]. The most reliable protein-ligand complexes were placed in the orthorhombic box $20 \times 20 \times 20$ Å (buffer) filled with TIP3P water molecules. During the 50 ns dynamic, the NPT ensemble was simulated, Nose-Hoover chain thermostat (coupling strength 20 ps) and Martyna-Tobias-Klein barostat were applied. The protein-ligand complexes were neutralized by 27 Na+ ions, additionally, the 0.15 M salt concentration was emulated by adding monovalent Na+ and Cl- ions. To increase calculation efficiency, all protein atoms beyond 10 Å around glutathione atoms were frozen. In every MD we analyzed protein-ligand contacts and RMSD fluctuation during the trajectory to examine the stability of the protein-ligand complex. Obtained MD trajectories were clusterized by glutathione and target cysteine (244 or 458) atom coordinates, up to 10 clusters for every trajectory (cluster sets) were separated. In all clusters sets the structure with the optimal cysteine-glutathione superposition (minimum distance between cysteine and glutathione sulfur atoms) was selected. Chosen best structures were used in the further transition state scanning calculations.

2.6 Quantum Chemistry Calculations

To analyze the availability of disulfide covalent bond formation between oxidized cysteine and reduced glutathione, we cut out the binding pocket 5 Å around the glutathione, full amino acid residues and water molecules were included. Additionally, we deleted water molecules 4 Å around the glutathione to reduce the number of calculated atoms. The distorted valence on the protein backbone atoms caused by fragment separation was restored by adding a proton to the NH group and by adding OH atoms to the CO group. To achieve better accuracy, all involved ions were deleted. For all obtained protein-ligand-water complexes were optimized both 'reagents' and 'products' geometries in the ground state and detected the transition state in MOPAC using the PM6-D3H4 method. Due to a large number of atoms COSMO solvation model was not applied. Despite the low accuracy, vibrational frequencies were calculated to examine the ground state achievement. Thus, we calculated the energy barrier (heat of formation) of studied reaction for all complexes to determine the availability of the reaction in principle.

3 Results

3.1 ATPase Conformations Clustering

In concordance to clustering results, four separate clusters were obtained. As expected, they reflect the tendency of ATPase geometry rearrangement during the ion channel action cycle, so the structures from the same cycle stage were grouped in the single cluster. Clusters were enumerated in agreement with the included conformation numbers. To obtain general preliminary results, one conformation from each cluster was randomly selected (conformations 7, 14, 21, and 28 respectively) for subsequent analysis. To examine the geometry differences between selected clusters, we studied the RMSD values between four chosen structures aligned to the conformation 2. The gradual RMSD growing was observed. For both full protein and target cysteine environment (amino acid residues around 5 Å) very low geometry rearrangement was observed. One of the explanations could rely on the significant movement of the extra-membrane domain only.

3.2 Glutathione Conformer Search

Our experimental conformer sampling assay allowed us to obtain glutathione geometries similar to both extended and folded conformations which were previously described by NMR data [6]. The closest to the experimental geometries water-ligand complexes are presented in Fig. 1. Geometry deviations (RMSD) of the other glutathione conformations were no more than 1.8 Å relative to the presented configurations. At the same time, explicit simulation of water molecules made it possible to calculate the glutathione - water bulk interaction energy that will be used for docking energy corrections assay. Since the docking process runs usually in a vacuum, it does not take into account the geometry transition barrier between docked conformer and docking pose, especially with solvent media corrections. Previously obtained energy of solvation will allow us to get a reliable evaluation of proper protein-ligand interaction energy for the most promising protein-ligand complexes. To achieve this goal, the short ligand geometry optimization in the corresponding approximation (PM6-D3H4 or B97-3c) is needed. There is no doubt that the considered pipeline could be used as the first iteration corrections: one ought to arrive at the final decision related to the recognition availability after the so-called "funnel MD" which simulates the docking algorithm as a stochastic process in explicit solvent media only.

3.3 Docking Assay and MD Relaxation

It has been known that 244, 458, and 459 cysteines play an important role in glutathiony-lation [8, 10]. Twenty glutathione conformations were docked in two sites centered on the 244 and 458, 459 cysteines respectively. We examined all possible types of protein-ligand interactions using Interaction fingerprints for docking poses clustering. In each cluster, the most reliable glutathione docking pose was selected in agreement with their Glide score and Emodel criterium assessment. For conformation 14 (cluster 2) no appropriate docking pose near Cys 458 was found - the distance between glutathione and cysteine sulfur atoms was too large. A similar situation was observed near Cys 244: only in conformations 21 and 28, the protein-ligand complex geometry close to the transition state was achieved. Thus, we discovered five complexes, in which the disulfide bond formation could take place in theory. Their stability and geometry rearrangement analysis in the presence of water were validated using MD simulations.

To examine the MD results both protein-ligand interactions analysis and trajectories clustering were performed. In concordance to the obtained results, the glutathione was washed out from the binding site in conformations 7, despite the large negative value of docking binding energy. For conformations 21 and 28, the significant glutathione relocation was discovered, although it was tightly bounded with the protein. In conformation 21 the distance between glutathione and Cys244 was increased relative to the docking pose, thus this complex was dropped out from the further analysis. Protein-ligand

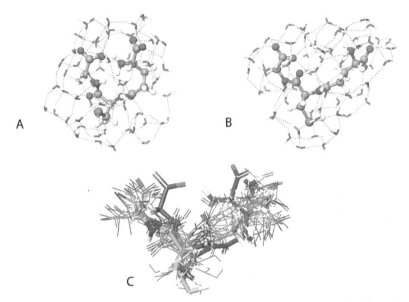

Fig. 1. Folded (A) and extended (B) conformers of glutathione in water bulk. Aligned confor-mations of glutathione conformers (C), folded and extended geometries are colored in green and violet correspondingly. Water molecules are hidden. (Color figure online)

interaction analysis for the last three complexes (conformation 21 - Cys458, confor-mation 28 - Cys458, and conformation 28 - Cys244) is presented below. Glutathione rearrangement in the binding pocket is reflected on the RMSD plots, the extremely low protein geometry changes are connected with the relatively small amount of non-frozen atoms. MD clusters with the smallest distance between sulfur atoms are presented as 3D structures, amino acid residues involved in protein-glutathione interactions are signed. Discovered optimum MD clusters were used as start configurations in transition state scanning assays.

3.4   Quantum Chemistry Calculations

Table 1 presents the data of thermodynamic calculations of the corresponding S-S bond formation reactions in the studied docking poses. Quantum chemical calculations for docking poses were carried out in parallel with molecular dynamics calculations.

The calculation data show that, in theory, glutathione binding is possible in all three considered complexes; however, the reaction in conformation 28 for Cys244 is the most profitable. The strong scatter in the dG values is most likely caused by the presence of water in the resulting system. Perhaps the water molecule stabilizes not the most profitable system geometries, and it is challenging to obtain more reasonable values of energy. Also, transition states for the examined docking poses were optimized, the TS method of the MOPAC program was used for optimization. The criterion for the correctness of the calculation results was the presence of the single imaginary frequency in the resulting structure.

Table 1. Thermodynamic calculations of the corresponding S-S bond formation reactions.

Complex	dG (kJ/mole)
7 - Cys458	−95
21 - Cys458	−7
28 - Cys458	−21
28 - Cys244	−318

Obtained results clearly show that the SN1 mechanism is the most probable for the studied reaction. I.e., at the first stage of the reaction, the proton of the SH group is attached to oxygen on the SOH group with subsequent separation of water, and at the second stage, a covalent S-S bond is formed. However, further refinement by more accurate methods is required.

Acknowledgments. This work (except conformational movement modeling) has been supported by the Russian Science Foundation (grant No. 19-14-00374). The conformational movement modeling has been performed under support of the Russian Science Foundation (grant No 19-71-30020).

References

1. Bowers, K.J., et al.: Molecular dynamics—Scalable algorithms for molecular dynamics simulations on commodity clusters. In: Proceedings of the 2006 ACM/IEEE Conference on Supercomputing - SC 2006 (2006). https://doi.org/10.1145/1188455.1188544
2. Brandenburg, J.G., et al.: B97-3c: a revised low-cost variant of the B97-D density functional method. J. Chem. Phys. (2018). https://doi.org/10.1063/1.5012601
3. Delfino, F., et al.: Structural transition states explored with minimalist coarse grained models: applications to Calmodulin. Front. Mol. Biosci. (2019). https://doi.org/10.3389/fmolb.2019.00104
4. Deng, Z., et al.: Structural interaction fingerprint (SIFt): a novel method for analyzing three-dimensional protein-ligand binding interactions. J. Med. Chem. (2004). https://doi.org/10.1021/jm030331x
5. Friesner, R.A., et al.: Extra precision glide: docking and scoring incorporating a model of hydrophobic enclosure for protein-ligand complexes. J. Med. Chem. (2006). https://doi.org/10.1021/jm0512560
6. Mandal, P.K., et al.: Glutathione conformations and its implications for in vivo magnetic resonance spectroscopy. J. Alzheimer's Dis. (2017). https://doi.org/10.3233/JAD-170350
7. Neese, F.: Software update: the ORCA program system, version 4.0. Wiley Interdiscip. Rev. Comput. Mol. Sci. **8**(1), e1327 (2018). https://doi.org/10.1002/wcms.1327
8. Petrushanko, I.Y., et al.: Cysteine residues 244 and 458–459 within the catalytic subunit of Na, K-ATPase control the enzyme's hydrolytic and signaling function under hypoxic conditions. Redox Biol. (2017). https://doi.org/10.1016/j.redox.2017.05.021
9. Petrushanko, I.Y., et al.: S-glutathionylation of the Na, K-ATPase catalytic α subunit is a determinant of the enzyme redox sensitivity. J. Biol. Chem. (2012). https://doi.org/10.1074/jbc.M112.391094

10. Poluektov, Y.M., et al.: Na, K-ATPase α-subunit conformation determines glutathionylation efficiency. Biochem. Biophys. Res. Commun. (2019). https://doi.org/10.1016/j.bbrc.2019.01.052

11. Řezáč, J., et al.: Semiempirical quantum chemical PM6 method augmented by dispersion and H-bonding correction terms reliably describes various types of noncovalent complexes. J. Chem. Theory Comput. (2009). https://doi.org/10.1021/ct9000922

12. Roos, K., et al.: OPLS3e: extending force field coverage for drug-like small molecules. J. Chem. Theory Comput. (2019). https://doi.org/10.1021/acs.jctc.8b01026

13. Schrödinger Release: Desmond Molecular Dynamics System

14. Stewart, J.J.P.: MOPAC2016 (2016). https://doi.org/10.2106/JBJS.G.00147

15. Tamazian, G., et al.: Modeling conformational redox-switch modulation of human succinic semialdehyde dehydrogenase. Proteins Struct. Funct. Bioinforma. (2015). https://doi.org/10.1002/prot.24937

HiChew: a Tool for TAD Clustering in Embryogenesis

Nikolai S. Bykov[1]([⊠])(iD), Olga M. Sigalova[2](iD), Mikhail S. Gelfand[3,4](iD), and Aleksandra A. Galitsyna[3,4,5](iD)

[1] Faculty of Computer Science, Higher School of Economics, Moscow, Russia
nickandroid1@gmail.com
[2] Genome Biology Unit, EMBL, Heidelberg, Germany
[3] Skolkovo Institute of Science and Technology, Moscow, Russia
[4] A.A. Kharkevich Institute for Information Transmission Problems, Moscow, Russia
[5] Institute of Gene Biology, Russian Academy of Sciences, Moscow, Russia

Abstract. The three-dimensional structure of the *Drosophila* chromatin has been shown to change at the early stages of embryogenesis from the state with no local structures to compartmentalized chromatin segregated into topologically associated domains (TADs). However, the dynamics of TAD formation and its association with the expression and epigenetics dynamics is not fully understood. As TAD calling and analysis of TAD dynamics have no standard, universally accepted solution, we have developed HiChew, a specialized tool for segmentation of Hi-C maps into TADs of a given expected size and subsequent clustering of TADs based on their dynamics during the embryogenesis. To validate the approach, we demonstrate that HiChew clusters correlate with genomic and epigenetic characteristics. Particularly, in accordance with previous findings, the maturation rate of TADs is positively correlated with the number of housekeeping genes per TAD and negatively correlated with the length of housekeeping genes. We also report a high positive correlation of the maturation rate of TADs with the growth rate of the associated ATAC-Seq signal.

Keywords: Chromatin 3D structure · Embryogenesis · Epigenetics · Housekeeping genes · TADs · Clustering

1   Introduction

Interconnection between the epigenetic regulation of biological processes, genes expression, and three-dimensional organization of chromatin during the development of an organism remains unclear [8,10,15]. The genome folding is established during early embryogenesis at different levels [2,10]. One of these levels, topologically associated domains (TADs), demonstrates diverse maturation rates during the embryo development [3,10]. The chromatin conformation in *Drosophila*

This work is supported by RFBR grant 19-34-90136, and by Skoltech Systems Biology Fellowship for Aleksandra Galitsyna.

© Springer Nature Switzerland AG 2020
Z. Cai et al. (Eds.): ISBRA 2020, LNBI 12304, pp. 381–388, 2020.
https://doi.org/10.1007/978-3-030-57821-3_37

melanogaster is established during the zygotic genome activation (ZGA) between the 13th and 14th nuclear cycles, and the boundaries of TADs appear at early expressed loci enriched in housekeeping genes and remain stable in later-stage embryos [14]. Thus, TADs are finally matured by the stage 3–4 hours [10]. To identify these structural units, dynamic programming algorithms like Armatus or modularity are used [3,6]. A plethora of methods could be applied to cluster TADs in the space of developmental stages by their maturation rate with subsequent analysis of functional annotation in genome regions forming clustered TADs [3,10].

Here, we introduce HiChew, a tool for TAD segmentation that fits the expected mean TAD size and subsequent clustering of these TADs. We apply HiChew to data from four time points in early embryogenesis of *Drosophila melanogaster* [10] in order to test the hypothesis that TADs differ in the maturation time, and this time is associated with functional annotation of the corresponding genomic segments, in particular, gene composition and chromatin accessibility measured by ATAC-Seq [2,10,16].

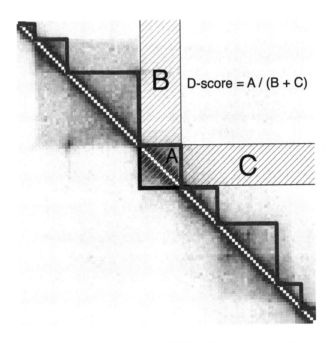

Fig. 1. *Drosophila melanogaster* embryo TAD calling: stage 3–4 h, chromosome X, 2400..2500 bins, method – modularity, expected TAD size – 60 Kb. (Color figure online)

2   Results

2.1   TAD Calling and Clustering with HiChew

HiChew Design. Since there are many methods for TAD calling but no clear consensus for the selection of their parameters [7,13], we have developed a console program that constructs TAD segmentation for a given organism fitting the expected mean size of TADs [6]. The program is based on existing algorithms for TAD calling, in particular, dynamic programming with modularity and Armatus scoring functions [12]. In usual cases, these algorithms both require a user to pre-set the parameter γ, which cannot be directly converted to the expected TAD size of the output segmentation. HiChew allows the user to avoid the need to pre-set this parameter γ, but to provide interpretable and straightforward expected size of TADs instead.

Next, the compactness of each TAD is measured by D-score [3] at each time point in the time-resolved Hi-C (Fig. 1). D-score equals to the ratio of the sum of chromatin contacts within a TAD to the sum of contacts of this TAD with the rest of the chromosome [3] (Fig. 1). Each TAD is characterized by a set of D-scores corresponding to stages of the embryo development. These sets of D-score series are further clustered to produce groups of TADs that have similar time dynamics of folding.

HiChew Pipeline. HiChew constructs the TAD segmentation that fits the given expected TAD size. The resulting TAD segmentation is clustered by D-scores using the HiChew clustering function (Fig. 2b). However, this general scheme is a complex pipeline consisting of several building blocks that are repeated for each chromosome arm. We provide the pipeline and its details in Fig. 2a.

HiChew Implementation. HiChew is based on two Python libraries for Hi-C data processing, Cooler [1] (for reading and converting cool files) and Lavaburst [12] (for TAD calling given parameters for Armatus or modularity). The following procedures are implemented using these libraries and used as building blocks in HiChew pipeline (Fig. 2):

1. **Filtering out noisy TADs.** For each Hi-C map iterative correction [11] is applied (using Cooler v. 0.8.5) to identify low-coverage bins. TADs that appear to be in the neighborhood of such bins might arise a technical artifact. Thus they are excluded from the downstream analyses.
2. **Refining interval of the grid of parameter γ.** The method is based on a binary search algorithm. To initialize it, we specify boundaries of the interval large enough to allow the search for the optimal value of γ. We do this because segmentation stops changing starting from some γ due to the nature of the Armatus and modularity scoring functions. Then the method reduces the upper boundary of the interval of the γ parameter using the binary search algorithm, until the segmentation obtained with the upper boundary begins to change.

3. **Identifying the optimal value of** γ. To find the optimal value of γ, the user specifies the expected average TAD size. For each γ in the refined grid, HiChew constructs the corresponding segmentation into TADs, calculates the average TAD size and selects the one closest to user-specified expectation. Subsequently, HiChew iteratively decreases the grid step in the neighborhood of the found optimum, until the calculated average TAD size becomes closer to the specified one than the user-defined error ε.
4. **Clustering of TADs by D-score profiles.** The constructed TADs are clustered by D-score profiles using one of the clustering algorithms that HiChew provides (K-means, MeanShift, Affinity Propagation, Hierarchical clustering and Spectral clustering).

TAD Calling for the 3–4 Hours Fruit Fly Embryo. As the first step of HiChew testing for *Drosophila melanogaster* embryogenesis, we segmented 5Kb-binned Hi-C map of the *Drosophila* embryo at the 3–4 h. Each map consists of five regions (chromosome arms 2R, 2L, 3R, 3L and chromosome X). We selected the modularity method [4,6] with the expected TAD size of 60 Kb [10] (Fig. 1) to produce the TAD segmentation. This TAD segmentation is used further for the analysis of D-scores.

TAD Clustering. The D-score profiles were clustered using K-means algorithm [3]. We selected 7 as the optimal number of clusters according to the Silhouette criterion value (0.25). The dynamics of D-scores in clusters is presented in Fig. 3b.

2.2 Housekeeping Genes Are Prevalent in Faster-Forming TADs

To identify trends in the D-score dynamics we have built a regression line for each cluster (Fig. 3b-c). In order to computationally address the type of pattern for each cluster, we introduce maturation rate of the cluster, which is calculated as the regression coefficient α for the corresponding D-scores series. Maturation rate α for each cluster correlates well with the ratio of tissue-specific over housekeeping genes (Fig. 3a,c), as supported by the Pearson correlation coefficient -0.95 (seven data points, $p < 0.001$). Thus, housekeeping genes are more prevalent in faster-forming TADs.

2.3 Shorter Housekeeping Genes Are More Prevalent in Faster-Forming TADs

In each cluster, the length of housekeeping genes significantly differs from the length of tissue-specific genes within the clusters (MWU-test, $p < 10^{-5}$ for each cluster, Fig. 3d). Maturation rate correlates well with the median length of both housekeeping and tissue-specific genes (Fig. 3c-d, Pearson's $r = -0.95$, $p = 0.001$ for housekeeping genes and $r = -0.76$, $p = 0.046$ for tissue-specific genes). Moreover, the length of housekeeping genes significantly differs for almost all pairs of clusters (MWU-test, $p < 0.02$, Fig. 3d), suggesting that shorter housekeeping genes might be prevalent in faster-forming TADs.

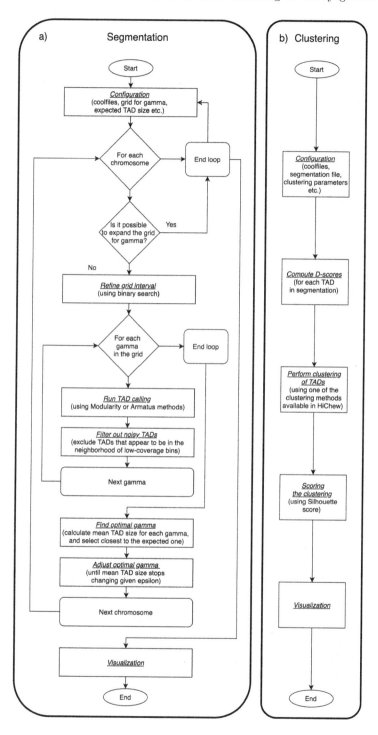

Fig. 2. HiChew flowchart for TAD segmentation (a) and clustering (b).

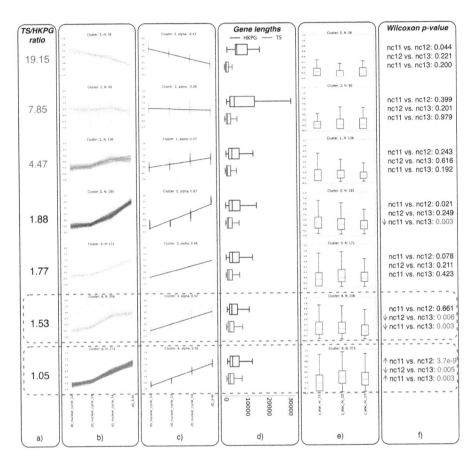

Fig. 3. a) The ratio of the numbers of tissue-specific to housekeeping genes. Higher values are highlighted in red. b) K-means clustering of seven clusters of D-score profiles in the space of four early embryogenesis stages. c) Linear regression lines computed for each cluster from (b). d) Whisker-box plots for the length of housekeeping and tissue-specific genes in each cluster; blue boxes for housekeeping genes, red boxes for tissue-specific genes. e) Whisker-box plots for the ATAC-Seq signal averaged for TADs in each considered stage (Fig. 3b). f) Wilcoxon test p-values for the differences in the ATAC-Seq signal distribution for TADs between stages (Fig. 3e). Significant p-values are highlighted in red. Arrows indicate significant changes (rise or fall) in ATAC-Seq signal between different stages (according to the Whisker-box plots from the Fig. 3e and Wilcoxon test p-values for these box plots). (Color figure online)

2.4   Faster-Forming TADs Demonstrate Distinct ATAC-Seq Changes

We noticed that the fastest-forming TADs (cluster 6 that highlighted with the purple dashed frame, Fig. 3) demonstrate a significant increase of the ATAC-Seq signal at nuclear cycle 12 and then decrease at nuclear cycle 13. Notably, the drop of the ATAC-Seq signal at nuclear cycle 13 can be detected for the second group of fast-forming TADs (sorted by the linear regression coefficient in clusters, cluster 4 – also highlighted with the purple dashed frame). No other cluster features a statistically different ATAC-Seq signal for pairwise comparisons of subsequent nuclear cycles. This might indicate that unbinding of regulatory factors [5] might cause rapid compaction of TADs during embryogenesis.

3   Discussion

Here we report the development of console program HiChew, implemented to segment the genome into TADs and cluster TADs by their compactness (D-score) for time-resolved Hi-C.

We observe that within the clusters, housekeeping genes are significantly longer than tissue-specific genes. Also, the length of housekeeping genes significantly differs among all clusters. These observations are consistent with the results of earlier studies [9,10], in particular, that TAD boundaries are positioned at the early and short transcription loci, enriched in housekeeping genes.

By applying the program to Hi-C data of *Drosophila melanogaster* embryo [10], we demonstrate that the resulting clusters have distinct chromatin openness characteristics and different housekeeping genes ratio. Taken together, this work might be considered as proof that the constructed clusters by HiChew are biologically relevant.

4   Data and Software Availability

Hi-C data of the early embryogenesis of *Drosophila melanogaster* for four stages nc12-14 and 3–4 h for chromosome arms 2R, 2L, 3R, 3L and chromosome X, was obtained from [10], accession: E-MTAB-4918. ATAC-Seq data (three stages nc11-13, accession: GSE83851) was obtained from [2]. Annotation of housekeeping genes was taken from [16]. HiChew is available in the GitHub repository https://github.com/encent/hichew.

References

1. Abdennur, N., Mirny, L.: Cooler: scalable storage for Hi-C data and other genomically labeled arrays. Bioinformatics **36**(1), 311–316 (2020). https://doi.org/10.1093/bioinformatics/btz540
2. Blythe, S.A., Wieschaus, E.F.: Zygotic genome activation triggers the DNA replication checkpoint at the midblastula transition. Cell **160**(6), 1169–1181 (2015). https://doi.org/10.1016/j.cell.2015.01.050

3. Bonev, B., et al.: Multiscale 3D genome rewiring during mouse neural development. Cell **171**(3), 557–572 (2017). https://doi.org/10.1016/j.cell.2017.09.043

4. Brandes, U., et al.: On finding graph clusterings with maximum modularity. In: Brandstädt, A., Kratsch, D., Müller, H. (eds.) WG 2007. LNCS, vol. 4769, pp. 121–132. Springer, Heidelberg (2007). https://doi.org/10.1007/978-3-540-74839-7_12

5. Buenrostro, J.D., Giresi, P.G., Zaba, L.C., Chang, H.Y., Greenleaf, W.J.: Transposition of native chromatin for fast and sensitive epigenomic profiling of open chromatin, DNA-binding proteins and nucleosome position. Nat. Methods **10**(12), 1213–1218 (2013). https://doi.org/10.1038/nmeth.2688

6. Flyamer, I.M., et al.: Single-nucleus Hi-C reveals unique chromatin reorganization at oocyte-to-zygote transition. Nature **544**(7648), 110–114 (2017). https://doi.org/10.1038/nature21711

7. Forcato, M., Nicoletti, C., Pal, K., Livi, C.M., Ferrari, F., Bicciato, S.: Comparison of computational methods for Hi-C data analysis. Nat. Methods **14**(7), 679–685 (2017). https://doi.org/10.1038/nmeth.4325

8. Fulco, C.P., et al.: Activity-by-contact model of enhancer-promoter regulation from thousands of CRISPR perturbations. Nat. Genet. **51**(12), 1664–1669 (2019). https://doi.org/10.1038/s41588-019-0538-0

9. Heyn, P., et al.: The earliest transcribed zygotic genes are short, newly evolved, and different across species. Cell Rep. **6**(2), 285–292 (2014). https://doi.org/10.1016/j.celrep.2013.12.030

10. Hug, C.B., Grimaldi, A.G., Kruse, K., Vaquerizas, J.M.: Chromatin architecture emerges during zygotic genome activation independent of transcription. Cell **169**(2), 216–228 (2017). https://doi.org/10.1016/j.cell.2017.03.024

11. Imakaev, M., et al.: Iterative correction of Hi-C data reveals hallmarks of chromosome organization. Nat. Methods **9**(10), 999–1003 (2012). https://doi.org/10.1038/nmeth.2148

12. Lavaburst. https://github.com/nvictus/lavaburst (Accessed 10 Feb 2020)

13. Sauerwald, N., Singhal, A., Kingsford, C.: Analysis of the structural variability of topologically associated domains as revealed by Hi-C. NAR Genom. Bioinform. **2**(1) (2020). https://doi.org/10.1093/nargab/lqz008

14. Schulz, K.N., Harrison, M.M.: Mechanisms regulating zygotic genome activation. Nat. Rev. Genet. **20**(4), 221–234 (2019). https://doi.org/10.1038/s41576-018-0087-x

15. Stadhouders, R., et al.: Transcription factors orchestrate dynamic interplay between genome topology and gene regulation during cell reprogramming. Nat. Genet. **50**(2), 238–249 (2018). https://doi.org/10.1038/s41588-017-0030-7

16. Ulianov, S.V., et al.: Active chromatin and transcription play a key role in chromosome partitioning into topologically associating domains. Genome Res. **26**(1), 70–84 (2016). https://doi.org/10.1101/gr.196006.115

SC1: A Tool for Interactive Web-Based Single Cell RNA-Seq Data Analysis

Marmar Moussa and Ion I. Măndoiu$^{(\boxtimes)}$

Computer Science and Engineering Department, University of Connecticut,
Storrs, CT 06269, USA
{marmar.moussa,ion.mandoiu}@uconn.edu

Abstract. Single cell RNA-seq (scRNA-Seq) is critical for studying cellular function and phenotypic heterogeneity as well as the development of tissues and tumors. Here, we present a web-based interactive scRNA-Seq data analysis tool publicly accessible at https://sc1.engr.uconn.edu. The tool implements a novel method of selecting informative genes based on Term-Frequency Inverse-Document-Frequency (TF-IDF) scores and provides a broad range of methods for cell clustering, differential expression, gene enrichment, interactive visualization, and cell cycle analysis. In just a few steps, researchers can generate a comprehensive initial analysis and gain powerful insights from their single cell RNA-seq data.

1 Introduction

Currently there are only few packages for comprehensive scRNA-Seq data analysis. Most of them are implemented using the R programming language, require considerable programming knowledge, and are not easy to use by researchers in life sciences.

In this work, we present a web-based, highly interactive scRNA-Seq data analysis tool publicly accessible at https://sc1.engr.uconn.edu. The tool includes several data quality control (QC) options, a novel method for gene selection based on *Term-Frequency Inverse-Document-Frequency (TF-IDF)* scores [9], followed by cell clustering and visualization tools as well as Differential Expression (DE) analysis and gene enrichment steps. Additional analyses include various 3D interactive visualizations based on t-SNE and UMAP dimensionality reduction algorithms as well as a novel approach to clustering and ordering cells according to their cell cycle phase [7]. With robust default parameter values SC1 empowers researchers to generate a comprehensive initial analysis of their scRNA-Seq data in just a few steps, while also allowing them to conduct in depth interactive data exploration and parameter tuning.

2 SC1 Workflow

The SC1 workflow is implemented in the R programming language, with an interactive web-based front-end built using the Shiny framework [1]. In the following we present details of the main analysis steps of the workflow.

© Springer Nature Switzerland AG 2020
Z. Cai et al. (Eds.): ISBRA 2020, LNBI 12304, pp. 389–397, 2020.
https://doi.org/10.1007/978-3-030-57821-3_39

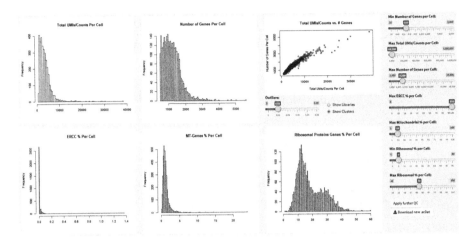

Fig. 1. SC1 QC dashboard.

Data Pre-processing. Before a detailed analysis of scRNA-Seq datasets (in 10X Genomics or csv format) can be performed, several pre-processing steps are carried out, starting with an initial quality control step in which cells with less than 500 detected genes and genes detected in less that 10 cells are excluded. Imputation is provided as an optional pre-processing step. Empirical experiments in [6] show that over-imputation is a concern for most existing methods. In SC1 we implemented the Locality Sensitive Imputation method (LSImpute) from [8], which was shown in [6] to yield high accuracy with minimum over-imputation. SC1 pre-processing also includes performing dimensionality reduction using three commonly used algorithms: Principal Component Analysis (PCA) [2], t-distributed Stochastic Neighborhood Embedding (t-SNE) projections [11], and Uniform Manifold Approximation and Projection (UMAP) [5].

scDat Upload. Pre-processed data is saved in SC1's ".scDat" file format that can then be uploaded for interactive analysis. Several publicly available datasets from [3,4,12] spanning different scRNA-Seq technologies are provided in SC1 as example datasets. Initial data exploration includes detecting the species (mouse or human), generating basic summary statistics including the number of expressed genes and the number of cells per library, and the ability to relabel the libraries. 'At-a-glance' two dimensional views of the data are also generated based on PCA, tSNE, and UMAP.

Quality Control Dashboard. Before further analyses, SC1 allows users to perform additional Quality Control (QC) checks as shown in Fig. 1, whereby poor quality cells and outlier cells and genes can be excluded from subsequent analysis. The tool implements widely used criteria for cell filtering: library size, number of detected genes, as well as the fraction of reads mapping to mitochondrial genes, ribosomal protein genes, or synthetic spike-ins. SC1 also allows outlier removal

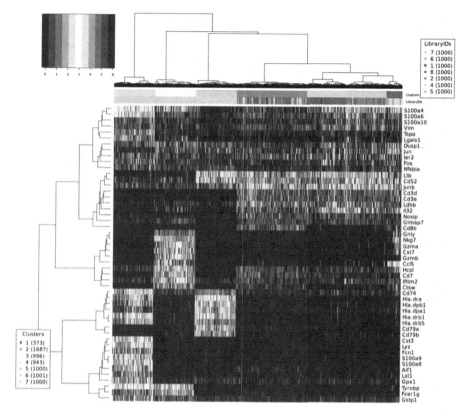

Fig. 2. Heat map of genes with top average TF-IDF scores for cells of the 7-class PBMC mixture from [9].

based on the ratio between the number of detected genes to total read/UMI count per cell.

Gene Selection. SC1 implements a novel method of selecting informative genes based on the average TF-IDF (*Term Frequency times Inverse Document Frequency*) scores, as detailed in [9]. TF-IDF scores are applied to scRNA-Seq data by considering the cells to be analogous to documents; in this analogy, genes correspond to words and UMI counts replace word counts. The TF-IDF scores can then be computed from UMI counts (or expression values). Similar to document analysis, the genes with highest TF-IDF scores in a cell are expected to provide most information about the cell type. Genes with highest average TF-IDF scores differentiate best between heterogeneous cell populations; visually this leads to a clear "chess-board" effect in the heat map constructed using the top average TF-IDF genes as shown in Fig. 2.

Clustering. By default, SC1 automatically infers the number of clusters using the Gap Statistics method as described in [9]. However, users can also manually

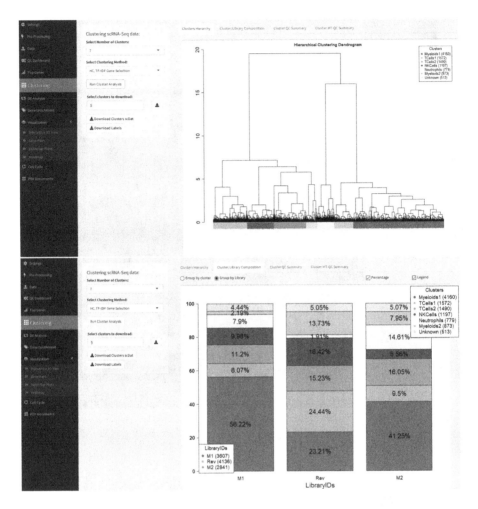

Fig. 3. SC1 clustering.

specify the number of clusters based on prior knowledge of the expected sample heterogeneity. Valuable insight into sample heterogeneity is also provided by inspecting the heat map generated using the top TF-IDF genes (Fig. 2) before clustering. Clustering can be performed using Ward's Hierarchical Agglomerative Clustering or Spherical K-means (both using the top average TF-IDF genes as features) or using Graph-based Clustering using binarized TF-IDF data as described in [9]. Several visualizations describe clustering details (see Fig. 3).

Differential Expression Analysis. Differential expression (DE) analysis is done by performing "One vs. the Rest" t-tests for each of the identified clusters. Results of the Log2 Fold Change and the p-value from the analysis are provided as a downloadable numeric matrix. A custom test of two selected groups of clusters or

Fig. 4. SC1 differential expression analysis.

libraries is also provided, with results provided both as a downloadable numeric table and as a Volcano plot visualizing the Log2 Fold Change and p-values for the tested groups (Fig. 4).

Enrichment Analysis. DE analysis is followed by cluster-based gene functional enrichment analysis performed using the 'gProfileR' R package [10] with results visualized as word clouds (Fig. 5) and provided as downloadable term significance values to help with cluster annotation. Labels assigned to the clusters at this step update throughout SC1 tool output and visualization plots.

Interactive Data Visualization. Many SC1 analysis steps generate visualizations of the results, including for instance the violin plots showing the probability density of gene expression values for each selected cluster/library and the bar-plots showing percentage of cells expressing selected genes by cluster or by library. Additional visualizations include:

– Clustering and gene co-expression visualization. SC1 includes multiple interactive visualization options; the interactive 3D t-SNE or UMAP visualization tabs include the ability to select genes individually, in pairs, or in groups as predefined gene sets. Cells are identified where all (AND) or any (OR) of the selected genes are detected. Identified cell populations can be selected or excluded to form a subset that can be downloaded and used to form a new sub-population for further analysis in SC1 (Fig. 6). Identifying various cell populations in SC1 and downloading relevant cells' expression profiles can be achieved in various ways in SC1: by selecting pre-defined libraries or conditions or selecting cell populations based on gene selection, also selecting specific cell types from clustering analysis results. Gene pair co-expression can also be visualized using interactive 3D plots as well as scatter plots Fig. 7)

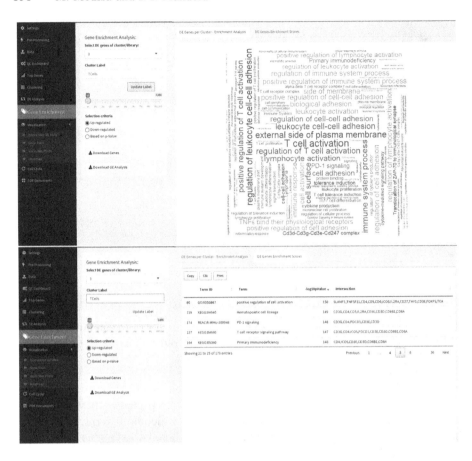

Fig. 5. SC1 cluster-based gene enrichment analysis.

– Detailed and summary heatmaps. SC1 provides several ways to select genes and cells visualized in configurable heat maps. Automatic identification of informative genes based on average TF-IDF allows the generation of exploratory heat maps to investigate the heterogeneity of the data. Also, a list of highly expressed/abundant genes can be downloaded from SC1 and used to construct a heat map. SC1 also supports custom gene selection by manually selecting or uploading a list of genes of interest to use for heat map construction. After the DE analysis step is concluded, the list of differentially expressed genes can also be visualized as a heat map. The expression/count values are by default log transformed in SC1 heat maps using the $log2(x+1)$ transformation. The summary heat map view in SC1 provides a "pseudo-bulk" view of the data, showing average expression profiles for selected genes by cluster or library (Fig. 8). The gene expression levels in summary heat maps are row-normalized, i.e., gene means expressions in libraries and clusters are normalized by dividing by the max mean expression of each gene

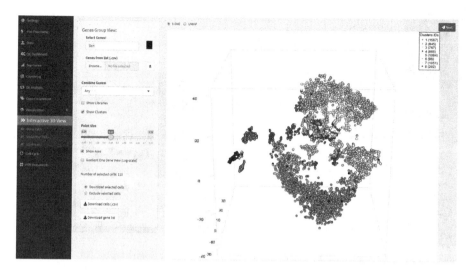

Fig. 6. SC1 3D visualization of clustering results and selected genes on data from [4].

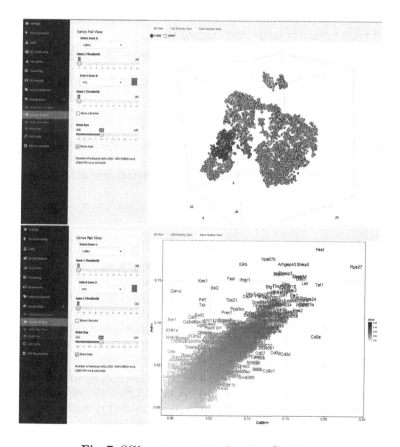

Fig. 7. SC1 gene co-expression visualization.

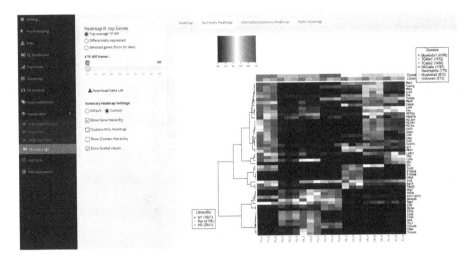

Fig. 8. Summary heat map showing cluster/library breakdown mean expression profiles of selected genes. (Color figure online)

over all libraries and clusters. This assigns a maximum value of 1 (red) to the groups for which the mean expression of the gene is the highest.

Cell Cycle Analysis. The variation in the gene expression profiles of single cells in different phases of the cell cycle can present a leading source of variance between cells and can interfere with cell type identification and functional analysis of scRNA-Seq data. In SC1, an orthogonal analysis of cell cycle effects can be performed at any stage of the analysis by clustering and ordering cells according to the expression levels of cell cycle genes, as described in [7].

3   Conclusion

SC1 provides a powerful tool for interactive web-based analysis of scRNA-Seq data. The SC1 workflow is implemented in the R programming language, with an interactive web-based front-end built using the Shiny framework [1]. SC1 employs a novel method for gene selection based on *Term-Frequency Inverse-Document-Frequency (TF-IDF)* scores [9], and provides a broad range of methods for cell clustering, differential expression analysis, gene enrichment, visualization, and cell cycle analysis. Future work includes integrating additional clustering methods, as well as other differential expression analysis methods and integrating methods for cell differentiation analysis. As the amount of scRNA-Seq data continues to grow at an accelerated pace, we hope that SC1 will help researchers to fully leverage the power of this technology to gain novel biological insights.

References

1. Chang, W., Cheng, J., Allaire, J., Xie, Y., McPherson, J.: Shiny: web application framework for R. http://CRAN.R-project.org/package=shiny (2017)
2. Erichson, N.B., Voronin, S., Brunton, S.L., Kutz, J.N.: Randomized matrix decompositions using R. arXiv preprint arXiv:1608.02148 (2016)
3. Gubin, M.M., et al.: High-dimensional analysis delineates myeloid and lymphoid compartment remodeling during successful immune-checkpoint cancer therapy. Cell **175**(4), 1014–1030 (2018)
4. Lukowski, S.W., et al.: Detection of HPV E7 transcription at single-cell resolution in epidermis. J. Investig. Dermatol. **138**(12), 2558–2567 (2018)
5. McInnes, L., Healy, J., Melville, J.: Umap: Uniform manifold approximation and projection for dimension reduction. arXiv preprint arXiv:1802.03426 (2018)
6. Moussa, M., Măndoiu, I.: Locality sensitive imputation for single-cell RNA-Seq data. J. Comput. Biol. **26** (2019). https://doi.org/10.1089/cmb.2018.0236
7. Moussa, M.: Computational cell cycle analysis of single cell RNA-Seq data. In: 2018 IEEE 8th International Conference on Computational Advances in Bio and Medical Sciences (ICCABS), p. 1. IEEE (2018)
8. Moussa, M., Măndoiu, I.: Locality sensitive imputation for single-cell RNA-Seq data. In: ISBRA2018 Proceedings (2018)
9. Moussa, M., Măndoiu, I.: Single cell RNA-Seq data clustering using TF-IDF based methods. BMC Genomics **19**(Suppl 6), 4922 (2018)
10. Reimand, J., Kull, M., Peterson, H., Hansen, J., Vilo, J.: g:Profiler-a web-based toolset for functional profiling of gene lists from large-scale experiments. Nucleic Acids Res. **35**(suppl2), W193–W200 (2007)
11. van der Maaten, L., Hinton, G.: Visualizing high-dimensional data using t-SNE. J. Mach. Learn. Res. **9**, 2579–2605 (2008)
12. Zheng, G.X., et al.: Massively parallel digital transcriptional profiling of single cells. bioRxiv p. 065912 (2016)

Quantitative Analysis of the Dynamics of Maternal Gradients in the Early Drosophila Embryo

Ekaterina M. Myasnikova[1,2], Victoria Yu. Samuta[1], and Alexander V. Spirov[1(✉)]

[1] I.M. Sechenov Institute of Evolutionary Physiology and Biochemistry of the Russian Academy of Sciences (IEPhB RAS), St-Petersburg, Russia
alexander.spirov@gmail.com
[2] Peter the Great St. Petersburg Polytechnical University, St-Petersburg, Russia

Abstract. Predetermination, formation and maintenance of the primary morphogenetic gradient (bicoid, bcd, gradient) of the early Drosophila embryo involves many interrelated processes. Here we focus on a systems biological analysis of the bcd mRNA redistribution in an early embryo. The results of the quantitative analysis of experimental data, together with the results of their dynamic modeling, substantiate the role of active transport in the redistribution of the bcd mRNA.

Keywords: Biomolecular imaging · Gene expression · Dynamic modeling · Systems biology

1 Introduction

In the 1960s and early 1970s, the concept of morphogenetic gradients in developmental biology acquired the form in which it is still known. Namely, substrates (morphogens) diffuse through arrays of cells, forming spatial concentration gradients (Crick 1970; Gierer and Meinhardt 1972). Simple molecular mechanisms, such as the localized protein synthesis, diffusion and degradation, can create a constant uneven distribution of morphogens in the cells of the embryo. Cells are able to recognize local concentrations of morphogen molecules and, in response, trigger activation of certain genes when a certain concentration threshold of these molecules is reached (Crick 1970). As a result, patterns of expression of genes controlling subsequent morphogenesis begin to form. Since then, a number of morphogenetic gradients have been extensively studied experimentally. The most thoroughly studied is the Bicoid (Bcd) protein gradient along the anteroposterior (AP) axis of the Drosophila embryo.

The gradient of Bcd protein emerges when maternal bcd mRNA accumulates (is deposited) at the anterior end of the embryo at the early stage of its formation (St Johnston et al. 1989). bcd mRNA begins to translate as soon as an egg is laid, and the protein gradient is formed within 3 h.

bcd mRNA is deposited in the head of the zygote as a conglomerate, forming a complex with several proteins (Weil et al. 2012). This forms an extremely specialized

© Springer Nature Switzerland AG 2020
Z. Cai et al. (Eds.): ISBRA 2020, LNBI 12304, pp. 398–405, 2020.
https://doi.org/10.1007/978-3-030-57821-3_40

structure (P-bodies (Weil et al. 2012)), which supports the translation stability and production of large amounts of Bcd protein. However, the mechanism of action of this structure has not yet been clarified. Therefore, the widely used working hypothesis considers this conglomerate as a point source (not extended spatially) for the synthesis of Bcd protein. This led to the formulation of the Synthesis-Degradation-Diffusion (SDD) model for the Bcd morphogenetic gradient (Grimm et al. 2010). The idea underlying the SDD model is that the Bcd protein synthesis begins at the anterior point of the embryo, then the protein diffuses from the head to the tail, while gradually degrading.

Recent studies of this primary morphogenetic gradient suggest the higher complexity of its regulation. In particular, it turned out that the bcd mRNA forms not a point source, but an extended spatial gradient, which is highly dynamic (Spirov et al. 2009; Cai et al. 2017). This leads to significant consequences for the mechanism of formation of the morphogenetic gradient of the Bcd protein.

The purpose of our report is to outline our systematic approach to analyzing the behavior of the bcd mRNA using dynamic modeling, based on the accumulated experimental data.

2   Biological Background

Below, we briefly describe what is known about the behavior of the bcd mRNA in the early Drosophila embryogenesis. In a mature zygote, the bcd mRNA is fixed in the anterior cortical layer of an egg. Following fertilization, mRNA-containing particles are released into the cytoplasm. Following this, complex processes of redistribution of this mRNA in the earliest embryogenesis are started (Alexandrov et al. 2018). There are at least three stages of redistribution of the mRNA bcd.

Firstly, it is the redistribution of particles containing mRNA during the very first cycles of cleavage division at the pre-blastoderm stage (Little et al. 2011). The detailed analysis of the data suggests a consistent full-scale reorganization of the early head region of an embryo (Spirov et al. 2009; Little et al. 2011). The second event is the further expansion of the mRNA gradient in the posterior direction (Spirov et al. 2009). It lasts from the late cleavage stage to the beginning of the 14th division. The third key event is the basal-apical redistribution of the bcd mRNA at the beginning of cycle 14 (Spirov et al. 2009).

We can consider these observed redistributions and reorganizations of the mRNA-containing material as the identification of some key stages in the formation of the source of the Bcd protein gradient (Bcd production site). These stages are: formation (1), sequential enlargement and amplification (2), and, finally, disassembly of the spatially distributed source of the Bcd protein (3).

3   Methods and Approaches

3.1   Quantitative Data on the Redistribution of the bcd mRNA in Zygote-Early Embryo

Embryos were heat-fixed and processed by the FISH method on the bcd mRNA modified to achieve the high sensitivity (as described in Spirov et al. (2009)).

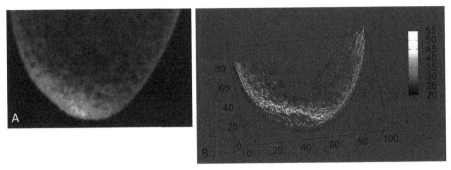

Fig. 1. An example of a sagittal image of a bcd mRNA pattern (FISH) in the head of an early embryo (A) and the result of its processing presented as a 3D plot (B). Plot (B) was obtained by digitizing the results of scanning by a small window sliding along a series of profiles parallel to the contour of the embryo.

Next, the embryos were subjected to layer-by-layer confocal scanning. Then, the obtained sagittal images were subjected to quantitative processing so as to obtain a series of expression profiles, as in Fig. 1.

3.2 Dynamic Modeling of bcd mRNA Redistribution

Now we will give a brief description of our dynamic model and the key results obtained by numerical experiments with the model. As mentioned, the second stage of the dynamics of the bcd mRNA consists in the propagation of the gradient of this RNA in the posterior direction.

We have reason to believe that usually any bcd mRNA profile consists of a short and sharp anterior gradient and a long flattened "tail" (Alexandrov et al. 2018). The tail may be flat, decreasing or increasing (in the anteroposterior direction). For the purposes of our dynamic modeling, we consider only profiles with a decreasing tail.

We assume here that the well-known tests with the bcd mRNA injection (RNA injection assay (Ferrandon et al. 1994; Ferrandon et al. 1997)) reflect some key features of the mechanisms of the bcd mRNA transport. Our main hypothesis here is that the mRNA released from the initial anterior conglomerate forms complexes with adapter proteins and molecular motors. These complexes, in turn, are able to move along the MT network of the syncytial embryo. The experimental data on which this hypothesis is based are presented in recent publications (Spirov et al. 2009; Cai et al. 2017).

Therefore, our first equation describing the large apical bcd complexes is:

$$\frac{d[bcd]}{dt} = D_{rnp}\Delta[bcd] - C[bcd] \tag{1}$$

Here [bcd] is the concentration of the bcd mRNA (in fact, RNA molecules are located in large partially immobilized macromolecular complexes, which we call RNP for simplicity); D_{rnp} is the diffusion coefficient for these large RNA-containing complexes; C is the coefficient corresponding to the release of the bcd mRNA from large slow complexes to smaller and/or significantly more mobile ones (we call them bcd').

Due to the apparently very low rate of mRNA propagation in the posterior direction, the diffusion coefficient must be very low. This would correspond to those large bcd mRNA complexes that are still probably linked to some cytoskeletal elements.

The second equation describes the dynamics of bcd in complexes that can be transported via microtubules (MT):

$$\frac{d[bcd']}{dt} = D_{rna}\Delta[bcd'] + C[bcd] - \Phi([bcd]) \tag{2}$$

Here [bcd'] is the concentration of bcd mRNA in complexes with motors such as Dynein that can move along MT; D_{rna} is the diffusion coefficient for RNA-containing complexes capable of active MT transport; $\Phi([bcd])$ is a term describing in a general way the degradation of a bcd mRNA.

We assume that the MT network is not oriented, so bcd transport can be described by the Fick's law. The diffusion coefficient in this case should be significantly higher than in the above case (Eq. 1).

We performed quite simple, but sufficiently detailed tests with one-dimensional models. We used 500 cells of 1 μm each. The duration of each run is 7200 s, ~120 min. The initial conditions were as follows: the first N cells contain X units of mRNA each, the rest cells – 0. We neglected the mRNA degradation during these first 2 h of embryo development. The model was fitted to real experimental data by the method of genetic algorithms.

4   Results and Discussion

4.1   Redistribution of bcd mRNA in a Syncytial Embryo

In this brief report, we present the results of modeling two of the three consecutive stages of the dynamics of the bcd mRNA gradient (second and third, final).

A typical result is shown in Fig. 2. As we can see, the simulation result is similar to real observations.

As we found out in these computational experiments, the diffusion coefficient for larger and very slow bcd containing particles is about 0.05 μm2/s. This is a very slow movement indeed, even if we compare with the Gregor's results for the Bcd protein, D = 0.3 μm2/s (Gregor et al. 2007). It is six times smaller and this corresponds to either large multimolecular complexes, or partial immobilization through the cytoskeleton, or both. On the contrary, as one would expect, the second transportation process looks really fast: D = 1.50 μm2/s.

It is appropriate to recall here how large complexes such as bcd2-Stau-Dynein are. So, we come to the conclusion that this numerical result really implies the processes of active molecular transport for the bcd mRNA.

4.2   Elimination of the bcd mRNA Gradient

In this subsection, we present the results of modeling the third stage of the bcd mRNA dynamics. At this stage, the mRNA gradient is being quickly eliminated. These simulations are illustrated in Fig. 3. This 1D model of transport was organized to simulate

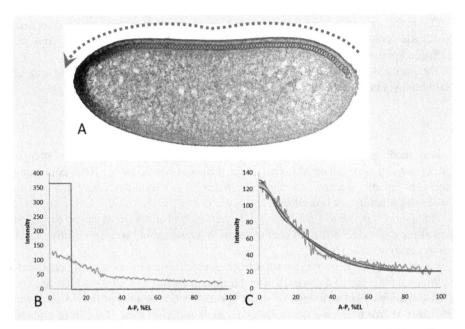

Fig. 2. Modeling the redistribution of bcd mRNA in a syncytial embryo. (A) Acquisition of intensity profiles for anteroposterior direction (red arrow). An example of extracting the anteroposterior profile of bcd mRNA from mid-cycle 14. (B) The initial distribution of mRNA in the model (rectangular, blue) and the data for fitting the model (real representative profile of the bcd mRNA in the 13th cycle embryo; red). (C) Simulation results (superposition of the model results, blue, and the real profile in red). (Color figure online)

the process in the apical part of the early embryo (as opposed to the model presented in Fig. 2). Here, our goal was to fit the simulation results to really observed changes in the bcd profile at the stages of the mature syncytial stage, the beginning of cellularization. As the initial conditions, a representative bcd profile for the 13th cycle embryo was used here.

The key point here was to study the effect of the mRNA degradation law. We were interested in the law and parameters of mRNA degradation (and dose dependence for bcd gene). We searched for the best value of the order of degradation rate (first order, or lower or higher). The goal was to obtain an embryo profile typical for the very beginning of the 14th cycle (~10 min), starting from a given profile of the end of the 13th cycle. Our modeling approach and typical result are presented in detail in Fig. 3. Randomized active transport was modeled according to the Fick's law; degradation was modeled by a 1st order reaction.

To our surprise, we found that the best value of the degradation rate of the bcd mRNA matching the experimental data was about 3/2, as illustrated in Fig. 4. That is, degradation is not just proportional to concentration, but the higher is the concentration of mRNA, the faster is its degradation (Fig. 4).

What is especially interesting, modeling with a degradation rate higher than 1 demonstrates such a significant feature as the robustness (insensitivity) of the solution to the

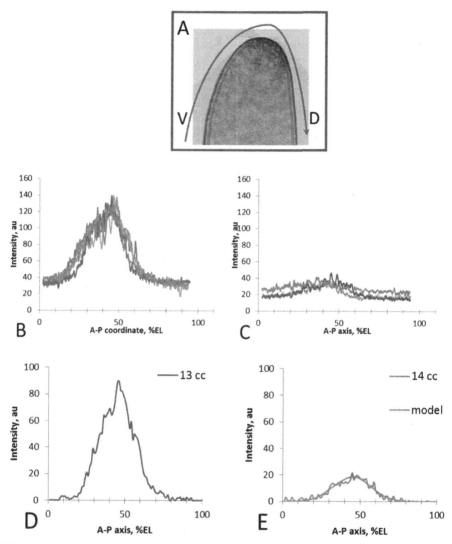

Fig. 3. Data and model of the bcd mRNA degradation organized to simulate the process in the apical part of the early embryo only. (A) Schematic of the profile extraction: from ventral (V) to dorsal (D), through the anterior pole (red arrow) (B) A set of representative profiles for cycle 13 embryos (profiles are extracted as in scheme (A). (C) A set of representative profiles for embryos from the beginning of the 14th cycle. (D–E) A typical simulation result: the initial profile (D) quickly decreases to (E). Degradation was modeled by a 1st order reaction. (Color figure online)

variability of the initial gradient. Such doses of bcd have been tested: 1 (haploid), 2 (WT) and higher doses (4 and 6). Simulation results demonstrated that higher-order degradation (e.g. 2/3) can compensate for the gene dosage at the mRNA level (Fig. 4).

Such compensation was reported at the level of accuracy of the Hb factor domain, but the mechanisms are still unclear (Houchmandzadeh et al. 2002). It is known that the

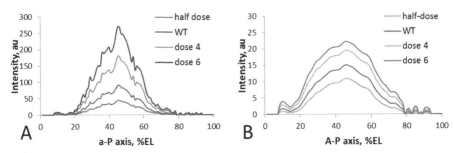

Fig. 4. The mRNA degradation model, as in Fig. 3, at various doses of the bcd gene. Typical simulation result: the initial profile (A) quickly decreases to (B). Randomized active transport was modeled according to the Fick's law; degradation was modeled by the reaction of order 3/2. Different doses of bcd have been tested: 1 (haploid), 2 (WT) and higher doses (4 and 6). As it turned out, higher-order degradation can compensate the gene dose at the mRNA level.

beginning of cycle 14A, immediately before the mid-blast transition, is characterized by a much more stable pattern of segmentation than in the early cycles (Holloway et al. 2006; Surkova et al. 2008). What we found here means that the robustness of pattern formation can be realized not only at the level of segmentation factors, but also at the level of some mRNAs.

Acknowledgement. This work was supported by the Russian Foundation for Basic Research (project no. 20-04-01015).

References

Crick, F.: Diffusion in embryogenesis. Nature **225**, 420–422 (1970)

Gierer, A., Meinhardt, H.: A theory of biological pattern formation. Kybernetik **12**, 30–39 (1972)

St Johnston, D., Driever, W., Berleth, T., Richstein, S., Nüsslein-Volhard, C.: Multiple steps in the localization of bicoid RNA to the anterior pole of the Drosophila oocyte. Development **107**(Suppl), 13–19 (1989)

Weil, T.T., Parton, R.M., Herpers, B., et al.: Drosophila patterning is established by differential association of mRNAs with P bodies. Nat. Cell Biol. **14**, 1305–1313 (2012)

Grimm, O., Coppey, M., Wieschaus, E.: Modelling the Bicoid gradient. Development **137**, 2253–2264 (2010)

Spirov, A., Fahmy, K., Schneider, M., Noll, M., Baumgartner, S.: Formation of the bicoid morphogen gradient: an mRNA gradient dictates the protein gradient. Development **136**, 605–614 (2009)

Cai, X., Spirov, A., Akber, M., Baumgartner, S.: Cortical movement of Bicoid in early Drosophila embryos is actin- and microtubule-dependent and disagrees with the SDD diffusion model. PLoS ONE **12**(10), e0185443 (2017)

Alexandrov, T., Golyandina, N., Holloway, D., Shlemov, A., Spirov, A.: Two-exponential models of gene expression patterns for noisy experimental data. J. Comput. Biol. **25**(11), 1220–1230 (2018)

Little, S.C., Tkacik, G., Kneeland, T.B., Wieschaus, E.F., Gregor, T.: The formation of the Bicoid morphogen gradient requires protein movement from anteriorly localized mRNA. PLoS Biol. **9**(3), e1000596 (2011)

Ferrandon, D., Elphick, L., Nusslein-Volhard, C., St Johnston, D.: Staufen protein associates with the 3UTR of bicoid mRNA to form particles that move in a microtubule-dependent manner. Cell **79**, 1221–1232 (1994)

Ferrandon, D., Koch, I., Westhof, E., Nüsslein-Volhard, C.: RNA–RNA interaction is required for the formation of specific bicoid mRNA 3' UTR–STAUFEN ribonucleoprotein particles. EMBO J. **16**, 1751–1758 (1997)

Gregor, T., Wieschaus, E.F., McGregor, A.P., Bialek, W., Tank, D.W.: Stability and nuclear dynamics of the Bicoid morphogen gradient. Cell **130**, 141–152 (2007)

Houchmandzadeh, B., Wieschaus, E., Leibler, S.: Establishment of developmental precision and proportions in the early Drosophila embryo. Nature **415**, 798–802 (2002)

Holloway, D.M., Harrison, L.G., Kosman, D., Vanario-Alonso, C.E., Spirov, A.V.: Analysis of pattern precision shows that Drosophila segmentation develops substantial independence from gradients of maternal gene products. Dev. Dyn. **235**(11), 2949–2960 (2006)

Surkova, S., et al.: Characterization of the Drosophila segment determination morphome. Dev. Biol. **313**(2), 844–862 (2008)

Atom Tracking Using Cayley Graphs

Marc Hellmuth[2], Daniel Merkle[1], and Nikolai Nøjgaard[1(✉)]

[1] Department of Mathematics and Computer Science,
University of Southern Denmark, Odense, Denmark
{daniel,nojgaard}@imada.sdu.dk
[2] School of Computing, University of Leeds, Leeds, UK
mhellmuth@mailbox.org

Abstract. While atom tracking with isotope-labeled compounds is an essential and sophisticated wet-lab tool in order to, e.g., illuminate reaction mechanisms, there exists only a limited amount of formal methods to approach the problem. Specifically when large (bio-)chemical networks are considered where reactions are stereo-specific, rigorous techniques are inevitable. We present an approach using the right Cayley graph of a monoid in order to track atoms concurrently through sequences of reactions and predict their potential location in product molecules. This can not only be used to systematically build hypothesis or reject reaction mechanisms (we will use the mechanism "Addition of the Nucleophile, Ring Opening, and Ring Closure" as an example), but also to infer naturally occurring subsystems of (bio-)chemical systems. We will exemplify the latter by analysing the carbon traces within the TCA cycle and infer subsystems based on projections of the right Cayley graph onto a set of relevant atoms.

1 Introduction

Traditionally, atom tracking is used in chemistry to understand the underlying reactions and interactions behind some chemical or biological system. In practice, atoms are usually tracked using isotopic labeling experiments. In a typical isotopic labeling experiment, one or several atoms of some educt molecule of the chemical system we wish to examine are replaced by an isotopic equivalent (e.g. ^{12}C is replaced with ^{13}C). These compounds are then introduced to the system of interest, and the resulting product compounds are examined, e.g. by mass spectrometry [5] or nuclear magnetic resonance [7]. By determining the positions of the isotopes in the product compounds, information about the underlying reactions might then be derived. From a theoretical perspective, characterizing a formal framework to track atoms through reactions is an important step to understand the possible behaviors of a chemical or biological system. In this contribution, we introduce such a framework based on concepts rooted in

D. Merkle—This work is supported by Novo Nordisk Foundation grant NNF19OC0057834 and by the Independent Research Fund Denmark, Natural Sciences, grant DFF-7014-00041.

© Springer Nature Switzerland AG 2020
Z. Cai et al. (Eds.): ISBRA 2020, LNBI 12304, pp. 406–415, 2020.
https://doi.org/10.1007/978-3-030-57821-3_41

semigroup theory. Semigroup theory can be used as a tool to analyze biological systems such as metabolic and gene regulatory networks [9,15]. In particular, Krohn-Rhodes theory [17] was used to analyze biological systems by decomposing a semigroup into simpler components. The networks are modeled as state automatas (or ensembles of automatas), and their characteristic semigroup, i.e., the semigroup that characterizes the transition function of the automata [14], is then decomposed using Krohn-Rhodes decompositions or, if computational not feasible, the holonomy decomposition variant [10]. The result is a set of symmetric natural subsystems and an associated hierarchy between them, that can then be used to reason about the system. In [4] algebraic structures were employed for modeling atom tracking: graph transformation rules are iteratively applied to sets of undirected graphs (molecules) in order to generate the hyper-edges (the chemical reactions) of a directed hypergraph (the chemical reactions network) [1,2]. A semigroup is defined by using the (partial) transformations that naturally arise from modeling chemical reactions as graph transformations. Utilizing this particular semigroup so-called pathway tables can be constructed, detailing the orbit of single atoms through different pathways to help with the design of isotopic labeling experiments.

In this work, we show that we can gain a much deeper understanding of the analyzed system by considering how atoms move in relation to each other. To this end, we show how the possible trajectories of a subset of atoms can be intuitively represented as the (right) Cayley graph [8] of the associated semigroup. Moreover, natural subsystems can be defined in terms of reversible atom configurations. We demonstrate the usefulness by differentiating chemical pathways, and we show how the (right) Cayley graph provides a natural handle for the analysis of cyclic chemical systems such as the TCA cycle. We note, that an extended version of this paper can found at https://cheminf.imada.sdu.dk/preprints/isbra-2020.pdf.

2   Chemical Networks and Their Algebraic Structures

We consider a chemical network CN modeled as a hypergraph and generated by a graph grammar as done in [2]. In this model, vertices of CN are molecules modeled as graphs and hyper-edges are reactions modeled as graph transformations. A graph transformation is modeled as an application of a Double-pushout rule transforming the educts of a reaction to its products. A rule application in CN defines a bijective map, called an atom map, specifying how atoms move through reactions of CN. We let $tr(CN)$ be all atoms maps that can be derived from CN. Where possible, we use common graph and semigroup notations and definitions, and refer to the original paper [13] for more details.

Characteristic Monoids. Assume we are given some chemical network CN that is some hypergraph modeling some chemistry. As we are interested in tracking the possible movements of atoms in CN, we are inherently interested in the reactions of CN, i.e., in its edge set $E(\text{CN})$. Indeed, atoms can only reconfigure to construct new molecules under the execution of some reaction. We will refer to the execution of a reaction as an *event*. The possible reconfigurations of atoms caused by a single event, is given by the set of atom maps $tr(\mathrm{CN})$ constituting a set of (partial) transformations on $X = \bigcup_{M \in V(\mathrm{CN})} V(M)$. Note, the vertex $M \in V(\mathrm{CN})$ corresponds to an entire molecule for which $V(M)$ denotes the set of atoms (=labeled vertices). A transformation t on X describes the position (i.e., in what molecule and where in the molecule the atom is found) of each atom in X when X is transformed by t. In what follows, we will sometimes refer to such transformations on X as *atom states*, as the transformations encapsulates the "state" of the network, i.e., the position of each atom. To track the possible movement of atoms through a chemical network, we must consider sequences of events.

Definition. *Let Σ be an alphabet containing a unique identifier t for each atom map in $tr(\mathrm{CN})$. An* event trace *is an element of the free monoid Σ^*.*

The free monoid Σ^* contains all possible sequences of events that can move the atoms of X. Note, Σ^* does not track the actual atoms through event traces. For this, we use the following structure:

Definition. *Let the* characteristic monoid *of CN be defined as the transformation monoid $\mathcal{S}(\mathrm{CN}) = (X, \langle tr(\mathrm{CN}) \cup 1_X \rangle)$. Moreover, given a set of edges $E \subseteq E(\mathrm{CN})$, and the set of atoms $Y \subseteq X$ found in E (that is $Y = \cup_{e \in E} Y_e$), we let the* characteristic monoid *of E be defined as $\mathcal{S}(E) = (Y, \langle tr(E) \cup 1_Y \rangle)$.*

Let $\sigma : \Sigma \to tr(\mathrm{CN})$ be the function, that maps all identifiers of Σ to their corresponding atom map in $tr(\mathrm{CN})$. Given an event trace $t = t_1 t_2 \ldots t_n \in \Sigma^*$, we let the events of t refer to their corresponding transformations in $tr(\mathrm{CN})$ when acting on an element $s \in \mathcal{S}(\mathrm{CN})$, i.e., $st = s\sigma(t_1)\sigma(t_2) \ldots \sigma(t_n) \in \mathcal{S}(\mathrm{CN})$. Every event trace $t \in \Sigma^*$ gives rise to a member $\mathcal{S}(\mathrm{CN})$, in particular the transformation $1_X t$, that represents the resulting atom state obtained from moving atoms according to t. Hence, there is a homomorphism from Σ^* to $\mathcal{S}(\mathrm{CN})$, meaning that $\mathcal{S}(\mathrm{CN})$ captures all possible movements of atoms through reactions of CN.

Often, we are only interested in tracking the movement of a small number of atoms. Let $\bar{z}$ be a tuple of distinct elements from X that we want to track. Then, there is again a homomorphism from Σ^* and $\mathcal{O}(\bar{z}, \mathcal{S}(\mathrm{CN}))$. Namely, for a given event trace $t \in \Sigma^*$, we can track the atoms of $\bar{z}$ as the atom state $1_{\{x \mid x \in \bar{z}\}} t$ corresponding to an element in the orbit $\mathcal{O}(\bar{z}, \mathcal{S}(\mathrm{CN}))$, if we treat the element as a (partial) transformation. As a result, $\mathcal{O}(\bar{z}, \mathcal{S}(\mathrm{CN}))$ characterizes the possible movements of the atoms in $\bar{z}$, and we will refer to its elements as atom states similarly to elements in $\mathcal{S}(\mathrm{CN})$ as they conceptually represent the same thing.

We note, the above definitions are not unlike some of the core definitions within algebraic automata theory [14]. Here, the possible inputs of an automata is often defined in terms of strings obtained from the free monoid on the alphabet of the automata. The characteristic semigroup is then defined as the semigroup that characterizes the possible state transitions. In the same vein, we can view our notion of event traces as the possible "inputs" to our chemical network CN that moves some initial configuration of atoms 1_X. The characteristic monoid of CN then characterize the possible movements of atoms through event traces.

In what follows we let Cay(CN) denote the Cayley graph Cay($\mathcal{S}$(CN), tr(CN)$\cup 1_X$). Similarly, given a tuple of atoms $\bar{z}$, we let PCay(CN, $\bar{z}$) denote the projected Cayley graph PCay($\mathcal{S}$(CN), tr(CN) $\cup 1_X, \bar{z}$). We note, that by Definition, $\mathcal{S}$(CN) is constructed from the generating set $\langle tr$(CN) $\cup 1_X\rangle$, and hence Cay(CN) and PCay(CN, $\bar{z}$) are well defined. Since the transformation 1_X will always result in a loop on every vertex of the (projected) Cayley graph, and conveys no meaningful information, we will refrain from including any edge arising from 1_X. We can illustrate the relation between atom states using the Cayley graph Cay(CN). More precisely, there exists an edge between two atom states $a, b \in \mathcal{S}$(CN) with label t, if it is possible to move the atoms in a to b using t. It is natural to relate Σ^* to Cay(CN). Namely, any path in Cay(CN) corresponds directly to an event trace in Σ^*. Hence, where Σ^* encapsulates the "inputs" of the chemical network and $\mathcal{S}$(CN) contains the possible atoms states derived from Σ^*, the Cayley graph Cay(CN) captures *how* atom states from $\mathcal{S}$(CN) can be created by event traces.

Natural Subsystems of Atom States. In the intersection between group theory and systems biology, attempts to formalize the notion of natural subsystems and hierarchical relations within such systems have been done by works such as [15]. Here, natural subsystems are defined as symmetric structures arising from a biological system. Such symmetries manifests as permutation groups of the associated semigroup representing said system. In such a model the Krohn-Rhodes decomposition or the holonomy decomposition [10] can be used to construct a hierarchical structure on such natural subsystems of the biological system. In terms of atom tracking, however, defining natural subsystems in terms of the permutation groups in $\mathcal{S}$(CN) does not have an immediately useful interpretation. Similarly, the hierarchical structure obtained from methods such as holonomy decomposition are not intuitive to interpret. Instead, when talking about natural subsystems in terms of atom tracking, we are interested in systems of reversible *event traces*, i.e., event traces that do not change the original configuration of atoms. To this end, it is natural to define natural subsystems of $\mathcal{S}$(CN) in terms of Green's relations [6]. For elements $s_1, s_2 \in \mathcal{S}$(CN), we define the reflexive transitive relation $\succeq_\mathcal{R}$ as $s_1 \succeq_\mathcal{R} s_2$, if there exists an event trace $t \in \Sigma^*$ such that $s_1 t = s_2$. In addition, we define an equivalence relation $\mathcal{R}$, where s_1 is equivalent to s_2, in symbols $s_1 \mathcal{R} s_2$ whenever $s_1 \succeq_\mathcal{R} s_2$ and $s_2 \succeq_\mathcal{R} s_1$.

Definition (Natural Subsystems). *The natural subsystems of $\mathcal{S}$(CN) is the set of equivalence classes induced by the $\mathcal{R}$-relation.*

The equivalence classes correspond to the strongly connected components of the Cayley graph Cay(CN) [11]. We note, that for a tuple of atoms $\bar{z}$, the natural extension to natural subsystems of the orbit $\mathcal{O}(\bar{z}, \mathcal{S}$(CN)) is simply the strongly connected components of its projected Cayley graph PCay(CN, $\bar{z}$). The $\mathcal{R}$ relation is interesting, as the equivalence classes on $\mathcal{S}$(CN) induced by the $\mathcal{R}$ relation forms pools of reversible event traces. More precisely, let $s_1 \mathcal{R} s_2$ for some $s_1, s_2 \in \mathcal{S}$(CN), where $s_1 \cdot t_{12} = s_2$ and $s_2 \cdot t_{21} = s_1$ for some $t_{12}, t_{21} \in \Sigma^*$. Then, the event traces t_{12} and t_{21} are reversible, i.e. we can re-obtain s_1 as

$s_1 t_{12} t_{21} = s_1$ and s_2 as $s_2 t_{21} t_{12} = s_2$. Additionally, the quotient graph of the equivalence classes of the $\mathcal{R}$ relation on the Cayley graph Cay(CN) naturally forms a hierarchical relation on the atom states of $\mathcal{S}$(CN) that has a useful interpretation from the point of view of chemistry as we will see in Sect. 3.

3   Results

Differentiating Pathways. In this section, we will explore the possibilities of using the characteristic monoids of chemical networks to determine if it is possible to distinguish between two pathways P_1 and P_2, based on their atom states of their respective characteristic monoids. The motivation stems from methods such as isotope labeling. Here, a "labeled" atom, is a detectable isotope whose position is known in some initial molecule and can then be detected, along with its exact position, in the product molecules of some pathway. In contrast to [4], we will not focus on the orbits of atoms in isolation, as we lose the ability to reason about atom positions in relation to each other. Moreover, as we will see here, the Cayley graph of the chemical network can be used to identify the exact event two pathways split.

Given a chemical network CN, a pathway P is a set of hyper-edges (i.e. reactions) from CN equipped with a set of input and output molecules. We think of a pathway as a process that consumes a set of input molecules to construct a set of output molecules, using the reactions specified by P. In our case, a "labeled" atom is a point in $\mathcal{S}$(CN). Given two pathways P_1 and P_2, we can characterize the possible movement of atoms as the characteristic monoids $\mathcal{S}(P_1)$ and $\mathcal{S}(P_2)$. In practice, it might not be feasible to track every atom in CN, e.g. we are only able to replace a few atoms with its corresponding detectable isotope, and hence it becomes useful to consider the orbits $\mathcal{O}(\bar{z}, \mathcal{S}(P_1))$ and $\mathcal{O}(\bar{z}, \mathcal{S}(P_2))$ where $\bar{z}$ is the atoms from the input molecules we can track. Clearly, of the atom states in $\mathcal{O}(\bar{z}, \mathcal{S}(P_1))$ and $\mathcal{O}(\bar{z}, \mathcal{S}(P_2))$, we can only expect to observe, e.g. in an isotope labeling experiment, the atom states that locates the tracked atoms in the output molecules. As a result, we arrive at the following observation:

Observation 1. *Let $Y_i \subseteq \mathcal{O}(\bar{z}, \mathcal{S}(P_i))$, $i \in \{1, 2\}$, be the atom states we can hope to observe after some isotope labeling experiment. Then, we can always distinguish between P_1 and P_2 if $Y_1 \cap Y_2 = \emptyset$.*

Example: Consider the network CN depicted in Fig. 1a modelling the creation of product 4-phenyl-6-aminopyrimidine (denoted P) from the educt 4-(benzyloxy)-6-bromopyrimidine (denoted E) using ammonia. This well investigated and widely used substitution mechanism (ANRORC) [16] was proven to non-trivially function via ring opening and ring closure (and an accompanied carbon replacement) via isotope labeling. Two possible pathways are modelled: the input molecules for the two pathways are the molecules E, NH_3, NH_2, while the output is the single molecule P. The first, seemingly correct but wrong, pathway $P_1 = \{r_3\}$ converts E and an NH_3 molecule directly into P, by replacing the Br

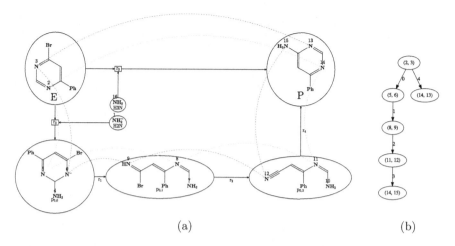

(a)   (b)

Fig. 1. (a) The chemical network for the creation of P from E using ammonia. The dotted red and blue lines shows the possible atom trajectories for the atoms 2 and 3 respectively. (b) The projected Cayley graph PCay(CN, (2, 3)). (Color figure online)

atom with NH_2. The second pathway consists of the reactions $P_2 = \{r_0, r_1, r_2, r_4\}$ and models the ANRORC mechanism.

Assume we wanted to device a strategy to decide what pathway is executed in reality. By replacing the nitrogen atoms of the E molecule with the isotope ^{13}N we would be able to observe where the atoms are positioned in the produced P molecule. Since we, by assumption, only label the nitrogen atoms of the E molecule, i.e., the atoms 3 and 2, we can look at the orbits of the characteristic monoids $\mathcal{O}((2,3), \mathcal{S}(P_1))$ and $\mathcal{O}((2,3), \mathcal{S}(P_2))$ with the order of 5 and 2 respectively. We see that both orbits only contains a single element locating $(2,3)$ in the P molecule, namely the element $(14, 15)$ for $\mathcal{O}((2,3), \mathcal{S}(P_1))$ and $(14, 13)$ for $\mathcal{O}((2,3), \mathcal{S}(P_2))$. As the possible configurations are different for P_1 and P_2, it is hence possible to always identify if the P molecule was created by P_1 or P_2.

This fact, also becomes immediately obvious by looking at the projected Cayley graph PCay(CN, (2, 3)) depicted in Fig. 1b, that shows the immediate divergence of atom states of the two pathways.

Natural Subsystems in the TCA Cycle. The citric acid cycle, also known as the tricarboxylic (TCA) cycle or the Krebs cycle, is at the heart of many metabolic systems. The cycle is used by aerobic organisms to release stored energy in the form of ATP by the oxidation of acetyl-CoA into water and CO_2. The details for the TCA cycle can be found in any standard chemistry text book, e.g. [12]. In [18], the trajectories of different carbon atoms in the TCA cycle was examined to explain the change of their oxidation states. It is well known that there is an enzymatic differentiation of the two carboxymethyl groups in citrate, which requires a rigorous stereochemical modeling of the graph grammar rules used [3]. Ignoring such stereochemical modeling would lead to atom mappings not occurring in nature. We will provide a formal handle to analyze theoretically possible carbon trajectories using the algebraic constructs provided in this paper. As we will see, such structures provides intuitive interpretations for the TCA cycle.

In our setting, the TCA cycle is a chemical network CN giving rise to transformations of the underlying monoid (depicted and discussed in the Appendix of [13]). To start the cycle, an Acetyl-CoA molecule is condensed with an oxaloacetate (OAA) molecule, executing a cycle of reactions that ends up regenerating the OAA molecule while expelling two CO_2 and water on the way. For the sake of demonstration, assume we are interested in answering the following questions: What are the possible trajectories of the carbons of an OAA within a TCA cycle while (i.) ignoring the enzymatic differentiation of the two carboxymethyl groups in citrate (denoted TCA-□), or (ii.) not ignoring (denoted TCA-△). To answer these questions, we will decompose the characteristic monoid of the TCA cycle into its natural subsystems and examine them using the projected Cayley graph. When an original atom is expelled from the cycle, we will consider it permanently lost. The carbon atoms of the OAA molecule that we are interested in tracking are annotated with the ids 4, 5, 6, and 7. Let $\bar{z} = (4, 5, 6, 7)$. The projected Cayley graph of $\text{PCay}(\text{CN}, \bar{z})$ wrt. TCA-□ (resp. TCA-△), consists of 213 (resp. 67) vertices. The full Cayley graphs are depicted in the Appendix of [13]. When a carbon atom leaves the TCA cycle we denote it by "_". E.g. the atom state $(\_, 7, 6, \_)$ should be read as the original carbon atoms with ids 4 and 7 has been expelled, while the carbon atoms with ids 5 and 6 are located at the atoms with id 7 and 6 respectively.

We can find the natural subsystems of CN as the strongly connected components of $\text{PCay}(\text{CN}, \bar{z})$. In TCA-□ (resp. TCA-△) we find 92 (resp. 51) strongly connected components of which 8 (resp. only 1) are non-trivial. Any non-trivial strongly connected component must invariably contain at least one tour around the TCA cycle. Moreover, any non-trivial strongly connected component represents a sequence(s) of reactions that uses (some of the) original atoms of the OAA molecule. To simplify $\text{PCay}(\text{CN}, \bar{z})$ such that only the information on carbon traces of the atoms of OAA are depicted, we will construct the simplified projected Cayley graph, denoted $\text{SCay}(\text{CN}, \bar{z})$, as follows: collapse any vertex in $\text{PCay}(\text{CN}, \bar{z})$ that is part of a trivial strongly connected component and whose atoms are not located in an OAA molecule. Moreover, for any non-trivial strongly connected component, hide the edges between atom states in the same strongly connected component, and finally only include atom states if the atoms are located in a OAA molecule. The resulting graphs for TCA-□ and TCA-△ are depicted in Fig. 2. Each box in the figure represents a natural subsystem that contains an atom state where every atom is either expelled or located in an OAA molecule. When ignoring the stereochemical formation of citrate, $(\_, 5, 6, 7)$ is a grey node in $\text{SCay}(\text{CN}, \bar{z})$ (i.e., a representative of a strongly connected component $\text{PCay}(\text{CN}, \bar{z})$), i.e., there is a trajectory where three of the four original carbons of OAA are re-used at the same location after a TCA-□ cycle turnover. However in TCA-△ only $(\_, 5, \_, \_)$ is a representative of a strongly connected

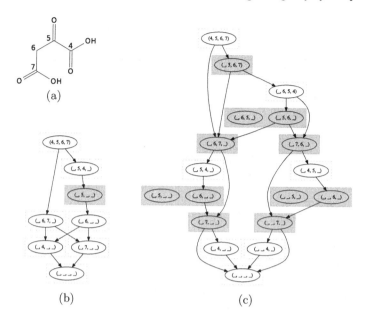

Fig. 2. (a) The oxaloacetate molecule. The carbon atoms are equipped with the ids 4, 5, 6, and 7. (b) The simplified projected Cayley graph SCay(CN, $(4,5,6,7)$), when adjusting for stereospecific citrate in $tr(\mathrm{CN})$. (c) The simplified projected Cayley graph SCay(CN, $(4,5,6,7)$) when not considering stereospecificity.

component, i.e., only the carbon with id 5 of OAA can be kept at the same location when a multitude of TCA-△ turnovers are executed. If that carbon changes location it will leave the TCA cycle after exactly two more turnovers (the natural subsystems reachable from $(\_,5,\_,\_)$ do not correspond to strongly connected components) via positions $5 \rightarrow 6 \rightarrow 4 \rightarrow \_$ or via $5 \rightarrow 6 \rightarrow 7 \rightarrow \_$. To the best of our knowledge such investigations have not been executed formally before.

4   Conclusion

In this work we have extended the insights provided by [4], by showing the natural relationship between event traces, the characteristic monoid and its corresponding Cayley graph. The projected Cayley graph provides valuable insights into local substructures of reversible event traces.

We see future steps for this approach to branch in at least two directions. On one hand, these methods shows obvious applications in isotopic labeling design. To this end, it is natural to extend the system to model the actual process of such experiments. E.g. when doing isotopic labeling experiments with mass spectrometry, molecules are broken into fragments and the weight of such fragments are deduced to determine the topology of the fragment. Using our model to track where the atoms might end up in such fragments and how it affects their weight seems like a natural next step. On the other hand, a more rigorous investigation of the fundamental properties derived from semigroup theory of the characteristic monoid seems appealing. As we have shown here, understanding such relations might grant insights into the nature of the examined system.

References

1. Andersen, J.L., Flamm, C., Merkle, D., Stadler, P.F.: Inferring chemical reaction patterns using rule composition in graph grammars. J. Syst. Chem. **4**(1), 4 (2013)
2. Andersen, J.L., Flamm, C., Merkle, D., Stadler, P.F.: A software package for chemically inspired graph transformation. In: Echahed, R., Minas, M. (eds.) ICGT 2016. LNCS, vol. 9761, pp. 73–88. Springer, Cham (2016). https://doi.org/10.1007/978-3-319-40530-8_5
3. Andersen, J.L., Flamm, C., Merkle, D., Stadler, P.F.: Chemical graph transformation with stereo-information. In: de Lara, J., Plump, D. (eds.) ICGT 2017. LNCS, vol. 10373, pp. 54–69. Springer, Cham (2017). https://doi.org/10.1007/978-3-319-61470-0_4
4. Andersen, J.L., Merkle, D., Rasmussen, P.S.: Graph transformations, semigroups, and isotopic labeling. In: Cai, Z., Skums, P., Li, M. (eds.) ISBRA 2019. LNCS, vol. 11490, pp. 196–207. Springer, Cham (2019). https://doi.org/10.1007/978-3-030-20242-2_17
5. Chahrour, O., Cobice, D., Malone, J.: Stable isotope labelling methods in mass spectrometry-based quantitative proteomics. J. Pharm. Biomed. Anal. **113**, 2–20 (2015)
6. Clifford, A., Preston, G.: The Algebraic Theory of Semigroups, vol. 2. American Mathematical Society, Providence (1967)
7. Deev, S.L., Khalymbadzha, I.A., Shestakova, T.S., Charushin, V.N., Chupakhin, O.N.: 15n labeling and analysis of 13c–15n and 1h–15n couplings in studies of the structures and chemical transformations of nitrogen heterocycles. RSC Adv. **9**, 26856–26879 (2019). https://doi.org/10.1039/C9RA04825A
8. Dénes, J.: Connections Between Transformation-semigroups and Graphs. Hungarian Academy of Sciences Central Research Institute for Physics (1966)
9. Egri-Nagy, A., Nehaniv, C.L.: Hierarchical coordinate systems for understanding complexity and its evolution, with applications to genetic regulatory networks. Artif. Life **14**(3), 299–312 (2008)
10. Egri-Nagy, A., Nehaniv, C.: Computational holonomy decomposition of transformation semigroups. arXiv preprint arXiv:1508.06345 (2015)
11. Froidure, V., Pin, J.E.: Algorithms for computing finite semigroups. In: Cucker, F., Shub, M. (eds.) Foundations of Computational Mathematics, pp. 112–126. Springer, Heidelberg (1997)
12. Harvey, R., Ferrier, D.: Biochemistry (Lippincott's Illustrated Reviews Series). Lippincott Williams & Wilkins, Baltimore (2010)
13. Hellmuth, M., Merkle, D., Nøjgaard, N.: Atom tracking using cayley graphs. https://cheminf.imada.sdu.dk/preprints/isbra-2020.pdf
14. Mikolajczak, B.: Algebraic and Structural Automata Theory. Elsevier, Amsterdam (1991)
15. Nehaniv, C.L., et al.: Symmetry structure in discrete models of biochemical systems: natural subsystems and the weak control hierarchy in a new model of computation driven by interactions. Philos. Trans. R. Soc. A Math. Phys. Eng. Sci. **373**(2046), 20140223 (2015)

16. Van der Plas, H.C.: The SN(ANRORC) mechanism: A new mechanism for nucle-ophilic substitution. Acc. Chem. Res. **11**(12), 462–468 (1978)
17. Rhodes, J., Nehaniv, C.: Applications of Automata Theory and Algebra. World Scientific, Singapore (2009)
18. Smith, E., Morowitz, H.J.: The Origin and Nature of Life on Earth: The Emergence of the Fourth Geosphere. Cambridge University Press, Cambridge (2016)

Author Index

Printed in the United States
By Bookmasters